PHYSICS RESEARCH AND TECHNOLOGY

SOLITONS

INTERACTIONS, THEORETICAL AND EXPERIMENTAL CHALLENGES AND PERSPECTIVES

PHYSICS RESEARCH AND TECHNOLOGY

Additional books in this series can be found on Nova's website under the Series tab.

Additional e-books in this series can be found on Nova's website under the e-book tab.

PHYSICS RESEARCH AND TECHNOLOGY

SOLITONS

INTERACTIONS, THEORETICAL AND EXPERIMENTAL CHALLENGES AND PERSPECTIVES

MATTHEW C. WILLIAMS
EDITOR

nova publishers
New York

For permission to use material from this book please contact us:
Telephone 631-231-7269; Fax 631-231-8175
Web Site: http://www.novapublishers.com

LIBRARY OF CONGRESS CATALOGING-IN-PUBLICATION DATA

ISBN: 978-1-62618-234-9

Published by Nova Science Publishers, Inc. † New York

CONTENTS

PREFACE

In mathematics and physics, a soliton is a self-reinforcing solitary wave (a wave packet or pulse) that maintains its shape while it travels at constant speed. Solitons are caused by a cancellation of nonlinear and dispersive effects in the medium. In this book, the authors discuss the interactions and theoretical and experimental challenges of solitons. Topics include soliton motion of electrons and its physical properties in coupled electron-phonon systems and ionic crystals; soliton excitations and its experimental evidence in molecular crystals; shapes and dynamics of semi-discrete solitons in arrayed and stacked waveguiding systems; ion-acoustic super solitons in plasma; diamond-controlled solitons and turbulence in extracellular matrix and lymphatic dynamics; and nonlinear waves in strongly interacting relativistic fluids.

The soliton motion of the electron and polaron and their properties arising from the nonlinear interactions between the electron and phonon are here studied using the nonlinear quantum mechanics in the coupled electron-phonon systems, superconductor, ionic crystals and semiconductors, respectively, in which the nonlinear interactions are due to the interactions between the electron and phonon as well as the electron and polarized lattice-field, respectively. In Chapter 1, the authors give first the Hamiltonians of these systems and corresponding dynamic equations of the solitons, which are some nonlinear Schrödinger equation. The soliton solutions of the dynamic equations, their properties of motion and some physical phenomena resulting from them are studied in detail. These results can be verified by some experiments. This means that the soliton excitations of the electrons and polarons arising from the nonlinear interactions are really existent in these systems, their motions result in the physical phenomena in these systems. Meanwhile, the authors demonstrated also that the macroscopic quantum phenomenon and Josephson effect occurred in superconductor and superconductive junctions are just caused by the soliton motions of electrons. Therefore, these results obtained have the important significances in physics.

In Chapter 2, the authors first establish Hamiltonian of molecular crystal and corresponding nonlinear dynamic equation of exciton through determining the mechanism of generation and forms of nonlinear interaction based on its molecular structure. Subsequently the authors find out the exciton-soliton and phonon-soliton solutions of the dynamic equations of exciton and phonon in acetanilide and study their properties, respectively. The authors further calculate the eigenenergy spectrum and specific heat of acetanilide and investigate theoretically the features of Raman scattering and infrared absorption of acetanilide arising from these soliton excitations, and obtain the rule of exponential decreases of infrared absorption strength of 1650cm^{-1} band, which corresponds to the soliton, with increasing the temperature of acetanilide using the soliton model. Finally, the authors measure the experimental results of Raman scattering and infrared absorption to verify the correctness of theoretical results mentioned above, in which the authors obtain that the frequencies of peaks in Raman and infrared spectra consist exactly with the theoretical result in eigenenergy levels of acetanilide and the experimental rule of change of strength of the peaks at 1650cm^{-1} with varying the temperature concord also the above theoretical result obtained by the soliton theory. Meanwhile, to use Mark free-electron laser captures truly the information of the soliton and obtain that the soliton lifetime is 15ps in this experiment and 35ps in pump-probe spectroscopy measures, respectively. These results obtained from theoretical and experimental studies represent clearly and confirm further that the exciton-soliton is really existent in the molecular crystal.

Chapter 3 studies the shapes and dynamics of semi-discrete solitons (SDSs) in the known model of the set of linearly-coupled waveguides with the self-focusing cubic nonlinearity. The model describes a planar array of optical fibers, as well as a stack of parallel planar waveguides, in the temporal and spatial domains, respectively. It also applies to the self-attractive Bose-Einstein condensate (BEC) loaded into an array of parallel tunnel-coupled "cigar-shaped" traps. It was known that the interplay of the group-velocity dispersion, discrete diffraction and nonlinearity gives rise to SDSs in this system. The authors here develop a variational approximation (VA) and additional analytical methods for the description of the SDSs, and study their mobility and collisions by means of direct simulations. The VA and another analytical method (an exact solution of the linearized equation in the core adjacent to the central one) produce an accurate description for the family of fundamental onsite-centered SDS solutions. The VA is developed too for transversely unstable intersite-centered and twisted solitons. In simulations, the solitons are not mobile in the discrete direction (while broader non-soliton modes may be mobile across the array). Collisions between the SDSs traveling in the longitudinal direction reveal a threshold separating the passage and merger, or mutual destruction of the colliding solitons. While the exact shape of the threshold, considered as a function of the solitons' energy, features irregularities, its average form is explained in an analytical form. Collisions between SDSs centered at two adjacent cores are studied too.

The role of the solitonic effects in the context of supercontinuum generation in optical fibers is reviewed in Chapter 4. Such effects become important whenever light propagates in the anomalous dispersion regime. When ultrashort pulses with large enough power are launched into highly nonlinear fibers, soliton fission gives origin to multiple fundamental solitons of different widths and peak powers. Among the host of soliton related effects, there are two which become particularly important: the soliton self-frequency shift and the emission of dispersive radiation in the normal dispersion region. If long duration pulses or a continuous wave are used to pump the supercontinuum, the modulation instability becomes the main initial mechanism, breaking the pump wave into multiple fundamental solitons. In this case, both the soliton self-frequency shift and inelastic collisions among such solitons play an important role.

New forms of nonlinear waves in plasmas – super-solitons are proposed in Chapter 5. They are different from ordinary solitons by nontrivial topology of their phase portraits. Phase trajectories of the super-nonlinear super-solitons envelope one or several separatrixes on the phase portrait.

A number of models of non-magnetized multicomponent plasmas are considered in the paper. These are: dusty electron-ion plasma, electron-positron-ion plasma, multi-ion plasma. Possibility of existence of the super-solitons in such plasmas is shown. Profiles of physical values in the specified waves are calculated. Examples of experimental observation of the super-solitons in the multi-ion plasmas are presented.

Chapter 6 presents local and global turbulence, shock-waves and solitons occurring in human lymphatic and matrix dynamics controlled by Diamond Biomesotherapy. The authors start with a basic description of the extracellular matrix (ECM), which is followed by an outline of Diamond ECM therapy. The next section presents modern research in lymphatic and ECM turbulent dynamics, from both a global-macroscopic perspective and local-microscopic (i.e., quantum) one. After that, the authors explore related Diamond therapies and fat-burning mechanics. The authors conclude with two biomedical applications: (i) a nonstandard Diamond Biomesotherapy for motor neuron diseases, and (ii) a Diamond fat-burning mechanism.

During the past decades the study of strongly interacting fluids experienced a tremendous progress. In the relativistic heavy ion accelerators, specially the RHIC and LHC colliders, it became possible to study not only fluids made of hadronic matter but also fluids of quarks and gluons. Part of the physics program of these machines is the observation of waves in this strongly interacting medium. From the theoretical point of view, these waves are often treated with li-nearized hydrodynamics. In Chapter 7 the authors review the attempts to go beyond linearization. The authors show how to use the Reductive Perturbation Method to expand the equations of (ideal and viscous) relativistic hydrodynamics to obtain nonlinear wave equations. These nonlinear wave equations govern the evolution of energy density perturbations (in hot quark gluon plasma) or baryon density perturbations (in cold quark gluon plasma and nuclear matter). Different nonlinear wave equations, such as the breaking wave, Korteweg-de Vries and Burgers equations, are obtained from different equations of state (EOS). In nuclear matter, the Walecka EOS may lead to a KdV equation. The authors explore equations of state such as those extracted from the MIT Bag Model and from QCD in the mean field theory approach. Some of these equations are integrable and have analytical solitonic solutions. The authors derive these equations also in spherical and cylindrical coordinates. The authors extend the analysis to two and three dimensions to obtain the Kadomtsev-Petviashvili (KP) equation, which is the generalization of the KdV. The KP is also integrable and presents analytical solitonic solutions. In viscous relativistic hydrodynamics the authors have second spatial derivatives which physically represent dissipation terms. The authors present numerical solutions and their corresponding algorithms for the cases where the equations are not integrable.

In: Solitons
Editor: Matthew C. Williams

ISBN: 978-1-62618-234-9

Chapter 1

THE SOLITON MOTIONS OF ELECTRONS AND ITS PHYSICAL PROPERTIES IN COUPLED ELECTRON-PHONON SYSTEMS AND IONIC CRYSTALS

Pang Xiao-Feng*

Institute of Life Science and Technology, University of Electronic Science and Technology of China, Chengdu, China

Abstract

The soliton motion of the electron and polaron and their properties arising from the nonlinear interactions between the electron and phonon are here studied using the nonlinear quantum mechanics in the coupled electron-phonon systems, superconductor, ionic crystals and semiconductors, respectively, in which the nonlinear interactions are due to the interactions between the electron and phonon as well as the electron and polarized lattice-field, respectively. We give first the Hamiltonians of these systems and corresponding dynamic equations of the solitons, which are some nonlinear Schrödinger equation. The soliton solutions of the dynamic equations, their properties of motion and some physical phenomena resulting from them are studied in detail. These results can be verified by some experiments. This means that the soliton excitations of the electrons and polarons arising from the nonlinear interactions are really existent in these systems, their motions result in the physical phenomena in these systems. Meanwhile, we demonstrated also that the macroscopic quantum phenomenon and Josephson effect occurred in superconductor and superconductive junctions are just caused by the soliton motions of electrons. Therefore, these results obtained have the important significances in physics.

* E-mail address: pangxf2006@yahoo.com.cn; Tel.:0086-2883202595; Address: Institute of Life Science and Technology, University of Electronic Science and Technology of China, Chengdu 610054, China. (Corresponding author)

1. The Soliton Motions of Electrons in Coupled Electron-Phonon Systems

As it is known, the investigations of property of coupled election-phonon systems have been a very important subject in the condensed matter physics because there are a great number of interesting and dramatic phenomena and effects in the systems, arising from the interaction between the electron density fluctuation and lattice deformation, for example, the charge density wave (CDW), polaron effect, superconductivity and dimerization of lattice and Peierls transition, and so on[1-7], which occur in some materials with different coupling strengths and are some nonlinear quantum mechanical phenomena. Therefore, it is quite necessary to study in-depth these problems.

In general, a coupled electron-phonon systems including electron-electron interaction are often described by the modified Hubbard model [4]. For example, one-dimensional modified Hubbard model[8] was studied in detail by Lindner and Fedyanin[9], and Makhankov et al. [10-12]. The model Hamiltonian is written in the form of

$$H = H_{ph} + H_e + H_{\text{int}} \tag{1}$$

where H_{ph} and H_e are the Hamiltonians of the phonon and electron subsystems, respectively, while H_{int} describes the interaction between them. H_e corresponds to the Hubbard model [8] in the nearest-neighbor approximation. We assume that the ground state is antiferromagnetic and that the lattice is divided into two equivalent ferromagnetic sublattices A and B, thus the magnetic moment of the system is zero in the absence of an external magnetic field. Based on this, Makhankov et al. [10-12] proposed the following for H_e,

$$H_e = \sum_{n\in A, j\in B;\sigma} J_{nj}(b^{A+}_{n,\sigma} b^{B}_{n,\sigma} + b^{B+}_{n,\sigma} b^{A}_{n,\sigma}) + \frac{V}{2}\sum_{n\in A,B;\sigma} N_{n,\sigma} N_{n,-\sigma} - \mu \sum_{n,\sigma} N_{n,\sigma} \tag{2}$$

where

$$J_{n,j} = \begin{cases} \bar{J}, (j = n \pm 1), \\ 0, (j \neq n \pm 1), \end{cases}$$

$b^{+}_{n,\sigma}(b_{n,\sigma})$ is the creation (annihilation) operator for an electron with spin σ in a Wannier state at site n, $N_{n,\sigma} = b^{+}_{n,\sigma} b_{n,\sigma}$ is the number operator, J is the transition amplitude between nearest neighbors, V is the repulsive interaction between electrons in the same atom, μ denotes the chemical potential. The lattice part of the Hamiltonian, H_{ph}, is given in the harmonic approximation by

$$H_{ph} = \frac{M}{2}\sum_n \left[(R_n)_t\right]^2 + \frac{W}{2}\sum_n (R_{n+1} - R_n)^2 \tag{3}$$

where M is the atomic mass, W denotes the force constant of the system, $R_n = R_n(t) = \overline{R}_n + u_n(t)$ is the position of the nth atom and $\overline{R}_n$ is its equilibrium position. In the linear approximation with respect to small deviations $u_n(t)$, the interaction between the electron subsystem and the lattice in Eq.(1) can be written as

$$H_{\text{int}} = I\sum_{n,\sigma}(u_{n,\sigma} - u_{n+1,\sigma})\left[b_{n.\sigma}^{B+}b_{n+1,\sigma}^{A} + b_{n+1,\sigma}^{A+}b_{n,\sigma}^{B} + b_{n+1,\sigma}^{B+}b_{n,\sigma}^{A} + b_{n,\sigma}^{A+}b_{n+1,\sigma}^{B}\right] \tag{4}$$

where

$$I = \left\langle (n+1)\left|\frac{\partial V\left(x - \tilde{R}_n\right)}{\partial x}\right| n\right\rangle$$

denotes the matrix element of the local "force" between Wannier states of neighboring lattice atoms, corresponding to the local potential $V = V(x - \overline{R}_n)$. Effects of anharmonicity in the lattice vibrations and nonlinearity in the expansion of electron integration force in this model have been studied, many interesting features of the soliton can be revealed.

Since the ground state of the system is antiferromagnetic, we are interested in the low-lying excitations, and in equations describing the amplitudes of spin up or down states $(\phi_{n\uparrow}, \phi_{n\downarrow})$ at the nth lattice site. We here eliminate the terms of the type $(u_{n+1} - u_n)b_{n,\sigma}^{B+}b_{n-1,\sigma}^{A}$, etc. To simplify the calculation, Makhankov et al. [10-12]. employed a method similar to the adiabatic approach related to the ground state, and examined the change of the state $\left|\left|0\right\rangle\right.$ under the action of the evolution operator, which is defined on single-particle states. Then Lindner et al. [9] and Makhankov et al. [10] defined

$$\phi_{n,\sigma}^{P}(t) = \langle\varphi(t)|b_{n,\sigma}^{P}(t)|\varphi(t)\rangle = \langle 0|b_{n,\sigma}^{P}(t)|0\rangle = \langle 0|Ub_{n,\sigma}U|0\rangle \tag{5}$$

Making use of (3) – (5), at $\hbar = 1$ we obtain the following equation for $b_{n,\sigma}^{A}(t)$

$$\begin{aligned} i\left(b_{n,\sigma}^{A}(t)\right)_t &= \left[b_{n,\sigma}^{A}, H\right]_- = Vb_{n,\sigma}^{A}(t)n_{n-\sigma} - \mu b_{n,\sigma}^{A}(t) + \\ &\left[\overline{J} + I(u_n - u_{n-1})\right]b_{n+1,\sigma}^{B}(t) + \left[\overline{J} + (u_{n-1} - u_n)\right]b_{n-1,\sigma}^{B}(t) \end{aligned} \tag{6}$$

The corresponding equation for $b_{n,\sigma}^{B}$ is the same as (6) except that A and B are interchanged. Applying (5) and (6), we can get

$$\begin{aligned} i\left(\phi_{n,\sigma}^{A}(t)\right)_t &= V\langle 0|b_{n,\sigma}^{A}(t)n_{n,-\sigma}|0\rangle - \mu\phi_{n,\sigma}^{A}(t) + \\ &\left[\overline{J} + I(u_n - u_{n-1})\right]\phi_{n+1,\sigma}^{B}(t) + \left[\overline{J} + I(u_{n-1} - u_n)\right]\phi_{n-1,\sigma}^{B}(t) \end{aligned} \tag{7}$$

and a similar equation for $\phi^{B}_{n,\sigma}(t)$, which is the same as the above except that A and B are interchanged. Applying the following decoupling procedure to (7),

$$\begin{aligned}&\langle 0|b^{p}_{n,\sigma}(t)b^{p+}_{n,-\sigma}(t)b^{p}_{\eta,-\sigma}(t)|0\rangle \to \langle 0|b^{p}_{n,\sigma}(t)|0\rangle\langle 0|b^{p+}_{n,-\sigma}(t)|0\rangle\langle 0|b^{p}_{\eta,-\sigma}(t)|0\rangle \\ &= \phi^{p}_{n,\sigma}(t)\phi^{p*}_{n,-\sigma}(t)\phi^{p}_{n,-\sigma}(t)\end{aligned} \quad (8)$$

where p = A,B, we finally get a system of nonlinear equations for the four complex function $\phi^{p}_{n,\sigma}(t)$ and the real function u_n. For the "pure" Hubbard model (I = 0), similar nonlinear equations can also be obtained for $\phi_{n,\sigma}(t)$ as well.

We now discuss the influence of spin ordering on the lattice vibrations. Makhankov et al. [10-12] constructed the functional,

$$F = \langle\phi(t)|H|\phi(t)\rangle = -\mu \sum_{n\in A\cup B;\sigma} |\phi_{n,\sigma}|^2 - \sum_{n,\sigma}\left[\bar{J} - I(u_n - u_{n+1})\right]\times(\phi^{A*}_{n,\sigma}\phi^{B}_{n+1,\sigma} + \phi^{B*}_{n+1,\sigma}\phi^{A}_{n,\sigma} + \phi^{B*}_{n,\sigma}\phi^{A}_{n+1,\sigma} + \phi^{A*}_{n+1,\sigma}\phi^{B}_{n,\sigma}) - \frac{1}{2}\sum_{n\in A\cup B}|\phi_{n,\sigma}|^2|\phi_{n,-\sigma}|^2 + \frac{M}{2}\sum_n\left[(R_n)_t\right]^2 + \frac{W}{2}\sum_n(R_{n+1} - R_n)^2$$

From the Hamilton equations,

$$M(R_n)_{tt} = P_n,\ P_n = -\frac{\delta F}{\delta R_n}$$

it follows that

$$\begin{aligned}M(R_n(t))_{tt} &= W[R_{n+1}(t) - 2R_n(t) + R_{n-1}(t)] - \\ I\sum_{\sigma}&\left\{\begin{matrix}\phi^{A*}_{n,\sigma}(t)\left[\phi^{B}_{n+1,\sigma}(t) - \phi^{B}_{n-1,\sigma}(t)\right] - \left[\phi^{B*}_{n+1,\sigma}(t) - \phi^{B*}_{n-1,\sigma}(t)\right]\phi^{A}_{n,\sigma}(t) - \\ \phi^{B*}_{n,\sigma}(t)\left[\phi^{A}_{n+1,\sigma}(t) - \phi^{A}_{n-1,\sigma}(t)\right] - \left[\phi^{A*}_{n+1,\sigma}(t) - \phi^{A*}_{n-1,\sigma}(t)\right]\phi^{B}_{n,\sigma}(t)\end{matrix}\right\}\end{aligned} \quad (9)$$

This equation is a discrete analogue of the harmonic oscillator equation under the action of an external force which does not vanish if variation in the initial spin ordering takes place. The usual condition for antiferromagnetism $\langle N^{A}_{n,\sigma}\rangle = \langle N^{B}_{n\pm 1,\sigma}\rangle$ now becomes $\phi^{A}_{n,\sigma} = \phi^{B}_{n\pm 1,\sigma}$ which causes the last term (−I) in Eq. (9) to vanish.

If we assume that the variation of magnetic order is described by a sufficiently smooth function of coordinate along the chain, we can use the continuum approximation

$$u_n(t) = R_n(t) - \bar{R}_n \to (\chi(\xi,t) - \xi)\gamma_0,\ \phi^{p}_{n,\sigma}(t) \to \phi^{p}_{\sigma}(\xi,t)$$

Then, expanding up to the second derivative and taking into account up to the cubic nonlinear terms, we have

$$R_{n+1}(t) \to \gamma_0\left(\chi + \chi_\xi + \frac{1}{2}\chi_{\xi\xi} + \cdots\right),$$

$$\phi^B_{n\pm1,\sigma}(t) \to \phi^A_{-\sigma}(\xi,t) \pm \phi^{A'}_{-\sigma}(\xi,t) + \frac{1}{2}\phi^{A''}_{-\sigma}(\xi,t) + \cdots,$$

$$\phi^A_{n\pm1,\sigma}(t) \to \phi^B_{-\sigma}(\xi,t) \pm \phi^{B'}_{-\sigma}(\xi,t) + \frac{1}{2}\phi^{B''}_{-\sigma}(\xi,t) + \cdots,$$

where $\phi^B_\sigma(\xi,t) = 0$ if $n \in A$, and similarly $\phi^A_\sigma(\xi,t) = 0$, if $n \in B$. Makhankov et al. [10-12] finally obtained the following equations for $\phi_\sigma(\xi,t)$ and $\chi(\xi,t)$

$$i\left(\phi_\sigma(\xi,t)\right)_t = \left[\left(\phi^n_{-\sigma}(\xi,t)\right)_{\xi\xi} + 2\phi_{-\sigma}(\xi,t)\right]\left[\tilde{J} + r_0 I\left(1 - \chi_\xi(\xi,t)\right)\right] + I\left|\phi_{-\sigma}\right|^2\phi_\sigma - \mu\phi_\sigma \tag{10}$$

$$\ddot{\chi}(\xi,t) = v_0^2\chi_{\xi\xi}(\xi,t) + \frac{2I}{r_0 M}\sum_\sigma \frac{\partial}{\partial\xi}\left(\phi^*_\sigma\phi_{-\sigma}\right), \tag{11}$$

respectively, where $v_0^2 = Wr_0^2 / M$ 。 Equation (10) is a discrete nonlinear Schrödinger equation.

In the quasi-static approximation ,or setting $\chi(\xi,t) = \chi(\xi - vt)$, we can integrate Equation (11) to obtain

$$\chi_\xi(\xi,t) = \frac{\tilde{I}}{2Ir_0}\sum_\sigma \phi^*_\sigma\phi_\sigma + K \tag{12}$$

where $\tilde{I} = (2I)^2 / M\left(v^2 - v_0^2\right)$ with $v^2 << v_0^2$ and the constant K depending on initial conditions. Using Equation (12), we can get from Equation (10)

$$\begin{aligned} i\left(\phi_\sigma(\xi,t)\right)_t &= T\left(\phi^n_{-\sigma}(\xi,t)\right)_{\xi\xi} + \left[2T - \tilde{I}\sum_\sigma \phi^*_\sigma(\xi,t)\phi_{-\sigma}(\xi,t)\right]\phi_{-\sigma}(\xi,t) - \\ &\mu\phi_\sigma(\xi,t) + V\left|\phi_{-\sigma}(\xi,t)\right|^2\phi_\sigma(\xi,t) \end{aligned} \tag{13}$$

where $T = \bar{J} - r_0 I(1 - K)$.

In terms of even and odd combinations of

$$\phi_\sigma = \{\phi_\uparrow, \phi_\downarrow\}, \quad \varphi^{(1)} = \phi_\uparrow + \phi_\downarrow, \quad \varphi^{(2)} = \phi_\uparrow - \phi_\downarrow,$$

where

$$\phi_\uparrow = \frac{\varphi^{(1)} + \varphi^{(2)}}{2}, \quad \phi_\downarrow = \frac{\varphi^{(1)} - \varphi^{(2)}}{2},$$

one gets

$$i\varphi_t^{(m)} = T\varphi_{\xi\xi}^{(m)} - \left\{ \delta_{m,1} 2T - \delta_{m,2} 2T - \mu - \frac{V - \hat{I}}{2} \left[\left|\varphi^{(1)}\right|^2 + \left|\varphi^{(2)}\right|^2 - 2\left|\varphi^{(m)}\right|^2 \right] \right\} \varphi^{(m)}$$

$$-\hat{I} \left\{ -\frac{1}{2} \left[\left|\varphi^{(1)}\right|^2 + \left|\varphi^{(2)}\right|^2 \right] \varphi^{(m)} + \frac{3}{4} \left|\varphi^{(m)}\right|^2 \varphi^{(m)} + \frac{1}{4} \left|\varphi^{(m)}\right|^2 \left[\delta_{m,1}\varphi^{*(2)} + \delta_{m,2}\varphi^{*(1)} \right] \right\}, \qquad (14)$$

where m= 1, 2. Assuming that the average occupation number $n = \left|\phi_\uparrow\right|^2 + \left|\phi_\downarrow\right|^2$ is constant and neglecting the term $\varphi_\sigma^* \varphi_{-\sigma} \varphi$ compared to $\left|\varphi_{-\sigma}\right|^2 \varphi_\sigma$, we get two decoupled nonlinear Schrödinger equations for '$\varphi^{(1)}$ and $\varphi^{(2)}$, respectively.

This demonstrates that the coupled electron-phonopn systems described by Equation (1) can be studied by using nonlinear quantum mechanics. In this case the soliton solutions of the two decoupled nonlinear Schrödinger equations can be represented by

$$\varphi^{(m)}(\xi, t) = \sqrt{2n} \sec h \left[\sqrt{\frac{n(V - \hat{I})}{T}} (\xi - vt) \right] \exp\left(\frac{iv\xi}{T} - i\omega_i t \right), (m = 1, 2)$$

where

$$\omega_1 = 2T - \mu - \frac{v^2}{4T}, \omega_2 = -2T - \mu - (v - \hat{I}) + \frac{v^2}{4T},$$

and v is the soliton velocity. Therefore, electron in the Hubbard model with electron-electron and electron-phonon interactions satisfies the nonlinear Schrödinger equation or the laws of nonlinear quantum mechanics. The soliton consists of a “drop” associated with a “bubble”, and is localized in the region

$$W_s \approx \sqrt{T(V - \bar{I}) / n}$$

which is accompanied by a lattice distortion wave that is also bounded in the same region, and $\chi(\xi, t)$ in Eq. (11) or Eq. (12) can be represented approximately by

$\chi \approx \tanh[(\xi - vt)/W_s]$ under appropriate initial conditions. The soliton possesses also the wave-corpuscle duality, which resumbles the microscopic particles described by nonlinear quantum mechanics[13].

However, for the pure Hubbard model, $\overline{I} = 0, T = \overline{J}$, and a half-filled band, i.e., n = 1, it follows that $\omega_1 = \omega_2$ for $\mu = V/2$. In the more general case, it is difficult to find the solution of Eq. (14). But if we neglect the last term on the right-hand side of Eq. (14), it becomes

$$i\varphi_t^{(m)} + \delta_{i,1} T\varphi_{xx}^{(m)} - \delta_{i,2} T\varphi_{xx}^{(m)} - \frac{1}{2} V\left(\eta\left|\varphi^{(m)}\right|^2 - \left|\varphi^{(m+1)}\right|^2\right)\varphi^{(m)} = 0, \quad (15)$$

where m= 1, 2 and

$$\eta = 1 - \hat{I}/2V, \varphi^{(1)} = \varphi^{(2)}$$

This is a vector nonlinear Schrödinger equation[13]. In the case of η = 1 (or $\overline{I}$ = 0) which corresponds to the situation when the electron-phonon interaction is "switched off", Makhankov et al. [10-12] proved that (15) is integrable. They were able to obtain the Lax pair, examine its symmetry property, and construct the soliton solutions for given values of η.

2. New State-Ansatz for the Particles and Corresponding Uncertainty Relationship in Strong Electron-Phonon Coupling Systems

2.1. New State-Ansatz and Its Features

As it is known, the concept of "small polaron" was introduced to describe the system in the weak coupled case. Then the ground state properties of coupled electron-phonon systems are obtained self-consistently from the minimization of total elastic plus electronic energies in mean-field theory by freezing out the lattice distortion. However, this method cannot generalize into the intermediately and strongly coupled regions, in which the motion of the electron is always accompanied by a phonon cloud, which follows instantaneously the electron to fluctuate. Thus the polarons occur in the systems. This is a nonadiabatic effect. The adiabatic and nonadiabatic dynamics in condensed matter were studied, for example, Q Niu et al. [14] studied the adiabatic quantum dynamics of local spin moment in itinerant magnets, H. Li et al. [15] studied the nonadiabatic quantum dynamics of a vortex in a Josephson junction. However, in a strongly coupled electron-phonon systems, such as Holstein model[2], a basic problem is how to treat the nondiabatic effect and polaronic behavior within the whole range of parameters. It is well known that the Holstein model[2] is a prototype model to discuss the electron- phonon dynamics and has usually been used to study the antagonistic tendencies (localization -delocalization) in the formation of the polarons and bipolarons in the systems. In this model the Hamiltonian of molecular crystal is usually represented[2] by

$$H=\sum_n\left(P_n^2/2M+\frac{W}{2}q_n^2\right)-t\sum_n\left(b_n^+b_{n+1}+b_{n+1}^+b_n\right)-\lambda\sum_n b_n^+b_nq_n \tag{16}$$

where M and W are the mass of one molecule and the force constant of the molecular crystals, respectively, t and λ are the hopping and transfer integral of the electron and the electron-phonon coupled constant, respectively, q_n and P_n are operators of ordinary coordinate and momentum of molecule at site n, respectively. $n_n=b_n^+b_n$ is number operator of the electron, $b_n^+(b_n)$ is creation (annihilation) operator of the electron at site n . If displacement q_n in Eq.(16) is replaced by relative displacement ($q_{n+1}-q_n$) for the phonons and the kinetic energy term of the electron. $\varepsilon_0 b_n^+b_n(\varepsilon_0=\hbar\omega_0)$,is added into Eq.(16) , then the Hamiltonian of the coupled electron -phonon systems in Eq.(16) is in essence similar with Eq.(2). Therefore, Eq.(16) can also give the soliton solution as mentioned above, thus the Holstein systems can be described by nonlinear quantum mechanics[13].

In this model, the motion and quantum tunneling of "heavy" polarons, arising from the postponement effect, result in a strong fluctuations and squeezing of multiphonon states, which resembles that in quantum optics. The squeezed state of the phonons is a two-phonon coherent state which is induced by the motions of the polarons, and it is deeply related with anomalous lattice fluctuation [16-46]. When the phonons are squeezed the state and properties of the polarons, including its binding energy, effective mass and band width, are simultaneously changed due to the counteraction of squeezed phonons on the latter according to the rule of action-counteraction in physics. The antisqueezed effects, thus, can also induce significantly changes of properties of the systems involving the ground state. This shows that there are the interaction and dependence between squeezed phonons and polarons. Therefore, the states of intermediately and strongly coupled electron-phonon systems are very complicated and are worth studying seriously. In such a case a difficult question is how we describe the above states in the systems.

The concept of squeezed coherent states of quantum optics is often used for a quantum treatment of the coherent two-phonon states in condensed matter problems with coupled electron-phonon systems. The resulting nonadiabatic effects imply consequences, e.g., for the formation of bipolarons [1], quantum fluctuations of polarons [2,5] or phonon-assisted tunneling [4]. The coherent two-phonon states are represented usually by a product of unitary operators generating displaced and squeezed states, e.g., for a coupled electron-phonon system, by

$$U=U_c(\alpha)U_s(\zeta)=\exp[\sum_n(\alpha a_n^+-\alpha^*a_n)b_n^+b_n)\times\exp(\sum_n(\zeta^*a_n^2-\zeta a_n^{+2})]=U_s(\zeta)U_c(\bar{\alpha}) \tag{17}$$

where $\bar{\alpha}=\alpha\cosh(2r)+\alpha^*e^{i\theta}\sinh(2r)$, $\zeta=re^{i\theta}$, $a_n^+(a_n)$ is creation (annihilation) operator of the phonon at site n, $U_c(\alpha)$ represents the Lang-Firsov ansatz for a unitary operator generating a linear displacement of atoms[15-18] due to their coupling to the electron density and $U_s(\zeta)$ represents a two-phonon (squeezing) unitary operator. However, in the coupled electron- phonon system, the effects of displacement and squeezing occur simultaneously, are mutually dependent and closely related, especially in the strong and intermediate coupling regimes.

Therefore, the state formed by this physical mechanism should be described more rigorously by a unified wave functional state ansatz, not being resolved into individual steps of factorized operators, which is represented by Eq. (17), and is used in many applications in condensed matter systems. Based on the above notions Pang et al. {37-44} proposed the state ansatz for the coherent two-phonon state related with the polaron effect , which is represented by

$$D = \exp\left(\sum_n (\zeta^* a_n^2 - \zeta a_n^{+2}) + (\alpha a_n^+ - \alpha^* a_n) b_n^+ b_n\right) \tag{18}$$

maintaining the number of parameters. The operators in the exponent do not commute, i.e., $D \neq U_c U_s$, because

$$[L_2, L_1] = -2r|\alpha|\left[a_n^+ e^{i(\theta-\gamma)} - a_n e^{-i(\theta-\gamma)}\right] b_n^+ b_n \text{ ,}$$

where

$$L_1 = (\alpha a_n^+ - \alpha^* a_n) b_n^+ b_n \text{ , } L_2 = (\zeta^* a_n^2 - \zeta a_n^{+2}) \text{ , } \alpha = |\alpha| e^{ir} \text{ , } \zeta = re^{i\theta} \text{ .}$$

Therefore, ansatz (18) accounts for an additional correlation of the Lang-Firsov type functional operator with the two-phonon (squeezed) state operator, i.e., it shows clearly the influence of the squeezed effect of the phonons on the polaron states in the coupling electron-phonon systems as above. The correlated effect really influences both the displacement and squeezing and the polaron characteristics, namely, (i)the Debye-Waller factor determining the extension of the polaron (parameter of tunneling) related with, e.g., the possibility of bipolaron formation,; (ii) quantum fluctuations of the ground state of a Holstein polaron[2] and of the dimerization order parameters [32].

The weaker the electron-phonon coupling, the weaker become the correlation effects and the quantum fluctuations can overcome the localizing influence of the electron-phonon coupling.

To understand clearly the physical relevance of the ansatz in Eq.(18) let us factorize it as follows,

$$D = \exp\sum_n (\zeta^* a_n^2 - \zeta a_n^{+2}) \times \exp\sum_n \left[\left(\Gamma a_n^+ - \Gamma^* a_n\right) + i\theta'\right] b_n^+ b_n \tag{19}$$

where

$$\Gamma = \frac{|\alpha|}{2r}\left[\left(\cosh 2r - 1\right) e^{i(\theta-r)} + \sinh 2r e^{ir}\right] \tag{20}$$

Γ^* is a complex conjugate of Γ and θ' is a c-number phase factor which cancels at unitary transformations and tends to zero for real α and ζ. In the last case, L_1 and L_2 become Lie operators,

$$[L_2, L_1] = 2r\alpha\left[a_n^\dagger - a_n\right] = 2rL_1 \text{ and } \Gamma = \Gamma_0 = \alpha\left(e^{2r} - 1\right)/2r$$

can be obtained easily [7]. Equation (19) show clearly that the correlated ansatz in Eq.(18) results in a renormalization of the displacement effect by squeezing and, consequently, of the related quantum effects on the polaron characteristics. The effective displacement Γ includes the correlation of the displacement α and of the squeezing ζ and, in view of $\left(e^{2r} - 1\right)/2r > 1$, it implies an effective enhancement of $|\alpha|$.

There are approaches including, e.g., correlations between different phonon modes [15] or a special type of squeezed coherent states in quantum optics [16], which introduce additional variational parameters.

We shall briefly summarize the properties of the unitary operator D, $D^{-1}D = I$. One can easily find

$$Da_nD^{-1} = a_n \cosh 2r + e^{i\theta}a_n^+ \sinh 2r - \Gamma b_n^+ b_n \tag{21}$$

and an analogous expression for Da_nD^{-1}. Here, r is given by Equation (20).

For comparison, in the case of the factorized displacement and squeezing generators, it reads

$$U_sU_ca_nU_c^{-1}U_s^{-1} = a_n \cosh 2r + a_n^+ \sinh 2r - \alpha b_n^+ b_n \tag{22}$$

and similarly for a Hermitian conjugate $U_sU_ca_n^+U_c^{-1}U_s^{-1}$. When comparing Eqs. (21) with Eq. (22) we see again that the difference of both transformations lies in replacing the displacement α in Eq. (22) by a joint correlation factor of the two effects mentioned above, i.e.,

$$\alpha \to \Gamma(\alpha, \zeta) \tag{23}$$

By other words, when going over from the correlated to the uncorrelated case one should set $\Gamma/\alpha \to 1$.

If the squeezed effect of the phonons is further taken into account in the strong coupling electron-phonon systems, then the squeezed and coherent states of the phonons and the polaron state are now represented by the following product of operators of respective displacement and squeezing and polaron parts:[16-17]

$$U = U_C(\zeta)U_l(\alpha)U_S(\xi) = \exp\left[\sum_n (-1)^l(\zeta a_n^+ - \zeta^* a_n)\right]\exp\left[\sum_n (\alpha a_n^+ - \alpha^* a_n)b_n^+ b_n\right]\exp\left(\sum_n (\xi^* a_n^2 - \xi a_n^{+2})\right) \quad (24)$$

for the coupled electron-phonon systems, where $a_n^+(a_n)$ and $b_n^+(b_n)$ are the creation (annihilation) operators of phonon and electron, respectively. Here ξ, ζ and α are some reales, which represent the squeezed and coherent effects of the phonons and polaronic effect of the electrons (α is related with the binding energy of the polaron, i.e., $\alpha^2=E_p/\hbar\omega$, E_p is the binding energy of polaron, ω is the phonon frequency), respectively. U_C and U_S are the ansatzs for the one-phonon coherent state and two-phonon squeezed state. respectively. U_l is an ansatz of Lang-Firsor small polaron state which associates the electron with the local lattice deformation.

Evidently, Eq.(24) is an approximate representation to the coupled electron-phonon systems because it denotes only the squeezed effect of the phonons induced by the polaron motion, but exhibits not its antisqueezed effect on the polaron state, or speaking, ignores the correlated effect between them. As a matter of fact, in practically physical processes the squeezing and antisqueezing effects occur simultaneously, influence mutually and relate closely, i.e., the motion of the polaron results in the squeezing of the phonons, but when the phonons are in the squeezed process the state of the polarons will also be changed due to the antisqueezed effect of the phonons on it according to the rule of action-counteraction in physics, specially in strongly and intermediately coupled regions. Thus, the binding energy, effective mass and band width of the polarons also are simultaneously changed due to the counteraction of squeezed phonons in this case as are mentioned above. Therefore, we cannot use Eq.(24), but should use an unified state- asatz to describe more rigorously the states of strongly coupled electron-phonon systems. However, an appropriate state-ansatz for the systems in this case occurs not up to now. Pang[37-44] proposed a new ansatz of state through introducing three variational parameters, r, ξ and α, which includes correlation effects between the squeezed-phononic and polaronic states to study the properties of the coupled electron-phonon systems[37-44], respectively. The new state-ansatz proposed by Pang[37-44] is represented by

$$D = \exp\left\{\sum_n \left[\left(\xi^* a_n^2 - \xi a_n^{+2}\right) - \left(\alpha a_n^+ - \alpha^* a_n\right)\left(b_n^+ b_n + (-)^n \zeta\right)\right]\right\} \quad (25)$$

It shows clearly the mutual correlations among the one-phonon coherent state, polaronic state and two- phonon squeezed state, or speaking, the squeezing-antisqueezed effects appearing in the systems, which was caused by the variations of electron density and motion of the polarons and nonadiabatic phonon fluctuation in strongly and intermediately coupling regions in the systems. Evidently, for the new ansatz it has the relation of $D^{-1}D=I$, but the operators in the exponents in Eq.(25) are not commute mutually, i.e.,

$$[L_1\ L_2] = 2r\,|\alpha|\,[a_n^+ e^{i(\theta-\gamma)} - a_n e^{-i(\theta-\gamma)}](b_n^+ b_n + (-1)^n \zeta)$$

where

$$L_1 = [\alpha a_n^+ - \alpha^* a_n](b_n^+ b_n + (-1)^n \zeta),$$
$$L_2 = \xi^* a_n^2 - \xi a_n^{+2}, \alpha = |\alpha| e^{i\gamma}, \xi = re^{i\theta}.$$

Thus $D \neq U = U_S U_l\, U_C$ or $U_C\, U_l U_S$. Therefore, equation (25) is a new wavevector of state, which is completely different from U in Eq.(24) and has
more deeply physical-meanings.

2.2. The Uncertainty Relationship of Position and Momentum

Let us now transform the coordinate and the momentum operators

$$q_i = \sqrt{\hbar / 2M\omega}\left(a_n^\dagger + a_n\right) \text{ and } P_n = i\sqrt{\hbar M\omega / 2}\left(a_n^\dagger - a_n\right), \; i = \sqrt{-1} \tag{26}$$

where $\omega = \sqrt{W / M}$. It yields from Equation (18) that [37-44]

$$Dq_n D^{-1} = \sqrt{\frac{\hbar}{2M\omega}}\left(\left(a_m^+ + a_n\right)\cosh 2r + \left(a_n^+ e^{i\theta} + a_n e^{-i\theta}\right)\sinh 2r - \frac{|\alpha|}{r}\left[\left(\cosh 2r - 1\right)\cos\left(\theta - r\right) + \sinh 2r \cos r\right] b_n^+ b_n\right), \tag{27}$$

$$DP_n D^{-1} = i\sqrt{\frac{\hbar M\omega}{2}}\left(\left(a_n^+ - a_n\right)\cosh 2r - \left(a_n^+ e^{i\theta} - a_n e^{-i\theta}\right)\sinh 2r + i\frac{|\alpha|}{r}\left[\left(\cosh 2r - 1\right)\sin\left(\theta - r\right) + \sinh 2r \cos r\right] b_n^+ b_n\right) \tag{28}$$

If $|0\rangle_D$, is the ground state of an electron-phonon system, then for the mean square fluctuations in our state generator ansatz $|D\rangle = D|0\rangle_D$ one obtains

$$\langle D|\Delta^2 q|D\rangle = \frac{\hbar}{2Mw} e^{4r}\left(1 - \left(1 - e^{-8r}\right)\sin^2\left(\frac{1}{2}\theta\right) + \frac{|\alpha|^2}{r^2} e^{-4r}\left[\left(\cosh 2r - 1\right)\cos\left(\theta - r\right) + \sinh 2r \cos r\right]^2 \langle \Delta^2 n\rangle\right) \tag{29}$$

$$\langle D|\Delta^2 P|D\rangle = \frac{\hbar M\omega}{2} e^{-4r}\left(1 + \left(e^{8r} - 1\right)\sin^2\left(\frac{1}{2}\theta\right) + \frac{|\alpha|^2}{r^2} e^{4r}\left[\left(\cosh 2r - 1\right)\sin\left(\theta - r\right) + \sinh 2r \sin r\right]^2 \langle \Delta^2 n\rangle\right) \tag{30}$$

where

$$\langle \Delta^2 n\rangle = \langle n^2\rangle - \langle n\rangle^2 \text{ and } \langle n\rangle = \frac{1}{N}\sum_n {}_D\langle 0|b_n^\dagger b_n|0\rangle_D$$

The minimum uncertainty principle then yields

$$\langle\Delta^2P\rangle\langle\Delta^2q\rangle=\frac{\hbar^2}{4}\left(1+(e8r-1)\sin 2\left(\frac{1}{2}\theta\right)+\frac{|\alpha|^2}{r^2}e^{4r}\left[(\cosh 2r-1)\sin(\theta-r)+\sinh 2r\sin r\right]^2\langle\Delta^2n\rangle\right)$$

$$\times\left(1-(1-e^{-8r})\sin^2\left(\frac{1}{2}\theta\right)+\frac{|\alpha|^2}{r^2}e^{-4r}\left[(\cosh 2r-1)\cos(\theta-r)+\sinh 2r\cos r\right]^2\langle\Delta^2n\rangle\right) \quad (31)$$

We compare Equations (29) - (8, 29) with the respective expressions for the uncorrelated case in Equation (17),

$$\langle U|\Delta^2q|U\rangle=\frac{\hbar}{2Mw}e^{4r}\left(1-(1-e^{-8r})\sin^2\left(\frac{1}{2}\theta\right)+4|\alpha|^2e^{-4r}\cos^2 r\langle\Delta^2n_0\rangle\right) \quad (32)$$

$$\langle U|\Delta^2P|U\rangle=\frac{\hbar Mw}{2}e^{-4r}\left(1+(e^{8r}-1)\sin^2\left(\frac{1}{2}\theta\right)+4|\alpha|^2e^{4r}\sin^2 r\langle\Delta^2n_0\rangle\right) \quad (33)$$

where

$$\langle\Delta^2n_0\rangle=\langle n_0^2\rangle-\langle n_0\rangle^2 \text{ and } \langle n\rangle=\frac{1}{N}\sum_n {}_U\langle 0|b_n^\dagger b_n|0\rangle_U, |U\rangle=U|0\rangle_U$$

where

$$\langle\Delta^2n_0\rangle=\langle n_0^2\rangle-\langle n_0\rangle^2 \text{ and } \langle n\rangle=\frac{1}{N}\sum_n {}_U\langle 0|b_n^\dagger b_n|0\rangle_U, |U\rangle=U|0\rangle_U,$$

respectively. Through comparing Eqs.(32)-(33) with Eqs.(29)-(30) we find that the correlations between the displacement and the squeezing of phonons enhance the quantum uncertainties of the phonon coordinate and momentum due to their coupling with the electron density fluctuations. The term related with the electron density correlations in Eq. (33) vanishes for real $\alpha(r=0)$, while in the correlated case in Eq.(30) the contribution of this term is larger due to the correlations except if $\theta=0$. The absolute minimum of the r.h.s. of Eq. (31) occurs if the density fluctuation of electrons $\Delta^2 n$ is zero, $\langle\Delta^2P\rangle\langle\Delta^2q\rangle\to\frac{1}{4}\hbar^2$ (a harmonic oscillator), which is consistent with the results in nonlinear quantum mechanics[13]. An enhancement of fluctuations beyond harmonic oscillator has been found by Borghi et al.'s exact numerical diagonalization of a one-dimensional model of electrons coupled to phonons [45].

We again calculate the uncertainty relation of the phonons in the systems represented by Eq.(25). In this case, the mean square fluctuations of coordinate and momentum of the phonon in the new state wavefunction $|D\rangle = D|0\rangle_D$ are, respectively, denoted by

$$\langle D|\Delta q^2|D\rangle = \frac{\hbar}{2M\omega}e^{4r}[1-(1-e^{-8r})\sin^2\frac{\theta}{2}+\frac{|\alpha|^2}{r^2}(1-e^{-2r})^2\cos^2(\gamma-\theta)(\langle\Delta n^2\rangle+\zeta m_e-(-1)^i\zeta\langle n\rangle)] \quad (34)$$

$$\langle D|\Delta P^2|D\rangle = \frac{\hbar M\omega}{2}e^{-4r}[1+(e^{8r}-1)\sin^2\frac{\theta}{2}+ \frac{|\alpha|^2}{r^2}(e^{2r}-1)^2\sin^2(\gamma-\theta)(\langle\Delta n^2\rangle+\zeta m_e-(-1)^i\zeta\langle n\rangle)] \quad (35)$$

Thus from Equations (34)-(35) we can obtain the sizes of the uncertainty between the position and momentum for the phonons arising from the strong electron-phonon coupling in the systems to be as follows

$$Q=\langle D|\Delta q^2|D\rangle\langle D|\Delta P^2|D\rangle = \frac{\hbar^2}{4}\{1+\sinh^2(4r)\sin^2\theta+2|\alpha|^2(\langle\Delta n^2\rangle+\zeta m_e- (-1)^i\zeta\langle n\rangle).[(e^{2r}-1)/2r]^2[2e^{-4r}+\sinh(4r)(\cos^2(\gamma-\theta)-\sin^2\frac{\theta}{2})]+ \frac{|\alpha|^4}{r^4}\sinh^4(r)e^{4r}\sin^2[2(\gamma-\theta)][\langle\Delta n^2\rangle+\zeta m_e-(-1)^i\zeta\langle n\rangle]^2\} \quad (36)$$

$$\langle\Delta n^2\rangle = \langle n^2\rangle-\langle n\rangle^2, \langle n\rangle = \frac{1}{N}\sum_i {}_D\langle 0|b_i^\dagger b_i|0\rangle_D,$$

$$m_e = \frac{1}{N}\sum_i(-1)^i\langle b_i^+ b_i\rangle = \frac{1}{N}\sum_i(-1)^i\langle n_i\rangle \quad (37)$$

From Eq.(36) we see that the term related to the electron density fluctuation vanishes for real a (here $\gamma=0$), but corresponding term in Eq.(35) vanishes not in such a condition, except for $\theta=0$. The last terms in Eqs.(34)-(35) show that the correlation effect enhances the fluctuations of coordinate and momentum, when compared with that of uncorrelated case. This conclusion is also consistent with the result of exact numerical diagonalization of one–dimensional model for a coupled electron-phonon systems by Borghi et al. [45].

However, the corresponding representations for the uncorrelated case in Eq.(24) are as follows

$$\langle U|\Delta q^2|U\rangle = \frac{\hbar}{2M\omega}e^{4r}[1-(1-e^{-8r})\sin^2\frac{\theta}{2}+4|\alpha|^2e^{-4r}\cos^2\gamma(\langle\Delta n^2\rangle+\zeta m_e-(-1)^i\zeta\langle n\rangle)] \quad (38)$$

$$\langle U|\Delta P^2|U\rangle = \frac{\hbar M\omega}{2}e^{-4r}[1+(e^{8r}-1)\sin^2\frac{\theta}{2}+4|\alpha|^2e^{4r}\sin^2\gamma(\langle\Delta n^2\rangle+\zeta m_e-(-1)^i\zeta\langle n\rangle)] \quad (39)$$

Thus

$$Q_o = \langle U|\Delta q^2|U\rangle\langle U|\Delta P^2|U\rangle = \frac{\hbar^2}{4}\{1+\sinh^2(4r)\sin^2\theta+2|\alpha|^2(\langle\Delta n_0^2\rangle+\zeta m_e^0-$$

$$(-1)^i\zeta\langle n_0\rangle)[2e^{-4r}+\sinh(4r)(\cos^2(\gamma-\theta)-\sin^2\frac{\theta}{2})]+$$

$$|\alpha|^4\sin^2[2(\gamma-\theta)][\langle\Delta n_0^2\rangle+\zeta m_e^0-(-1)^i\zeta\langle n_0\rangle]^2\} \quad (40)$$

$$\langle\Delta n_0^2\rangle = \langle n_0^2\rangle - \langle n_0\rangle^2, \langle n_0\rangle = \frac{1}{N}\sum_i {}_U\langle 0|b_i^\dagger b_i|0\rangle_U,$$

$$m_e = \frac{1}{N}\sum_i(-1)^i\langle b_i^+ b_i\rangle = \frac{1}{N}\sum_i(-1)^i\langle n_0\rangle \quad (41)$$

where m_e in Eqs,(37) and (41) are the order parameter of CDW, $\langle\cdots\cdots\rangle$ represent the average over the ground state of the systems in correlated and uncorrelated cases, respectively. Obviously, Eqs.(29)-(31) and Eqs.(34)-(37) differ from Eqs.(32))-(33) and Eqs.(38)- (41), respectively.

Thus we can obtain from the above comparison in them that the uncertainty representations of coordinate and momentum of the phonons are enhanced due to the mutual correlation and squeezing-antisqueezing effects between the phonons and polarons. The difference of the uncertainty between the correlated and uncorrelated case can be obtained from Eqs.(36) and (40), which is

$$\Delta Q = Q - Q_0 = \frac{\hbar^2}{4}\{|\alpha|^2[2e^{-4r}+\sinh(4r)][\langle\Delta n^2\rangle[(e^{2r}-1)/2r]^2-\langle\Delta n_0^2\rangle]+ \quad (41)$$
$$\zeta[m_e[(e^{2r}-1)/2r]^2-m_m^0]-(-1)^i\zeta[\langle n\rangle((e^{2r}-1)/2r)^2-\langle n_0\rangle]\}$$

Therefore, $\Delta Q = Q - Q_0 \neq 0$, i.e., the sizes of the uncertainty are different between correlated and uncorrelated state-ansatzs due to the squeezing-antisqueezing effect $(e^{2r}-1)/2r > 1$.This manifests that the squeezing-antisqueezing effect results in quantum fluctuations of the phonons.

From the above studies we can gain the following properties of the uncertainty relation of the phonons for both the correlated and uncorrelated state ansatzs in Eqs. (17)-(18) and Eqs.(24)-(25) in strong electron-phonon coupling systems.

(1) Their sizes are not a constant, but changed with varying characteristic parameters of the systems, i.e., they are increased with decreasing nonlinear interaction λ^2 and increasing phonon energy $\hbar\omega$, but approach the minimum value $\hbar^2/4$, when λ^2 is quite large and

$\hbar\omega$ is very small. This means that the strong coupling between the electron and phonon can suppress the quantum uncertainty and fluctuations of the phonons, thus the localized features of the phonons are enhanced in this case.

(2) Only if the density fluctuation of the electron, average values of electron density and order parameter of CDW vanish simultaneously as well as ζ and α are all real, then the absolute minimum of the r. h. side of Eqs.(34)-(35) appears. Thus we can obtain the minimum uncertainty relation, $<\Delta q^2><\Delta P^2> = \hbar^2/4$, in this case.

(3) For the case of one-phonon coherent state (i.e., r=0, $\theta = 0, \gamma = 0$ and $\alpha = 0$), we can obtain from Eq.(36) orEq.(40) that $<\Delta q^2><\Delta P^2> = \hbar^2/4$. For the pure squeezing state of phonon (i.e.,, $\theta = 0, \gamma = 0, \zeta = 0$ and $\alpha = 0$), we got $\langle \Delta q^2 \rangle = (\hbar / 2M\omega)\exp[4r]$ and $\langle \Delta P^2 \rangle = (\hbar M\omega / 2)\exp[-4r]$, thus $<\Delta q^2><\Delta P^2> = \hbar^2/4$. Meanwhile, when the one-phonon coherent and two-phonon coherent states occur simultaneously in the systems, then the minimum uncertainty relation exists also at $\theta = 0$.

(4) From Eqs.(34)-(41) we can obtain that the values of uncertainty for the phonons in the systems, in which the electrons have not spin, are large than that for spin -1/2, when λ^2 and $\hbar\omega$ are same. This means that the electrons with spins can cause reduction of the uncertainty of the phonons, thus the localized feature of the phonons is enhanced in this case.

(5) From Eqs. (34)-(41) we see that the squeezing-antisqueezing effect changes the anormalous fluctuation of the coordinate and momentum of lattice oscillator. However, for real ξ and α, the electron density fluctuations do not contribute to the momentum uncertainty. This is favourable for the physical well-founding of the present ansatz.

In order to understand clearly the physical meanings of new state-ansatz in Eq.(25) we , through calculation, factorize it as following form.

$$D = \exp\sum_n (\xi^* a_n^2 - \xi a_n^{+2}) \exp\sum_n -\{\frac{|\alpha|}{2r}(e^{2r}-1)[\cos(\gamma-\theta) + ie^{2r}\sin(\gamma-\theta)](a_n^+ - a_n) b_n^+ b_n\}.$$
$$\exp\sum_n -(-1)^n \{\frac{|\alpha|}{2r}(e^{2r}-1)[\cos(\gamma-\theta) + ie^{2r}\sin(\gamma-\theta)](a_n^+ - a_n)\zeta\} \tag{42}$$

If Eq.(42) is compared with Eq.(24) we find that the parameter denoting the feature of one- phonon coherent state, ζ, and the parameter exhibiting the characteristic of the polaron effect, $a\zeta$, are all related with the squeezed factor of the phonons, ξ, which are replaced by

$$\Gamma = \frac{|\alpha|}{2r}(e^{2r}-1)[\mathrm{con}(\gamma-\theta) + ie^{2r}\sin(\gamma-\theta)]$$

and $\Gamma\zeta$, respectively. In this case the polaronic and one-phonon coherent states are all affected by the antisqueezed effect of the phonons. Therefore, Eq.(25) represents exactly the squeezing-antisqueezed effect of the phonons and polarons in the coupled electron-phonon systems. Thus we refer to Eq.(25) as the correlated ansatz of state. A straightforward result, arising from this ansatz, is that the binding energy of the polaron is changed to $\bar{E}_p = -|\alpha|^2 \Gamma^2 \hbar\omega$ from $E_p = -|\alpha|^2 \hbar\omega$ in the uncorrelated case. When *a and* ζ are real, then $\bar{E}_p = -|\alpha|^2 (\frac{e^{2r}-1}{2r})^2 \hbar\omega$, hence $|\bar{E}_p| > |E_p|$ due to r>0 and $(\frac{e^{2r}-1}{2r}) > 1$. Thus the stability of the polaron is significantly enhanced in such a case, when compared it with the uncorrelated case. Therefore, we can conclude that the new ansatz mentioned above result in considerable changes of property of the systems.

3. Properties of Nonadiabatic Quantum Fluctuations for Phonon and Electron

3.1. Changes of Properties of Ground State of the Systems

As same as the squeezed effect of the phonons, its antisqueezed effect can also result in a lot of new phenomena to occur in the systems. We now study the squeezing-antisqueezed effect on the properties of the ground state of the systems by the Pang's ansatz[37-44].

The Holstein model in Eq.(16) describes a one-dimensionally and linearly coupled electron- phonon systems, in which the total number of electrons with spinless is N/2 for the half-filling band and, where N is the total number of lattice sites. It is obvious that in Eq.(16) the equilibrium position of every oscillator depends on whether there is an electron on it or not. If n=1, the equilibrium position of q_i is $\lambda/2N$, otherwise it is $-\lambda/2N$. Under the standard transformation in Eq.(26), the Hamiltonian in Eq.(16) is replaced by

$$H = \frac{N}{2}\hbar\omega + \sum_n \hbar\omega a_n^+ a_n - \sum_n \lambda\left(\frac{\hbar}{2M\omega}\right)^{1/2}(a_n^+ + a_n)(b_n^+ b_n - \frac{1}{2}) - t\sum_n (b_n^+ b_{n+1} + b_{n+1}^+ b_n) \quad (43)$$

Pang introduce three new parameters, r, f and m_0 into the Equation (25) [40], then Equation (25) becomes as

$$D = \exp\left\{\sum_n\left[(\xi^* a_n^2 - \xi a_n^{+2}) - (\alpha a_n^+ - \alpha^* a_n)(f(b_n^+ b_n - \frac{1}{2}) + (-1)^i \zeta m_0)\right]\right\} \quad (44)$$

The variational parameters *r*, *f*, m_0 represent the degree of squeezing of the lattice phonons, the size of phonon cloud brought by the electron, or the degree of polaron effect, the magnitude of phonon-staggered ordering or staggered lattice deformation in the squeezing-antisqueezed cases, respectively.

We can find out the change of energy of ground state of the systems in the this case, which is as follows

$$E_g(f,\rho)/N = {}_{ph}\langle 0|DHD^{-1}|0\rangle_{ph}/N = \frac{\hbar\omega}{4}(\rho^2+\rho^{-2}) - \frac{\lambda^2}{8W}f(\frac{e^{2r}-1}{2r})[2-f(\frac{e^{2r}-1}{2r})] +$$
$$\frac{\lambda^2}{2}(\frac{e^{2r}-1}{2r})^2 Wm_o^2 - \frac{1}{\pi}\int_0^{\pi/2}\{4\beta^2 t^2\cos^2 k + (\frac{e^{2r}-1}{2r})^2[2-f(\frac{e^{2r}-1}{2r})]^2 m_0^2\}^{1/2} \quad (45)$$

where

$$\beta(f,\rho) = \exp\left[-\frac{\lambda^2}{2W\hbar\omega}f^2\rho^2\left(\frac{e^{2r}-1}{2r}\right)^2\right] \quad (46)$$

is Debye-Waller factor, $|0\rangle_{ph}$ is the vacuum state of the phonon, the variational parameters $\rho=e^{-2r}$, f and m_0 can be determined self-consistently from

$$\frac{\partial E_g}{\partial\rho} = \frac{\partial E_g}{\partial f} = \frac{\partial E_g}{\partial m_o} = 0, \quad (47)$$

with,

$$m_e = Wm_o[(e^{2r}-1)/2r]\{[1-f((e^{2r}-1)/2r)]\lambda\}^{-1}, \quad (48)$$

where $0\le f\le 1$ and $0\le\rho\le 1$. We used the relation, $(n_i-\frac{1}{2})(n_i-\frac{1}{2})=\frac{1}{4}$, in calculation in the spinless case. Then, the properties of ground state of the systems are determined by the phonon energy $\hbar\omega$ and polaron's binding energy $E_p=\lambda^2/2W$ and time t.

As far as the energy of stable ground state in the variational method is concerned, we first should determine the minimum value of energy of ground state by adjusting the variational parameters, f and ρ, to their optimum values. For each f in the ranges of $0\le f\le 1$ we can determine a local minimum values of $E_g(f,\rho)$ by adjusting ρ to its local optimum value. Thus we can also get the curves of f-dependence of ρ. Figure1 shows the relation of $E_g(f,\rho)/N$ in Eq.(45) versus f and ρ. From this figure we know that the energy of ground state is negative in the regions of $0.2\le f\le 1$and $0.8\le f\le 1$. This is an interesting result. Through again the comparison of these local optimum values for the energies of ground state, we can determine the absolute minimum values of energy of ground state, $E_g(f_m, \rho_m)/N$, and corresponding optimum values of these variational parameters, f_m and ρ_m. Using this way we find finally out the absolute minimum value of energy of stable ground-state, which is $E_g(f_m,\rho_m)/N = -0.2659$ at $f_m=0.42$ and $\rho_m= 0.913$ in the cases of $J=\lambda/2Wt=4$, $\hbar\omega=2$, $W=0.25$ and $t=1$. In Fig.1 we give also the dependences of energy of ground state, $E'_g(f', \rho')/N$, in uncorrelated case in Eq.(2) *on* f and ρ in the dashed lines. From this figure we see clearly that the absolute minimum value is $E'_m(f'_m,\rho'_m)/N = -0.208$ at $f'_m=0.56$, $\rho'_m=0.951$. Obviously, this value is larger than $E_g(f_m,\rho_m)/N$. This means that the correlation and antisqueezed effects of one-phonon coherent state, polaronic state and two-phonon squeezed state represented by the new ansatz in Eq. (44) or Eq.(25) result exactly in the dropping of energy of ground state of the systems.

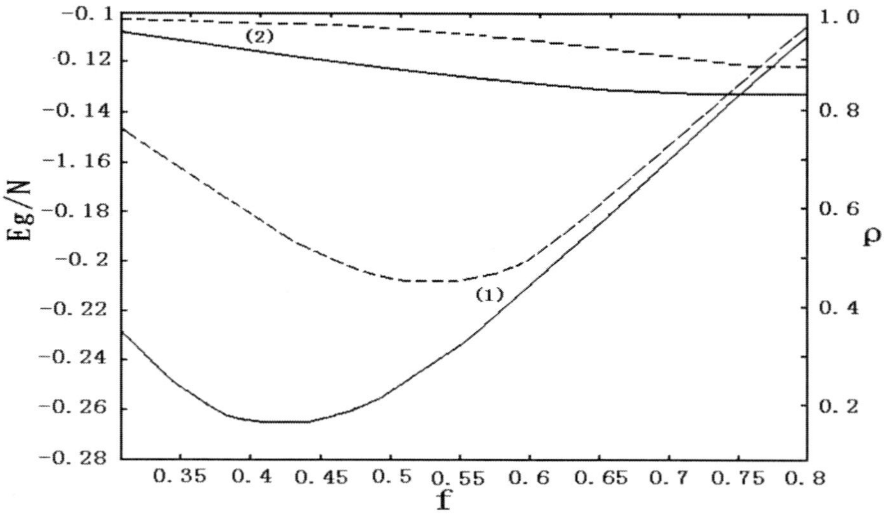

Figure 1. the dependences of Eg(fm , ρm)/N (real line) and E'g(f,' ρ')/N (dashed lines) on f and ρ, respective

In such a case we can also find out the dependence of $E_g(f_m , \rho_m)$/N on the $\hbar\omega$ and J, which is shown in Fig.2, where W=0.25, t=1. We see from this figure that $E_g(f_m , \rho_m)$/N increases with increasing $\hbar\omega$ and decreasing J. In order to understand further the features of correlation and squeezing- antisqueezed effects, we gave the changes of energy of ground state, resulting from these effects, $\Delta E_g / N = E'_g(f_m^0, \rho_m^0)/N - E_g(f_m, \rho_m)/N$ versus the $\hbar\omega$ and J relation, which is shown in Fig.3, where $E'_g(f'_m, \rho'_m)/N = \langle 0|UHU|0\rangle / N$ is the energy of ground state in the uncorrelated case. From this figure we see that there are always ΔE_g /N >0 for all values of J and $\hbar\omega$. This shows that the squeezing -antisqueezed effect results always in the decreases of energy of ground state of the systems as compared with that in the squeezed case due to the competition between the grown quantum fluctuations of the phonons and reduction of binding-energy of localized polaron. Thus we can conclude that the correlation and squeezing- antisqueezed effects can enhance the stability of the system. Then we can decide that the new ansatz is relevant and appropriate to the systems.

Also, we see from the above result that the polaron binding-energy, E_p, increases to

$$\frac{-\lambda^2}{8W}\left(\frac{e^{2r}-1}{2r}\right)f\left[2-\left(\frac{e^{2r}-1}{2r}\right)f\right]$$

from $\frac{-\lambda^2}{8W} f'(2-f')$ due to the squeezing-antisqueezed effect. Thus, the stability of the polaron is also enhanced because [(exp(2r)-1)/2r]>1.

However, the Debye-Waller factor, β, measuring the polaron narrowing of electron band, is increased due to the squeezing-antisqueezed effects, i.e., $\Delta\beta = \beta(f_m, \rho_m) - \beta'(f'_m, \rho'_m) > 0$, in such a case, where $\beta^{'}(f_m^{'}, \rho_m^{'}) = \exp\left[\frac{-\lambda^2 \rho_m^{'\,2} f_m^{'\,2}}{2\hbar\omega\, W}\right]$ is the Debye-Waller factor in uncorrelated case in Eq.(24).

In Fig.4 we plot the $\Delta\beta$ versus J and $\hbar\omega$ functional relation, This result shows that the increase of classical polaron narrowing of electron band, arising from the squeezed effect of the phonons, are also suppressed due to the antisqueezed effect.

Thus, the quantum tunneled effect of the polarons is also weakened. This shows that the correlation and squeezing- antisqueezed effects are not favorable to increases of quantum tunneling effect and dimerization of the polarons. This is an interesting result in physics.

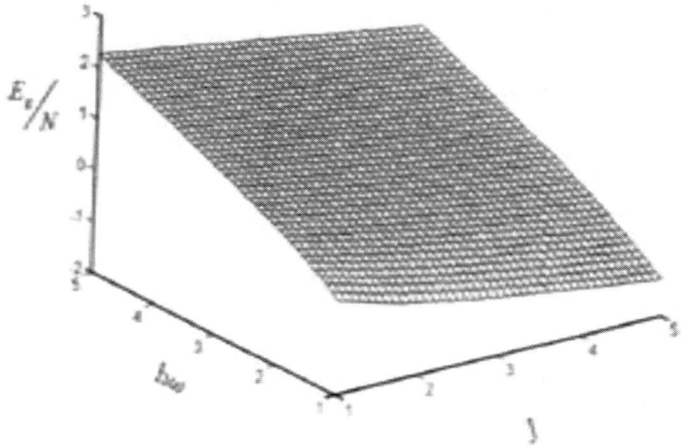

Figure 2. Changes of Eg(fm , ρm)/N with $\hbar\omega$ and J.

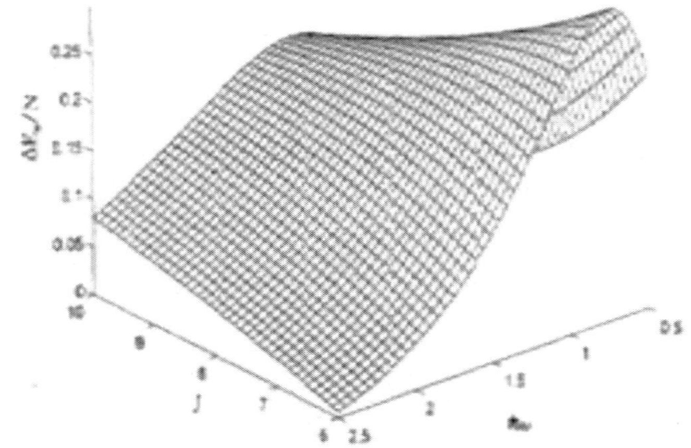

Figure 3. Dependence of ΔEg /N on J and $\hbar\omega$.

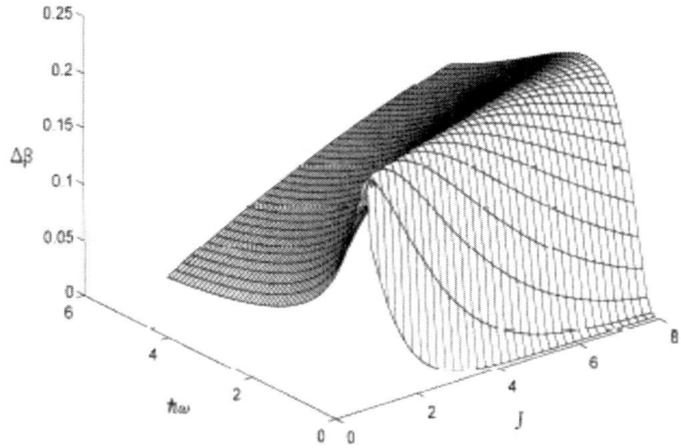

Figure 4. Dependences of Δβ on J and $\hbar\omega$.

3.2. Quantum Fluctuations of Phonons and CDW Order Effects

We now determine the phonon-staggered ordering parameter m_p, which can be determined in experiments. It is defined as

$$m_p = \frac{1}{N}\sum_n (-1)^n \langle q_n \rangle ,$$

where q_n is the position operator of the phonon mode and $\langle \cdots\cdots \rangle$ represents an average over the ground state of the systems. After finishing an unitary transformation of the new ansatz and making an average over the vacuum state of transformed phonon subsystem we can easily obtain $m_p = \lambda m_e / W$. Since $m_0 = \lambda m_e \{1 - f[(e^{2r}-1)/2r]/[(e^{2r}-1)/2r]/W$, then $m_p > m_0$. This means that m_p contains the contribution of polaron effect, arising from the antisqueezed effect, apart from the static staggered ordering parameter m_0. We find that the relation between the phonon- staggered ordering parameter m_p and CDW ordering parameter m_e, which is the same with that in the adiabatic limit and uncorrelated case. Meanwhile, we see from the above representations that m_p and m_e increase simultaneously with $\hbar\omega$, but the m_p increases and m_e decreases, when λ increases.

Since the lattice deformation which follows the electronic motion is proportional to the size of charge density, then it fluctuates as the electrons move. Thus, the anomalous fluctuation appeared are more larger than the zero-point fluctuations if a or $J > 1$. The fluctuation in every phonon mode q_n can be represented by $\langle \Delta q_n^{\ 2} \rangle = \langle q_n^{\ 2} \rangle - \langle q_n \rangle^2$. If the average is done over the ground state of harmonic oscillator with $\omega = \sqrt{W/M}$, then $\langle \Delta q_n^{\ 2} \rangle_h = \dfrac{\hbar}{2M\omega}$. When its average is found over the new ansatz in Eq.(44) in the coupled electron-phonon system at $\theta = 0$ and $\gamma = 0$, then the quantum fluctuation of phonon mode can be represented by

$$R = \langle \Delta q_n^2 \rangle - \langle \Delta q_n^2 \rangle_h = \frac{\hbar}{2M\omega}(\rho^{-2} - 1) + (\frac{\lambda}{2W})^2 f_m^2 (\frac{1-\rho_m^{-1}}{\ln \rho_m})^2 (1 - 4m_e^{\ 2}) \quad (49)$$

where first term $(\rho^{-2} - 1)$ in Eq .(49) exhibits the squeezed effect, the second term shows the anomalous fluctuation due to the polaron motion, electronic density fluctuation and antisqueezed effect. From the above representations we know that the m_e decreases with decreasing λ^2 or J and increasing $\hbar\omega$, but the m_p increases with increasing λ^2 or J and $\hbar\omega$. However, there is a nonmonotonic up-down-up tendency in the $\hbar\omega$-dependences of quantum fluctuation of phonon mode, R, i.e., R decreases with increasing $\hbar\omega$, and reaches a minimum at mid value of $\hbar\omega$, subsequently it is risen again with increasing $\hbar\omega$. On the other hand, R increases slowly with increasing λ, subsequently it is risen fast with increasing λ. This means that the weakened effect of larger ordering of the system is now suppressed by the fluctuation of phonons. This is an new result, which differs from uncorrelated case. In order to verify this conclusion, we have to find out the differences of the above physical quantities between the correlated ansatz, Eq.(25), and uncorrelated ansatz, Eq.(24), i.e.,

$$\Delta m_e = m_e(f_m, \rho_m) - m_e'(f_m', \rho_m'),$$

$$\Delta m_e = m_e(f_m, \rho_m) - m_e'(f_m', \rho_m'),$$

$$\Delta R = R(f_m, \rho_m) - R'(f_m', \rho_m'),$$

where the $m_e(f_m', \vartheta'_m), m_p'(f_m', \rho_m'), R'(f_m', \rho_m')$ are corresponding values in uncorrelated case in Eq.(24). These results are shown in Figs. 5, 6 and 7, respectively. From these figures we see clearly that there are always the relations of $\Delta m_e>0$, $\Delta m_p>0$ and $\Delta R>0$ in the correlation and squeezing-antisqueezed case, when compared with those in the squeezed and uncorrelated case due to the increases of CDW ordering and quantum fluctuations of phonons, resulting from the correlation and squeezing- antisqueezed effects. This shows that the increases of quantum fluctuation of the lattice phonons, arising from the squeezed effect of the phonons, are now suppressed due to the antisqueezed effect. This is a very interesting result in physics, which is worth studying deeply and further.

We also study the influences of correlation and squeezing-antisqueezed effects on the properties of the coupled electron-phonon systems with a half-filled band by using a new state ansatz and variational method[37-44]. In this case we introduce successively three variational parameters, m_0, f and r (or ρ), which describe the staggered lattice deformation without any polaron effect, the size of lattice deformation following instantaneously the electron (polaron effect) and the squeezing effect of multiphonons due to the electron band motion and its density fluctuation, respectively. The f indeed measures the on-site correlation

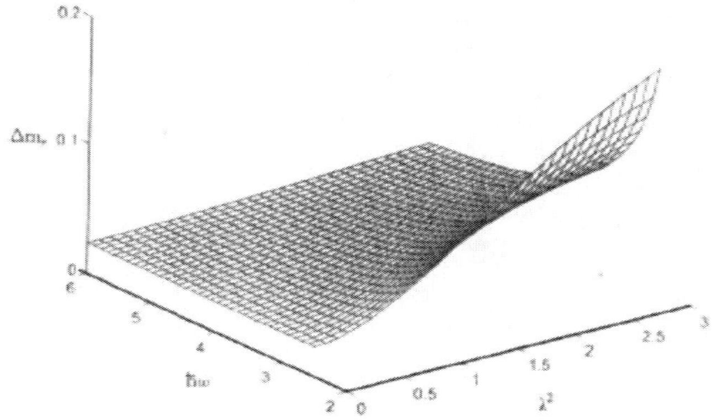

Figure 5. Dependences of Δme on λ2 and $\hbar$ ω.

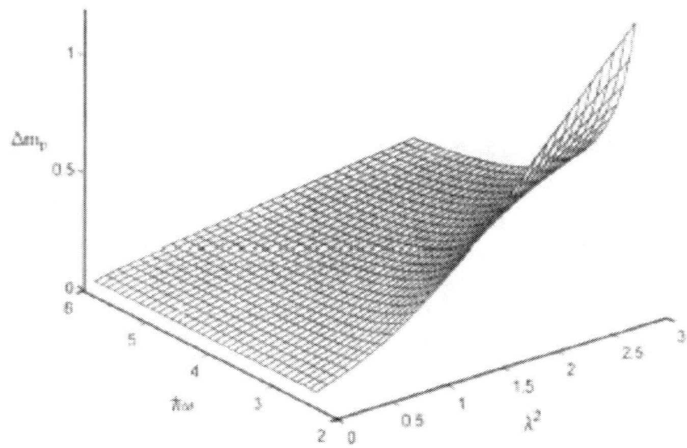

Figure 6. Changes of Δmp with λ2 and $\hbar$ ω.

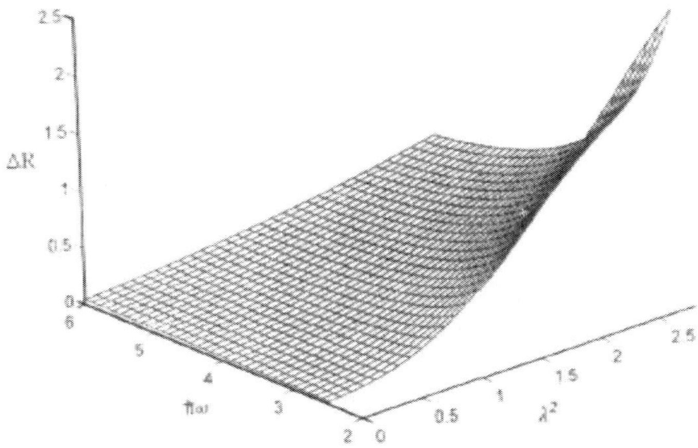

Figure 7. The influences of λ2 and $\hbar$ ω on ΔR.

between the electronic density and lattice deformation. While the squeezed and antisqueezed effects are purely a dynamical effect associated with the nonadiabatic and anomalous quantum fluctuations, resulting from the changes of electronic density and motions of the polarons. The new state-ansatz proposed by Pang shows accurately the influences of electron motion on the multiphonons (which causes the squeezed effect of multiphonons) and of antisqueezed effect of the phonons on the polaron band and one-phonon coherent state. Thus some new phenomena are obtained, which are as follows:

(1) The energy of ground state of the systems drops greatly. Then the enhancement of phonon displacement due to the correlation is responsible for the stabilization of strongly coupled polarons, when compared with the uncorrelated case. The increase of quantum zero fluctuation in Fig.3 is overcome by the growth of negative contribution of polaron binding energy. These contributions become more relevant for strong electron-phonon coupled systems with large λ.

(2) The binding energy of the polaron is increased, but the increases of band narrowing and tunneling effect of the polarons determined by the Debye-Waller factor, $\beta'(f'_m, \rho'_m) = \exp\left[-\lambda^2 \rho'^2_m f'^2_m / 2\hbar\omega W\right]$, are suppressed due to the antisqueezed effect, when compared with the uncorrelated case. This implies that the following phenomena occur. (a)The correlation effect induces the increase of quantum enhancement of the polaron pairing due to the squeezing-antisqueezed effect. Thus, the probability of the bipolarons, which are formed in this process, increases, then the two-polaron resonance effect is also enhanced in such a case. (b) The additional quantum tunneling effect of the systems increases[31-32]. (c) the dimerization order parameters[18-20,5-6], for example, the phonon staggered parameter m_p,.is enhanced.

(3) The correlation effect enhances the fluctuations of coordinate and momentum , or speaking, the quantum uncertainty of the phonons is enhanced, when compared with that in the uncorrelated case due to the enhancement of quantum fluctuations of the phonons. Thus a new uncertainty relation occurs.

(4) The CDW ordering parameter m_e is reduced, the staggered ordering parameter m_p are suppressed in this case.

4. The Properties of Soliton Motion of Electrons in Superconductors

As it is known, the superconductivity of material is caused by the electron-phonon interaction in the condition of low temperature [4,46-49]. Therefore, superconductor is a electron-phonon coupled system, in which the electrons become as Cooper pairs or superconductive electrons. In this section we discuss the nonlinear quantum-mechanical properties of motion of superconductive electrons in superconductors[13]. The study presented shows that the superconductive electrons move in the form of a soliton, which can just result in the superconductivity without electric resistance and a series of macroscopic quantum effects in the superconductors. Therefore, the properties and motions of the electron-soliton are not only important for understanding the essences and rule of superconductivity and macroscopic quantum effects, but also can verify the correctness of nonlinear quantum mechanics[13]. Thus it is worth to study the properties of the electron-soliton.

4.1. The Properties of Soliton Motion of Superconducting Electrons in Steady States without Electromagnetic-Field

It is known from BCS theory of superconductivity [4, 46-49], the states of the electrons in the superconductor are often represented by a macroscopic wave function,

$$\phi(\vec{r},t) = f(\vec{r},t)\phi_0 e^{i\theta(\vec{r},t)}, \text{ or } \phi = \sqrt{\rho}e^{i\theta},$$

which is same with Equation (1.21), where $\phi_0^2 = \alpha/2\lambda$. Landau et al. [50, 51] used the wave function to give the free energy density function, *f*, of a superconducting system, which is represented by

$$f_s = f_n - \frac{\hbar^2}{2m}|\nabla\phi|^2 - \alpha|\phi|^2 + \lambda|\phi|^4 \tag{50}$$

in the absence of any external field. If the system is subjected to an electromagnetic field specified by a vector potential $\vec{A}$, the free energy density of the system is of the form:

$$f_s = f_n - \frac{\hbar^2}{2m}\left|(\nabla - \frac{ie^*}{c\hbar}\vec{A})\phi\right|^2 - \alpha|\phi|^2 + \lambda|\phi|^4 + \frac{1}{8\pi}\vec{\mathrm{H}}^2 \tag{51}$$

where $e^*=2e$, $\vec{\mathrm{H}} = \nabla\times\vec{A}$, α and λ are some interactional constants related to the features of superconductor, m is the mass of electron, e^* is the charge of superconductive electron, c is the velocity of light, h is Planck constant, $\hbar = h/2\pi$, f_n is the free energy of normal state. The free energy of the system is $F_s = \int f_s d^3x$. In terms of the conventional field, $F_{jl} = \partial_j A_l - \partial_l A^j$, ($j, l$=1, 2, 3), the term $\vec{\mathrm{H}}^2/8\pi$ can be written as $F_{jl}F^{jl}/4$.

Equations (50) - (51) show the nonlinear features of the free energy of the systems because it is the nonlinear function of the wave function of the particles, $\phi(\vec{r},t)$. Thus we can predict that the superconductive electrons have many new properties relative to the normal electrons. From $\delta F_s/\delta\phi = 0$ we get

$$\frac{\hbar^2}{2m}\nabla^2\phi - \alpha\phi + 2\lambda\phi^3 = 0 \tag{52}$$

and

$$\frac{\hbar^2}{2m}(\nabla - \frac{ie^*}{c\hbar}\vec{A})^2\phi - \alpha\phi + 2\lambda\phi^3 = 0 \tag{53}$$

in the absence and presence of an external fields respectively, and the number current is of form

$$\vec{J} = +\frac{e^*\hbar}{2mi}(\phi^*\nabla\phi - \phi\nabla\phi^*) - \frac{e^*}{mc}|\phi|^2\vec{A} \tag{54}$$

Equations (52) - (54) are just well-known the Ginzburg-Landau (GL) equation [52-58] in a steady state, and in essence a time-independently nonlinear Schrödinger equation. Thus the superconductors are nonlinear quantum-mechanical systems, in which the properties of superconductive electrons can be completely described by nonlinear quantum-mechanics[13].

The above Ginzburg-Landau equations are fundamental equations of the superconductors, describing the motion of the superconductive electrons, in which there is the nonlinear interaction term of $2\lambda\phi^3$, arising from the electron-phonon interaction. However, the equations contain two unknown functions ϕ and $\vec{A}$ which make them extremely difficult to resolve.

We first study the properties of motion of superconductive electrons in the case of no external field. Then, we consider only a one-dimensional pure superconductor [62-63], where

$$\phi = \phi_0\varphi(x,t),\ \xi'^2(T) = \hbar^2/2m|\alpha|,\ x' = x/\xi'(T) \tag{55}$$

and where $\xi'(T)$ is the coherent length of the superconductor, which depends on temperature. For a uniform superconductor, $\xi'(T) = 0.94\xi_0[T_c/(T_c - T)]^2$, where T_c is the critical temperature and ξ_0 is the coherent length of superconductive electrons at T=0. In boundary conditions of $\varphi(x'=0)=1$, and $\varphi(x' \to \pm\infty)=0$, from Eqs. (52) and (54) we find easily its solution of

$$\varphi = \pm\sqrt{2}\sec h\left[\frac{x - x_0}{\xi'(T)}\right] \tag{56}$$

or

$$\phi = \pm\sqrt{\frac{\alpha}{\lambda}}\sec h[\frac{x - x_0}{\xi'(T)}] = \pm\sqrt{\frac{\alpha}{\lambda}}\sec h[\frac{\sqrt{2m\alpha}}{\hbar}(x - x_0)] \tag{57}$$

This is a well-known wave packet-type soliton solution. Therefore, the superconductive electrons have all properties of microscopic particles described by nonlinear quantum mechanics[13].

It can be used to represent the bright soliton occurred in the Bose-Einstein condensate found by Perez-Garcia et al. [59].

The energy of the soliton, (57), is given by

$$E_{so1} = \int_{-\infty}^{\infty} H'dx = \int_{-\infty}^{\infty}\left[\frac{\hbar^2}{2m}(\frac{d\phi}{dx})^2 - \alpha\phi^2 - \lambda\phi^4\right]dx = \frac{4\hbar\alpha^{3/2}}{3\lambda\sqrt{2m}} \quad (58)$$

We assume here that the lattice constant, $r_0=1$. The above soliton energy can be compared with the ground state energy of the superconducting state, $E_{\text{ground}}=-\alpha^2/4\lambda$. Their difference is

$$E_{so1} - E_{\text{ground}} = \alpha^{3/2}\left(\sqrt{\alpha} + \frac{16\hbar}{3\sqrt{2m}}\right)/2\lambda > 0$$

This indicates clearly that the soliton is not in the ground state, but in an excited state of the system, therefore, the soliton is a quasiparticle.

From the above discussion, we can see that, in the absence of external fields, the motion of superconductive electrons can be described by a nonlinear Schrödinger equation, its essence is a soliton in a uniform system.

The soliton is formed by a nonlinear interaction arising from the electron-phonon interaction, which suppresses the dispersive behavior of electrons.

The solitons with certain charges can move over a macroscopic distances retaining its shape, amplitude and energy along the superconductors, thus a superconducting phenomenon without resistance occurs in the systems.

4.2. The Features of Soliton Motion of Superconducting Electrons in Steady States with Electromagnetic-Field and Its Comparison with Experiments

We now consider the features of motion of superconductive electrons in the presence of an electromagnetic field $\vec{A}$; its equation of motion is denoted by Eqs. (53)-(54) in this case. Assuming now that the field $\vec{A}$ satisfies the London gauge $\nabla\cdot\vec{A}=0$. If the substitution of $\phi(\vec{r},t)=\varphi(\vec{r},t)\phi_0 e^{i\theta(\vec{r})}$ into Eqs. (53) and (54) we can get [61-62]:

$$J=\frac{e^*\phi_0^2}{m}(\hbar\nabla\theta-\frac{e^*}{c}\vec{A})\varphi^2 \quad (59)$$

and

$$\nabla^2\varphi-[(\nabla\theta-\frac{e^*}{\hbar c}\vec{\mathbf{A}})^2\varphi]-\frac{2m}{\hbar^2}(\alpha-2\lambda\phi_0^2\varphi^2)\varphi=0 \quad (60)$$

For bulk superconductors, J is a constant (permanent current) for a certain value of $\vec{A}$, and it thus can be taken as a parameter. Let $B^2=m^2J^2/\hbar^2(e^*)^2\phi_0^4$, $b=2m\alpha/\hbar^2=\xi'^{-2}$, from Equations (59) and (60), we can obtain [66-67]:

$$(\hbar\nabla\theta-\frac{e^*}{c}\vec{\mathbf{A}})=\frac{Jm}{e^*\phi_0^2\varphi^2} \quad (61)$$

$$\frac{d^2\varphi}{dx^2}=-\frac{d}{d\varphi}U_{\text{eff}}(\varphi),U_{\text{eff}}(\varphi)=\frac{B^2}{2\varphi^2}-\frac{1}{2}b\varphi^2+\frac{1}{4}b\varphi^4 \quad (62)$$

where U_{eff} is the effective potential of the superconductive electron in this case and it is schematically shown in Fig.8. Comparing Eqs.(61)-(62) with those in the absence of external fields we found that the equations of the superconductor have the same form and the electromagnetic field changes only the effective potential of the superconductive electron. When $\vec{A}=0$, the effective potential well is characterized by double wells. In the presence of an electromagnetic field, there are still two minima in the effective potential, corresponding to the two ground states of the superconductor in this condition. This shows that the spontaneous breakdown of symmetry occurs in the superconductor, thus the superconductive electrons also move in the form of solitons. To obtain the soliton solution, we integrate Eq. (62) and can get:

$$x=\int_{\varphi_1}^{\varphi}\frac{d\varphi}{\sqrt{2[E-U_{\text{eff}}(\varphi)}} \quad (63)$$

where E is a constant of integration which is equivalent to the energy, the lower limit of the integral, φ_1, is determined by the value of φ at x=0, i.e. $E=U_{\text{eff}}(\varphi_0)=U_{\text{eff}}(\varphi_1)$. Introduce the following dimensionless quantities $\varphi^2=u$, $E=\frac{b}{2}\varepsilon$, $2\tilde{d}=\frac{4J^2m\lambda}{(e^*)^2\alpha^2}$, and equation (63) can be written as the following upon performing the transformation $u\to-u$,

$$-\sqrt{2b}x=\int_{u_1}^{u}\frac{du}{\sqrt{u^3-2u^2-3\varepsilon u-2\tilde{d}^2}} \quad (64)$$

It can be seen from Figure 8 that the denominator in the integrand in Equation (64) approaches zero linearly when $u=u_1=\varphi_1^2$, but approaches zero gradually when $u=u_2=\varphi_0^2$. Thus we give [61-62]

$$u(x)=\varphi^2(x)=u_0-g\sec h^2\left(\sqrt{\frac{1}{2}}gbx\right)=u_1+g\tan h^2\left(\sqrt{\frac{1}{2}}gbx\right) \quad (65)$$

where $g=u_0-u_1$ and satisfies

$$(2+g)^2(1-g)=27\tilde{d}^2\text{, }2u_0+u_1=2\text{, }u_0^2+2u_0u_1=-2\varepsilon\text{, }u_1u_0^2=2\tilde{d}^2 \quad (66)$$

It can be seen from Equation (65) that for a large part of sample, u_1 is very small and may be neglected; the solution u is very close to u_0. We then get from Equation (65) that

$$\varphi(x) = \varphi_0 \tan h\left(\sqrt{\frac{1}{2}}gbx\right) \tag{67}$$

Substituting the above into Equation (61), the electromagnetic field $\vec{A}$ in the superconductors can be obtained

$$\vec{A} = -\frac{\vec{J}mc}{(e^*)^2\phi_0^2}\frac{1}{\varphi^2} - \frac{\hbar c}{e^*}\nabla\theta = \frac{\vec{J}mc}{(e^*)^2\phi_0^2\varphi_0^2}\cot h\left(\sqrt{\frac{1}{2}}gbx\right) - \frac{\hbar c}{e^*}\nabla\theta$$

For a large portion of the superconductor, the phase change is very small. Using $\vec{H} = \nabla\times\vec{A}$ the magnetic field can be determined and is given by [60-61]

$$\vec{H} = \frac{\vec{J}mc\sqrt{2gb}}{(e^*)^2\phi_0^2\varphi_0^2}[\cot h^3\left(\sqrt{\frac{1}{2}}gbx\right) + \cot h\left(\sqrt{\frac{1}{2}}gbx\right)] \tag{68}$$

Equations (67) and (68) are analytical solutions of the GL equation.(63) and (64) in the one-dimensional case, which are shown in Fig. 9. Equation (67) or (65) shows that the superconductive electron in the presence of an electromagnetic field is still a soliton.

However, its amplitude, phase and shape are changed, when compared with those in a uniform superconductor *and* in the absence of external fields, equation (65).

The soliton here is obviously influenced by the electromagnetic field, as reflected by the change in the form of solitary wave. This is why a permanent superconducting current can be established by the motion of superconductive electrons along certain direction in such a superconductor, because solitons have the ability to maintain their shape and velocity while in motion. Therefore, we can obtain that the superconductivity of materials is caused by soliton motion of superconductive electrons.

It is clear from Fig. 9 that $\tilde{H}(x)$ is larger where $\phi(x)$ is small, and vice versa. When $x\rightarrow 0$, $\tilde{H}(x)$ reaches a maximum, while ϕ approaches to zero. On the other hand, when $x\rightarrow\infty$, ϕ becomes very large, while $\tilde{H}(x)$ approaches to zero. These are exactly the well-known behaviors of vortex lines-magnetic flux lines in type-II superconductors [60-61]. Experiments discover that the external magnetic-field can partially penetrates inside type-II superconductors (here the Ginzburg-Landau parameter is greater than 1/2), such as the NbTi alloy and Nb_3Sn compounds, when the external field $\tilde{H}(x)$ is greater than the lower critical

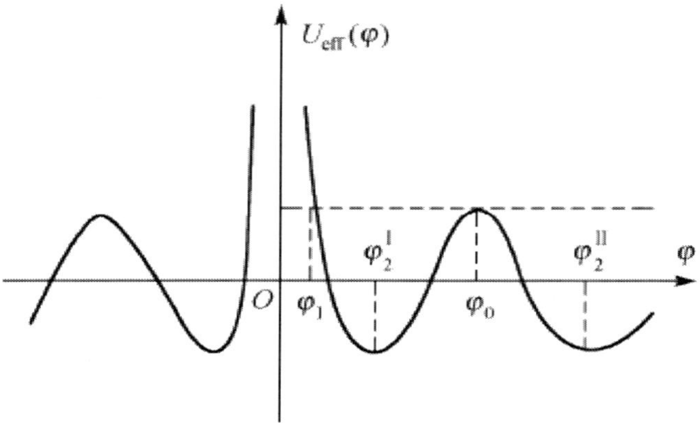

Figure 8. The effective potential energy in Equation (67).

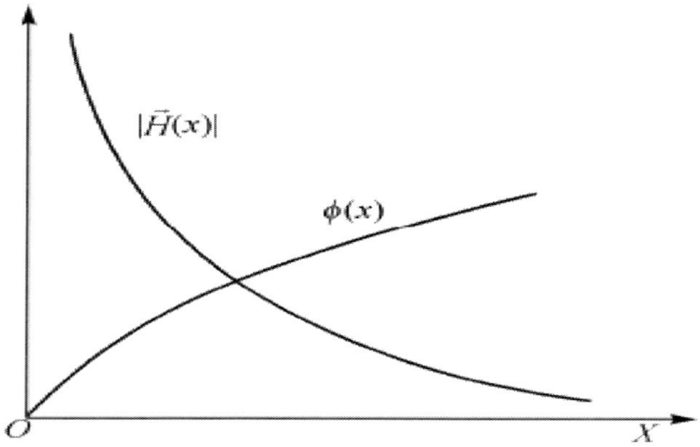

Figure 9. Changes of $\phi(x)$ and $\tilde{H}(x)$ with x in Equations (67)-(68).

magnetic field $\vec{H}_{c1}$, but less than the upper critical field $\vec{H}_{c2}$ [62-67]. The penetration of a magnetic field results in some small regions changing from superconductive to normal states. These small regions having the feature of normal state are of cylindrical shape and regularly arranged in the superconductor. Its physical image is same with Fig. 9. Each cylindrical region is called a vortex (or magnetic field line)[62-72]. The vortex lines are similar to the vortex structure formed in a turbulent flow of fluid. Both theoretical analysis and experimental measurements have shown that the magnetic flux associated with one vortex is exactly equal to one magnetic flux quantum, ϕ_0 when the applied field $\vec{H} \geq \vec{H}_{c1}$, the magnetic field penetrates into the superconductor in the form of vortex lines, increased one by one. This is just a macroscopic quantum effect occurred in type-II superconductors. The above results in Fig.9 give the structure of vortex lines. This means that nonlinear quantum mechanics is correct.

Recently, Garadoc-Daries et al. [73], Matthews et al. [74] and Madison et al. [75] observed vertex solitons in the Boson-Einstein condensates. Tonomure [76] observed experimentally magnetic vortexes in superconductors. These vortex lines in the type-II-superconductors are quantized. Thus, we can determine that the macroscopic quantum effects is in essence a nonlinear quantum effect and can be well described by the nonlinear Schrodinger equation[13].

We now proceed to determine the energy of the soliton given by (67). From the earlier discussion, the energy of the soliton is given by:

$$E=\int_{-\infty}^{+\infty}\left[\frac{1}{2}(\frac{d\varphi}{dx})^2+\frac{b}{2}\varphi^2-\frac{b}{4}\varphi^4-\frac{B^2}{2\varphi^2}\right]dx\approx\varphi_0{}^2\left[\frac{2b\varphi_0^2}{3}-1+\frac{b}{2}(1-\frac{\varphi_0^2}{2})\right]-\frac{B^2}{2\varphi_0^2}$$

which depends on the interaction between superconductive electrons and electromagnetic field.

From the above discussion, we understand that for a bulk superconductor, the superconductive electrons behave as solitons, regardless of the presence of external fields. Thus, the superconductive electrons are a special type of soliton. Obviously, the solitons are formed due to the fact that the nonlinear interaction $\lambda|\phi|^2\phi$, which is generated by the electron-phonon interaction, suppresses the dispersive effect of the kinetic energy in Eqs. (52) and (53). They move in the form of solitary wave in the superconducting state. In the presence of external electromagnetic fields, we demonstrate theoretically that a permanent superconductive current is established and that the vortex lines or magnetic flux lines also occur in type-II superconductors. These results are consistent with the experimental data.

5. The Features of Soliton Motion of the Electrons in Superconductive Junctions and Josephson Effect

5.1. Proximity Effect in Superconductive Junctions

Proximity effect refers to the phenomenon that a normal conductor (N) in contact with a superconductor (S) can be superconductive, which was observed experimentally. This is obviously the result of long- range coherent property of superconductive electrons. It can be regarded as the penetration of electron pairs from the superconductor into the normal conductor or a result of diffraction and transmission of superconductive electron wave. In this phenomenon superconductive electrons can occur in the normal conductor, but their amplitudes are much small compare to that in the superconductive region, thus the nonlinear term $\lambda|\phi|^2\phi$ in GL equations (53)-(54) can be neglected. Because of these, GL equations in the normal and superconductive regions have different forms. On the S side of the S-N junction, the GL equation is [77]

$$\frac{\hbar^2}{2m}(\nabla-\frac{ie^*}{ch}\vec{A})\phi-\alpha\phi+2\lambda\phi^3=0 \qquad (69)$$

while that on the N side of the junction is

$$\frac{\hbar^2}{2m}(\nabla - \frac{ie^*}{ch}\vec{A})\phi - \alpha'\phi = 0 \tag{70}$$

Thus, the expression for $\vec{J}$ remains the same on both sides.

$$\vec{J} = \frac{e^*\hbar}{2mi}(\phi^*\nabla\phi - \phi\nabla\phi^*) - \frac{(e^*)^2}{mc}|\phi|^2\vec{A} \tag{71}$$

In the S region, we have obtained solution of (69) in the previous section, and it is given by (65) or (67) and (68). In the N region, from Equations (70)- (71) we can easily obtain

$$\begin{cases} \varphi^2 = \frac{1}{2}\sqrt{(\varepsilon')^2 - 4d^2}\sin(2\sqrt{b'}x) + \frac{\varepsilon'}{2} \\ \phi_N^2 = \varphi^2\phi_0^2 e^{-2i\theta} = \frac{1}{2}\sqrt{(\varepsilon')^2 - 4d^2}\sin(2\sqrt{b'}x)e^{-i2\theta} + \frac{\varepsilon'}{2}\phi_0^2 e^{-i2\theta} \end{cases} \tag{72}$$

where

$$b' = \frac{2m\alpha'}{\hbar^2} = \frac{1}{\bar{\xi}'^2},\quad 2\tilde{d}^2 = \frac{4\vec{J}^2 m\lambda}{(e^*)^2\alpha'^2}\quad E' = \frac{b'}{2}\varepsilon'$$

here ε' is an integral constant. A graph of ϕ vs. x in both the S and the N regions, as shown in Fig.10, coincides with that obtained by Blackbunu [78]. The solution given in Eq. (72) is the analytical solution in this case. On the other hand, Blackbunu's result was obtained by expressing the solution in terms of elliptic integrals and then integrating numerically. From this, we see that the proximity effect is caused by diffraction or transmission of the superconductive electrons.

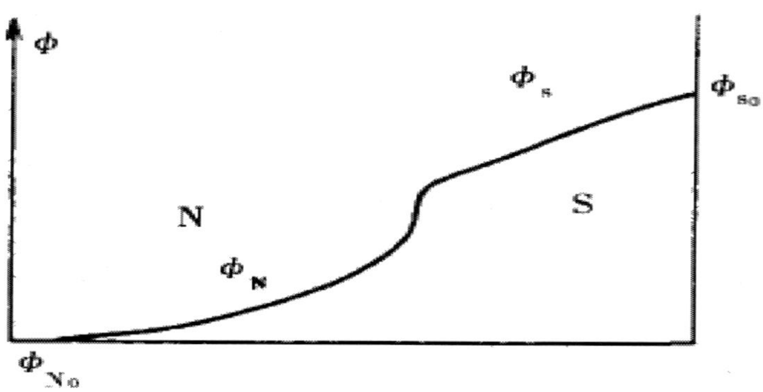

Figure 10. Proximity effect in S-N junction.

5.2. The Josephson Effect in S-J-S and S-N-S Junctions

A superconductor-normal conductor-superconductor junction (S-N-S) or a superconductor -insulator-superconductor junction (S-I-S) consists of a normal conductor or an insulator sandwiched between two superconductors as is schematically shown in Fig.11a.The thickness of the normal conductor or the insulator layer is assumed to be L and we choose the z coordinate such that the normal conductor or the insulator layer is located at $-L/2 \le x \le L/2$. The features of S-I-S junctions were studied by Jacobson et al. [79]. We will treat this problem using the above idea and method [80-81].

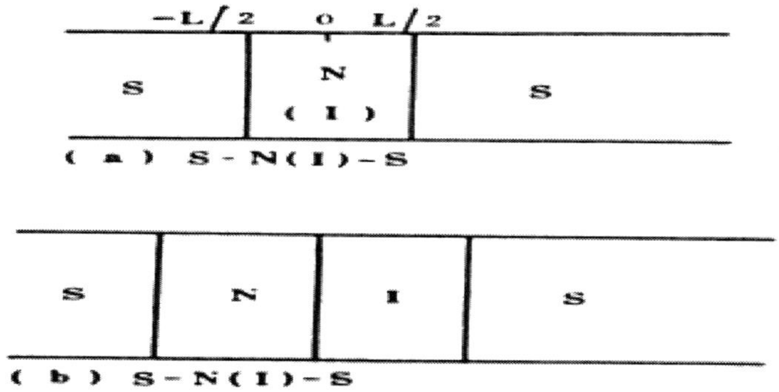

Figure 11. Superconductive junction for S-N-S and S-I-S.

The electrons in the superconducting regions ($|x| \ge L/2$) are depicted by GL equation (69). Its' solution was given earlier in Equation (67). After eliminating u_1 from Equation (66), we have [78-79]

$$J=\frac{1}{2}e^*\alpha u_0\sqrt{\frac{\alpha}{m\lambda}}(1-u_0).$$

Setting $dJ/du_0=0$, we get the maximum current $J_c=\frac{e^*\alpha}{3}\sqrt{\frac{\alpha}{3m\lambda}}$. This is the critical current of a perfect superconductor, corresponding to the three-fold degenerate solution of (66), *i.e.*,$u_1=u_0$.

From (71), we have

$$\vec{A}=-\frac{m\vec{J}c}{(e^*)^2\phi_0^2\varphi^2}+\frac{hc}{e^*}\nabla\theta.$$

Using the London gauge, $\nabla.\vec{A}=0$, we can get [75-76]

$$\frac{d^2\theta}{dx^2}=\frac{\mathrm{mJ}}{\mathrm{e}^*\phi_0^2\hbar}\frac{d}{dx}(\frac{1}{\varphi^2}).$$

Integrating the above equation twice , we get the change of the phase to be

$$\Delta\theta=\frac{\mathrm{mJ}}{\mathrm{e}^*\phi_0^2\hbar}\int(\frac{1}{\varphi^2}-\frac{1}{\varphi_\infty^2})dx \tag{73}$$

where $\varphi^2=\mathrm{u}$, and $\varphi_\infty^2=\mathrm{u}_0$. Here we have used the following de Gennes boundary conditions in obtaining Equation (73)

$$\left.\frac{d\phi}{dx}\right|_{|\mathrm{x}|\to\infty}=0,\left.\frac{d\theta}{dx}\right|_{|\mathrm{x}|\to\infty}=0,\ \ \phi(|\mathrm{x}|\to\infty)=\phi_\infty \tag{74}$$

If we substitute (64) - (67) into (73), the phase shift of wave function from an arbitrary point x to infinite can be obtained directly from the above integral, and takes the form of:

$$\Delta\theta_{\mathrm{L}}(\mathrm{x}\to\infty)=-\tan^{-1}\sqrt{\frac{\mathrm{u}_1}{\mathrm{u}_0-\mathrm{u}_1}}+\tan^{-1}\sqrt{\frac{\mathrm{u}_1}{\mathrm{u}-\mathrm{u}_1}} \tag{75}$$

For the S-N-S or S-I-S junction, the superconducting regions are located at $|\mathrm{x}|\geq\mathrm{L}/2$ and the phase shift in the S region is thus

$$\Delta\theta_{\mathrm{s}}=2\Delta\theta_{\mathrm{L}}(\frac{\mathrm{L}}{2}\to\infty)\approx2\tan^{-1}\sqrt{\frac{\mathrm{u}_1}{\mathrm{u}_{\mathrm{s}}-\mathrm{u}_1}} \tag{76}$$

According to the results in (70) - (71) and the above similar method, the change of the phase in the I or N region of the S-N-S or S-I-S junction may be expressed as [75-76]

$$\Delta\theta_{\mathrm{N}}=-2\tan^{-1}[\frac{2\mathrm{e}^*\mathrm{h}'}{\dot{\mathrm{J}}}\sqrt{\frac{\alpha^2}{8\mathrm{m}\lambda}}\tan(\frac{\sqrt{\mathrm{b}'}\mathrm{L}}{2})]+\frac{\mathrm{m}\dot{\mathrm{J}}\mathrm{L}}{2\mathrm{e}^*\mathrm{h}'\mu_0} \tag{77}$$

where

$$\mathrm{h}'=\sqrt{\frac{8\mathrm{m}\lambda}{\alpha^2}}\frac{\dot{\mathrm{J}}}{2\mathrm{e}^*}\frac{\tan(\Delta\theta_{\mathrm{N}}/2)}{\tan(\sqrt{\mathrm{b}'}\mathrm{L}/2)},\ \frac{\mathrm{m}\dot{\mathrm{J}}\mathrm{L}}{2\mathrm{e}^*\mathrm{h}'\mu_0}$$

is an additional term to satisfy the boundary conditions (74), and may be neglected in the case being studied.

Near the critical temperature ($T<T_c$), the current passing through a weakly linked superconductive junction is very small ($\dot{\mathrm{J}}<<1$), we then have

$$\mu_1' = \frac{4\dot{J}^2 m\lambda}{(e^*)^2\alpha^2} = 2\overline{A}^{-2}, \text{ and } g'=1$$

Since $\eta\varphi^2$ and $d\varphi^2/dx$ are continuous at the boundary x=L/2, we have

$$\frac{d\mu_S}{dx}\Big|_{x=L/2} = \frac{d\mu_N}{dx}\Big|_{x=L/2} \quad \eta_S\mu_S\Big|_{x=L/2} = \eta_N\mu_N\Big|_{x=L/2},$$

where η_s and η_N are the constants related to features of superconductive and normal phases in the junction, respectively. These give [80-81]

$$2\sqrt{b'}\,\overline{A}\sin(2\Delta\theta_N) = \varepsilon_1[1-\cos(2\Delta\theta_s)]\sin(\sqrt{b'}L)$$

$$\cos(\sqrt{b'}L)\sin(2\Delta\theta_s) = \varepsilon\sin(2\Delta\theta_N) + \sin(2\Delta\theta_s + \Delta\theta_N)$$

where $\varepsilon_1 = \eta_N/\eta_S$. From the two equations, we can get

$$\sin(\Delta\theta_s + \Delta\theta_N) = \frac{2\sqrt{2m\lambda}\,\dot{J}}{e^*\alpha}\sqrt{b'}\sin(\sqrt{b'}L)$$

Thus the Josephson current in superconductive junctions can be represented as

$$\dot{J} = \dot{J}_{max}\sin(\Delta\theta_s + \Delta\theta_N) = \dot{J}_{max}\sin(\Delta\theta) \tag{78}$$

where

$$\dot{J}_{max} = \frac{e^*\alpha_s}{2\sqrt{2m\lambda b'}}\cdot\frac{1}{\sin(\sqrt{b'}L)}, \Delta\theta = \Delta\theta_s + \Delta\theta_N \tag{79}$$

Equation (78) is the well-known example of the Josephson current, which was obtained from The Josephson effect of superconductivity junctions by Josephson[82-83]., that is a macroscopic quantum effect. At present, this effect can be explained based on the nonlinear quantum theory[13]. This again shows that this effect is a nonlinear quantum phenomenon and can be described by nonlinear quantum mechanics[13].

Also, from Eq.(79) we can see that the Josephson critical current[82-83] is inversely proportional to sin ($\sqrt{b'}L$), which means that the current increases suddenly whenever $\sqrt{b'}L$ approaches to $n\pi$, suggesting some resonant phenomena occurs in the system. This has not been observed before. Moreover $\dot{J}_{max}$ is proportional to $e^*\alpha_s/2\sqrt{2m\lambda b'} = (e^*\hbar\alpha_S/4m\sqrt{\lambda\alpha_N})$, which is related to $(T - T_c)^2$. Finally, it is worthwhile to mention that no explicit assumption was made in the above on whether the junction is a potential well (α<0) or a potential barrier (α>0). The results are thus valid and the Josephson effect in Eq. (78), occurs all for both potential wells and for potential barriers.

5.3. The Josephson Effect in SNIS Junctions

A superconductor-normal conductor-insulator-superconductor junction (SNIS) is shown schematically in Figure 11 (b). It can be regarded as a multilayer junction consist of S-N-S and S-I-S junctions. If appropriate thicknesses for the N and I layers are used (approximately 20 °A– 30 °A), the Josephson effect similar to that discussed above can occur in the SNIS junction. Since the derivations are similar to that in the previous sections, we will skip much of the details and give the results in the following. The Josephson current in the SNIS junction is still given by

$$\mathrm{j} = \mathrm{j}_{\max} \sin(\Delta\theta)$$

but, where $\Delta\theta = \Delta\theta_{s1} + \Delta\theta_{N} + \Delta\theta_{I} + \Delta\theta_{s2}$ and

$$\mathrm{j}_{\max} = \frac{1}{\sqrt{b'_N}}\{\frac{\varepsilon_1 \sinh(\sqrt{b'_N}L)}{2[\cosh(\sqrt{b'_N}L) - \cos(2\Delta\theta_N)]}\}\times \frac{1}{\sqrt{[1+\cos(2\Delta\theta_N)][1+\cos(2\Delta\theta_I)]} - \sqrt{[1-\cos(2\Delta\theta_N)][1-\cos(2\Delta\theta_I)]}} - \frac{1}{\sqrt{b'_N}}\{\frac{\varepsilon_1\sqrt{[1-\cos^2(2\Delta\theta_N)]}\sinh(\sqrt{b'_N}L)}{2[\cosh(\sqrt{b'_N}L) - \cos(2\Delta\theta_N)]^2 - 1 + \cos^2(2\Delta\theta_N)}\}\times \frac{1}{\sqrt{[1-\cos(2\Delta\theta_N)][1-\cos(2\Delta\theta_I)]} + \sqrt{[1+\cos(2\Delta\theta_N)][1+\cos(2\Delta\theta_I)]}},$$

The phase shifts $\Delta\theta_{s1}, \Delta\theta_{N}, \Delta\theta_{I}$ and $\Delta\theta_{s2}$ satisfy the following relations

$$4\sqrt{b'_N A}\sin(2\Delta\theta_N)] = \varepsilon_1[1-\cos(2\Delta\theta_{s1})\sinh(\sqrt{b'_N}L)]$$

$$4\sqrt{b'_I A}\sin(2\Delta\theta_I)] = \varepsilon_1[1-\cos(2\Delta\theta_{s2})\sinh(\sqrt{b'_I}L)]$$

$$\tan(\Delta\theta_I)] = \varepsilon_2 \frac{\tan(\sqrt{b'_N}L/2)}{\tanh(\sqrt{b'_N}L/2)}\tan(2\Delta\theta_N)(\sqrt{b'_N/b'_I}$$

$$\cosh(\sqrt{b'_N}L)\sin(2\Delta\theta_{s1}) = \sin(2\Delta\theta_N) + \sin(2\Delta\theta_{s1} + 2\Delta\theta_N)$$

$$\cos(\sqrt{b'_I}L)\sin(2\Delta\theta_{s2}) = \sin(2\Delta\theta_I) + \sin(2\Delta\theta_{s2} + 2\Delta\theta_I)]$$

It can be shown that the temperature dependence of J_{max} is $J_{max} \propto (T_c - T_0)^2$, which is quite similar to the results obtained by Blackburm et al. [84]. for the SNIS junction and those by Romagnan et al. [85]. using the Pb-PbO-Sn-Pb junction. Here, we obtained the same results using a complete different approach. This indicates again that we can obtain some results, which agree with the experimental data, by nonlinear quantum mechanics[13], thus it is correct.

6. Time- Dependent Features of Soliton Motion of the Electrons in Nonequilibrum Superconductive Systems

6.1. Soliton Solutions of Time-Dependent Dynamic Equation of the Electrons

We studied only the properties of motion of superconductive electrons in steady states in superconductors in section 4, and which are described by the time-independent GL equation[52-54]. In such a case, the superconductive electrons move as solitons. We ask, "What are the features of a time-dependent motion in non-equilibrium states of a superconductor?" Naturally, this motion should be described by the time-dependent Ginzburg-Landau (TDGL) equation [52-58, 86] in this case. Unfortunately, there are many different forms of the TDGL equation under different conditions. The one given in the following is commonly used when an electromagnetic field $\vec{A}$ is involved

$$\Gamma\left[\hbar\frac{\partial}{\partial t} - 2ie\mu(r)\right]\phi = \frac{-1}{2m}\left(\hbar\nabla - \frac{2ie}{c}\vec{A}\right)^2 + \alpha\phi - \lambda|\phi|^2\phi \tag{80}$$

$$\vec{J} = \sigma\left[-\frac{1}{c}\frac{\partial\vec{A}}{\partial t} - \nabla\mu(r)\right] + \frac{ie\hbar}{m}(\phi^*\nabla\phi - \phi\nabla\phi^*) - \frac{4e^2}{mc}\vec{A}|\phi|^2 \tag{81}$$

here $i = \sqrt{-1}$, $\nabla\times\nabla\times\vec{A} = \frac{1}{c}\frac{\partial}{\partial t}\left(-\frac{1}{c}\frac{\partial\vec{A}}{\partial t} - \nabla\mu\right) + \frac{4\pi\vec{J}}{c}$. and σ is the conductivity in the normal state, Γ is an arbitrary constant, and μ is the chemical potential of the system. In practice, Equation (80) is simply a time-dependent Schrödinger equation with a damping effect.

In certain situations, the following forms of the TDGL equation are also used.

$$i\hbar\frac{\partial\phi}{\partial t} = -\frac{\hbar^2}{2m}\left(\nabla - \frac{2ie}{\hbar c}\vec{A}\right)^2\phi + \alpha\phi - \lambda|\phi|^2\phi \tag{82}$$

or

$$i\left(\hbar\frac{\partial}{\partial t}-i2e\mu\right)\phi=\frac{1}{\Gamma}(\alpha-\lambda|\phi|^2)\phi+\frac{\xi'^2}{\Gamma}\left(\nabla-\frac{2ie}{\hbar c}\vec{A}\right)^2\phi \tag{83}$$

here $\xi'=\hbar/\sqrt{2m}$, and equation (82) is also a nonlinear Schrödinger equation under an electromagnetic field having soliton solutions. However, these solutions are very difficult to find, and no analytic solutions have been obtained. An approximate solution was obtained by Kusayanage et al. [87] by neglecting the ϕ^3 term in Eq. (80) or Eq. (82), in the case of $\vec{A}=(0,\bar{H}x,0)$, $\mu=-K\bar{E}x$, $\vec{H}=(0,\,0,\,\bar{H})$ and $\vec{E}=(\bar{E},0,0)$, where $\vec{H}$ is the magnetic field, while $\vec{E}$ is the electric field .We will solve the TDGL equation in the case of weak fields in the following.

TDGL equation (83) can be written in the following form when $\vec{A}$ is very small[88-89]

$$i\hbar\frac{\partial\phi}{\partial t}+\frac{\hbar^2}{2m\Gamma}\nabla^2\phi+\frac{\lambda}{\Gamma}|\phi|^2\phi=\left(\frac{\alpha}{\Gamma}-2e\mu\right)\phi \tag{84}$$

where α and Γ are material dependent parameters, λ is the nonlinear coefficient, m is the mass of the superconductive electron. Equation (84) is actually a nonlinear Schrödinger equation in a potential field $\alpha/\Gamma-2e\mu$. Cai, Bhattacharjee et al. [90], and Davydov [51] used it in their studies of superconductivity. However, this equation is also difficult to solve. In the following, Pang[91-95] solves the equation only in the one-dimensional case.

For convenience, let $t'=t/\hbar$, $x'=x\sqrt{2m\Gamma}/\hbar$, then Eq. (84) becomes

$$i\frac{\partial\phi}{\partial t'}+\frac{\partial^2\phi}{\partial x'^2}+\frac{\lambda}{\Gamma}|\phi|^2\phi=\left[\frac{\alpha}{\Gamma}-2e\mu(x')\right]\phi \tag{85}$$

If we let $\frac{\alpha}{\Gamma}-2e\mu=0$, then Equation (85) is the usual nonlinear Schrödinger equation whose solution is of the form [89-93]

$$\phi_s^0=\varphi_0(x',t')e^{i\theta_0(x',t')}, \tag{86}$$

here

$$\varphi_0(x',t')=\sqrt{\frac{\Gamma(\upsilon_e^2-2\upsilon_c\upsilon_e)}{2\lambda}}\times\sec h\left[\sqrt{\frac{(\upsilon_e^2-2\upsilon_c\upsilon_e)}{4}}(x'-\upsilon_e t')\right],$$

$$\theta_0(x',t')=\frac{1}{2}\upsilon_e(x'-\upsilon_c t') \tag{87}$$

In the case of $\frac{\alpha}{\Gamma} - 2e\mu \neq 0$, let $\mu = -K\bar{E}x'$, where K is a constant, and assume that the solution is of the form [91-95]

$$\phi = \varphi'(x',t')e^{i\theta(x',t')} \tag{88}$$

Substituting Equation (88) into Equation (86), we get:

$$-\varphi'\frac{\partial\theta}{\partial t'} - \varphi'\left(\frac{\partial\theta}{\partial t'}\right)^2 + \frac{\partial^2\varphi'}{\partial(x')^2} + \frac{\lambda}{\Gamma}(\varphi')^3 = \left(2Ke\bar{E}x' + \frac{\alpha}{\Gamma}\right)\varphi' \tag{89}$$

$$\frac{\partial\varphi'}{\partial t'} + 2\frac{\partial\varphi'}{\partial x'}\frac{\partial\theta}{\partial x'} + \varphi'\frac{\partial^2\theta}{\partial(x')^2} = 0 \tag{90}$$

Now let $\varphi'(x',t') = \varphi(\xi)$, $\xi = x' - u(t')$, $u(t') = -2\bar{E}Ke(t')^2 + \upsilon t' + d$, where $u(t')$ describes the accelerated motion of $\varphi'(x',t')$. The boundary condition at $\xi' \to \infty$ requires $\varphi(\xi)$ to approach zero rapidly. When $2\,\partial\theta/\partial\xi - \dot{u} \neq 0$, equation (90) can be written as:

$$\varphi'^2 = \frac{g(t')}{(\partial\theta/\partial\xi - \dot{u}/2)}$$

or

$$\frac{\partial\theta}{\partial x'} = \frac{g(t')}{\varphi'^2} + \frac{\dot{u}}{2} \tag{91}$$

where $\dot{u} = du/dt'$. Integration of (91) yields:

$$\theta(x',t') = g(t')\int_0^{x'}\frac{dx''}{\varphi'^2} + \dot{u}x'/2 + h(t') \tag{92}$$

and where $h(t')$ is an undetermined constant of integration. From Equation (92) we can get:

$$\frac{\partial\theta}{\partial t'} = \dot{g}(t')\int_0^{x'}\frac{dx''}{\varphi'^2} - \frac{g\dot{u}}{\varphi'^2} + \frac{g\dot{u}}{\varphi'^2}\Big|_{x'=0} + \frac{\ddot{u}}{2}x' + \dot{h}(t') \tag{93}$$

Substituting Equations (92) and (93) into Equation (89), we have:

$$\frac{\partial^2\varphi'}{\partial(x')^2}=\left[\left(2KEex'+\frac{\alpha}{\Gamma}\right)+\frac{u}{2}x'+h(t')+\frac{u^2}{4}+g\int_0^{x'}\frac{dx''}{\varphi'^2}+\frac{gu}{\varphi'^2}\Big|_{x'=0}\right]\varphi'-\frac{\lambda}{\Gamma}\varphi'^3+\frac{g^2}{\varphi'^3} \tag{94}$$

Since $\frac{\partial^2\varphi'}{\partial(x')^2}=\frac{d^2\varphi'}{d\xi^2}$, which is a function of ξ only, the right-hand side of Equation (94) is also a function of ξ only, so it is necessary that $g(t')=g_0=\text{constant}$, and

$$(2K\tilde{E}ex'+\frac{\alpha}{\Gamma})+\frac{\ddot{u}}{2}x'+\dot{h}(t')+\frac{\dot{u}^2}{4}+\frac{g\dot{u}}{f^2}\Big|_{x'=0}=\overline{V}(\xi) \tag{95}$$

Next, we assume that $V_0(\xi)=\overline{V}(\xi)-\beta$, where β is real and arbitrary, then

$$2K\tilde{E}ex'+\frac{\alpha}{\Gamma}=V_0(\xi)-\frac{\ddot{u}}{2}x'+\left[\beta-\frac{g\dot{u}}{\varphi^2}\Big|_{x'=0}-\dot{h}(t')-\frac{\dot{u}^2}{4}\right] \tag{96}$$

Clearly in the case discussed, $V_0(\xi)=0$, and the function in the brackets in Equation (95) is a function of t′. Substituting Equation (96) into Equation (94), we can get [91-95]:

$$\frac{\partial^2\varphi'}{\partial\xi^2}=\beta\varphi'-\frac{\lambda}{\Gamma}\varphi'^3+\frac{g_0^2}{\varphi'^3} \tag{97}$$

This shows that φ' is the solution of Eq. (96) when β and g are constant. For large $|\xi|$, we may assume that $|\varphi'|\le\beta/|\xi|^{1+\Delta}$, when Δ is a small constant. To ensure that φ' and $d^2\varphi'/d\xi^2$ approach zero when $|\xi|\to\infty$, only the solution corresponding to $g_0=0$ in Eq. (97) is kept, and it can be shown that this soliton solution is stable in such a case. Therefore, we choose $g_0=0$ and obtain the following from Eq. (91):

$$\frac{\partial\theta}{\partial x'}=\dot{u}/2 \tag{98}$$

Thus, we obtain from Equation (96) that

$$2K\tilde{E}ex'+\frac{\alpha}{\Gamma}=-\frac{\ddot{u}}{2}x'+\beta-\dot{h}(t')-\frac{\dot{u}^2}{4},$$

$$h(t')=\left(\beta-\frac{\alpha}{\Gamma}-\frac{1}{4}\upsilon^2\right)t'-\frac{4}{3}(K\overline{E}e)^2(t')^3+e\upsilon K\overline{E}(t')^2 \tag{99}$$

Substituting Equation (99) into Equations (92) - (93), we obtain:

$$\theta=\left(-2K\overline{E}et'+\frac{1}{2}\upsilon\right)x'+\left(\beta-\frac{\alpha}{\Gamma}-\frac{1}{4}\upsilon^2\right)t'-\frac{4}{3}(K\overline{E}e)^2(t')^3+e\upsilon K\overline{E}(t')^2 \tag{100}$$

Finally, substituting the Equation (100) into Equation (97), we can get

$$\frac{\partial^2\varphi'}{\partial\xi^2}-\beta\varphi'+\frac{\lambda}{\Gamma}\varphi'^3=0. \tag{101}$$

When $\beta>0$, the solution of. Equation (101) is of the form

$$\varphi'=\sqrt{\frac{2\beta\Gamma}{\lambda}}\sec h\left(\sqrt{\beta}\xi\right) \tag{102}$$

Thus [89-91]

$$\phi=\sqrt{\frac{2\beta\Gamma}{\lambda}}\sec h\left[\sqrt{\beta}\left(\sqrt{\frac{2m\Gamma}{\hbar^2}}x+\frac{2eK\overline{E}t^2-\upsilon t-d}{\hbar}\right)\right]\times\exp\left\{i\left[\left(\frac{-2eK\overline{E}t}{\hbar}+\frac{\upsilon}{2}\right)\sqrt{\frac{2m\Gamma}{\hbar^2}}x+\left(\beta-\frac{\alpha}{\Gamma}-\frac{1}{4}\upsilon^2\right)\frac{t}{\hbar}-\frac{4(eK\vec{E})^2t^3}{3\hbar^3}+\frac{\upsilon Ke\overline{E}t^2}{\hbar}\right]\right\} \tag{103}$$

6.2. The Time-Dependent Dynamic Features of Soliton Motion of the Electrons

Equation (103) is also a soliton solution, but its shape, amplitude and velocity have been changed relatively as compared to that of Eq. (87). It can be shown that Eq. (103) does indeed satisfy Eq. (85). Thus, equation (85) has a soliton solution. It can also be verified that this solition solution is stable.

For the solution of Eq. (103), we may define a generalized time-dependent wave number,

$$k=\frac{\partial\theta}{\partial x'}=\frac{\upsilon}{2}-2K\overline{E}et' \tag{104}$$

and a frequency

$$\omega = -\frac{\partial\theta}{\partial t'} = 2K\bar{E}ex' - \left(\beta - \frac{\alpha}{\Gamma} - \frac{1}{4}\upsilon^2\right) + e(K\bar{E}e)^2(t')^2 - 2K\bar{E}e\upsilon t' = 2K\bar{E}ex' - \beta - \frac{\alpha}{\Gamma} + k^2 \qquad (105)$$

The usual Hamilton equations for the superconductive electron (soliton) in the macroscopic quantum systems are still valid here and can be written as [91-95]

$$\frac{dk}{dt'} = -\frac{\partial\omega}{\partial x'}\Big|_k = -2K\bar{E}e\,, \qquad (106)$$

then the group velocity of the superconductive electron is

$$\upsilon_g = \frac{dx'}{dt'} = \frac{\partial\omega}{\partial k}\Big|_{x'} = 2\left(\frac{\upsilon}{2} - 2K\bar{E}et'\right) = \upsilon - 4K\bar{E}et' \qquad (107)$$

This means that the frequency ω still represents the meaning of Hamiltonian in the case of nonlinear quantum systems in Chapter 2. Hence,

$$\frac{d\omega}{dt'} = \frac{d\omega}{dk}\Big|_{x'}\frac{dk}{dt'} + \frac{\partial\omega}{\partial x'}\Big|_k\frac{dx'}{dt'} = 0\,, \qquad (108)$$

as seen in the usual stationary linear medium.

These relations, as shown in Eqs. (104) - (108) show that the superconductive electrons move as if they were classical particles moving with a constant acceleration in the invariant electric-field, and that the acceleration is given by $-4K\bar{E}e$. If $\upsilon>0$, the soliton initially travels toward the overdense region, it then suffers a deceleration and its velocity changes sign. The soliton is then reflected and accelerated toward the underdense region. The penetration distance into the overdense region depends on the initial velocity υ.

From the above studies we see that the time-dependent motion of superconductive electrons still behaves like a soliton in non-equilibrium state of superconductor. Therefore, we can conclude that the electrons in the superconductors are essentially a soliton in both time-independent steady state and time-dependent dynamic state systems. This means that the soliton motion of the superconductive electrons causes the superconductivity of material[96]. Then the superconductors have a complete conductivity and nonresistance property because the solitons can move over a macroscopic distances retaining its amplitude, velocity, energy and other quasi- particle features.

In such a case the motions of the electrons in the superconductors are described by a nonlinear Schrödinger equations (52), or (53) or (80) or (82) or (84). According to the soliton theory, the electrons in the superconductors are localized and have a wave-corpuscle duality due to the nonlinear interaction, which is completely different from those in the quantum mechanics concepts[13]. Therefore, we have sought that nonlinear quantum mechanics[13] can describe very well the superconductivity phenomenon occurred in low temperature.

7. Soliton Excitation of Polarons in Ionic Crystals and Its Properties

7.1. Soliton Excitations of Polarons on Ionic Crystals

The concept of polaron was first proposed by Landau to describe the motion of an additional electron in an ionic crystals. It was suggested that the electron would polarize the crystal through attracting positive ions and repulsing negative ions, thus the crystal energy would be lowered and the state of the electron is also changed in the process. The localized electron and the polarized lattice field, which surrounds the electron through electron-phonon interaction, incorporate together to form a polaron. In essence, the polaron is a quasiparticle formed by the combination of the electron with longitudinal-optics (LO) phonon[1-7,97], its scale is determined by the size of domain of polarized field of lattice around the electron using the formula of $L'=1/\Delta k=\sqrt{(\hbar/2m\omega_L)}$, where m and k are the mass and wavevector of the electron, respectively, ω_L is the frequency of LO phonon.

If the polarized domain are larger than the lattice constant of the crystal, then the quasiparticle is referred as to great polaron, on the contrary, is called small polaron. For general ionic-crystals, such as the halides of alkaline metals, L' is about 1nm; for polarized semiconductors containing the compounds formed by elements between II-VI and III-V periods, L' is about 10nm. Thus the quasiparticles in these materials are great polarons.

However, the semiconductors with narrow energy-band, such as NiO the quasiparticles are small polarons because their sizes are smaller than the constant of lattice.

Therefore, the properties of great and small polarons are completely different because of differences of their effective masses, moved velocities, bonding strengths and formed conditions. These problems were discussed in details by Pekar[98], Frohlech[99-100] and Holstein[101], and many others[102]. In this sections we first discuss the nonlinear quantum –mechanical properties of great polarons by new method.

We now take a look at the behavior of an additional electron in an one-dimensional polar crystal in the framework of Holstein model[101] which takes into consideration the appearance of soliton-like excitations of the additional electron in such systems. Assume in this case that the system consists of N identical diatomic molecules, with atomic masses of m_1 and m_2, respectively, and located at their equilibrium positions, $x_n^{(1)}$ and $x_n^{(2)}$ of the nth molecular cell. We here will consider only the variations of the interatomic distance due to the longitudinal optical oscillation with frequency ω_0, and ignore the longitudinal acoustic oscillations. The Hamiltonian operator for vibration of lattice molecules is given by

$$H_{ph}=-\frac{\hbar}{2M}\sum_n\frac{\partial^2}{\partial u_n^2}+\frac{1}{2}M\omega_0^2\sum_n u_n^2+\frac{1}{2}M\omega_1^2\sum_n u_n u_{n+1} \quad (109)$$

with $M=\dfrac{m_1m_2}{m_1+m_2}$, $u_n=x_n^{(2)}-x_n^{(1)}=a$

The last term in Equation(109) represents the nondiagonal dispersion effect of the optical phonons, with $\omega_1^2 << \omega_0^2$ .According to Fedyanin et al. [103-104], the Hamiltonian of an electron interacting with the lattice in the systems is given by

$$H_e + H_{\text{int}} = (\varepsilon_0 + W')\sum_n N_n - J\sum_n (b_n^+ b_{n+1} + b_{n+1}^+ b_n) + V\sum_n N_n N_{n+1} \qquad (110)$$

in the second quantization representation, where *W*' is the shift of the energy level ε_0 of the electron at the *n*th site due to the influence of the electron at other sites (we assume there is only one electron level '' associated with a possible bound state). Also in Eq.(110), *J* is the overlap integral ($J > 0$) and *V* is the strength of the electron-phonon interaction (having the dimensions [energy] × [length]$^{-1}$), $b_n^+(b_n)$ is the creation (annihilation) operator of the additional electron, $N_n = b_n^+ b_n$.

For small *V* , the Hamiltonian $H_{ph} + H_e + H_{\text{int}}$ can be transformed to the well-known Pekar- Fröhlich type which allows us to take into account the long-range interaction between the electron and the polarization field caused by the lattice deformation, whereas the Holstein Hamiltonian contains only short-range interaction. Applying the Bogolyubov variational principle, Fedyanin et al. [103-104] transformed the Hamiltonian $H_{ph} + H_e + H_{\text{int}}$ into the following form, with "separation" of the phonon and electron subsystems

$$H = \widetilde{H}_{ph} + \widetilde{H}_e$$

where

$$\widetilde{H}_{ph} = \hbar\sum_q \omega_q\left(a_q^+ a_q + \frac{1}{2}\right),$$

$$\tilde{H}_e = \left(\varepsilon_0 + W' - \frac{V^2}{2M\omega_0^2}\right)\sum_n N_n - Je^{-F_\theta}\sum_n\left(b_n^+ b_{n+1} + b_{n+1}^+ b_n\right) + \frac{V^2}{2M\omega_0^2}\sum_n N_n N_{n+1}, \qquad (111)$$

with

$$\left[a_q, a_{q'}^+\right] = \delta_{qq'},\ \left[b_n, b_{n'}^+\right]_+ = \delta_{nn'},\ \left[\widetilde{H}_e, \widetilde{H}_{ph}\right] = 0,$$

$$N_n = b_n^+ b_n,\ \omega_q^2 = \omega_0^2 + \omega_1^2 \cos q r_0,$$

$$F_\theta = \frac{V^2}{4MN\hbar\omega_0^4}\sum_q \omega_q |S_q|^2 \coth\left(\frac{\hbar\omega_q}{2K_BT}\right), (\theta = K_BT).$$

In the above,

$$S_q = \omega_0^2 \sum_n \frac{\Delta_{n+1}^m - \Delta_n^m}{\Delta} e^{iqmr_0}, \quad \Delta_n^m = \left(-\frac{\omega_1^2}{2\omega_0^2}\right)^{|n-m|} \frac{\Delta}{\omega_1^2}\left(1 - 2\frac{\omega_1^2}{\omega_0^2}\right)^{-2}$$

here Δ is the determinant of the matrix constituted of ω_1^2 and ω_o^2, Δ_n^m is its minor obtained by eliminating the *m*th column and the *n*th row.

For $\theta \to 0$, the basic contribution to Δ_n^m comes from the diagonal terms ($n = m$) of

$$S_\theta = \frac{V^2\theta}{2M\hbar^2\omega_0^4} + O\left(\frac{\omega_1^2}{\omega_0^2}\right)$$

Thus, the parameters corresponding to the phonon subsystem, ω_q and *V,* renormalize the Hamiltonian H_e due to the application of the Bogolyubov variational principle. Obviously, the Hamiltonian given in Eq. (111) is a nonlinear operator. Therefore, the polaron undergoes a change from a linear elementary excitation to a nonlinear soliton-excitation through the nonlinear interaction arising from the electron-phonon interaction in this system.

Using the Heisenberg equation[13], Makhankov et al. [11-12] showed that

$$i\hbar(\phi_n)_t = \left(\varepsilon_0 + W' - \frac{V^2}{2M\omega_0^2}\right)\phi_n(t) - Je^{-F_\theta}\left[\phi_{n+1}(t) + \phi_{n-1}(t)\right] + \frac{V^2}{M\omega_0^2}\left[1 - \phi_{n+1}^*(t)\phi_{n+1}(t) + \phi_{n-1}^*(t)\phi_{n-1}(t)\right]\phi_n(t) \tag{112}$$

In this case, the trial wave function can be written as

$$|\varphi(t)\rangle = U|0\rangle = \sum_n \varphi_n(t) b_n^+ |0\rangle_{po}, \tag{113}$$

where $|0\rangle_{P0}$ is the ground state of the polaron. In the continuum approximation, Makhankov and Fedyanin[12] obtained the following from Eq. (112) by neglecting terms involving derivatives higher than the second

$$i\hbar\phi_t = \left(\varepsilon_0 + W' - \frac{V^2}{2M\omega_0^2}\right)\phi(x,t) - Je^{-F_\theta}\left[2\phi + r_0^2\phi_{xx}\right] + \frac{V^2}{M\omega_0^2}\left[1 - |\phi|^2 - r_0^2|\phi_x|^2 - \frac{r_0^2}{4}|\phi_{xx}|^2 - \frac{r_0^2}{2}\left(\phi_{xx}^*\phi + \phi_{xx}\phi_x^*\right)\right]\phi(x,t). \tag{114}$$

If we further retain only terms with orders less than the fourth with respect to dispersion and nonlinearity in Equation (114), we can get

$$i\hbar\phi_t(x,t)=\left(\varepsilon_0+W'-\frac{V^2}{2M\omega_0^2}-2Je^{-F_\theta}+\frac{V^2}{M\omega_0^2}\right)\phi(x,t)-Jr_0^2e^{-F_\theta}\phi_{xx}-\frac{V^2}{M\omega_0^2}|\phi|^2\phi(x,t) \quad (115)$$

This is a standard nonlinear Schrödinger equation[13]. Therefore, the properties of the polarons can be described by nonlinear quantum mechanics[13]. Thus the polarons in ionic crystals have all properties of microscopic particles in nonlinear quantum mechanics [13]. The soliton solution of Eq.(115) is given by

$$\varphi(x,t)=\frac{r_0\exp[i(kx-\omega t+\theta_0)]}{2W_S\cosh[(x-vt-x_0)/W_S]} \quad (116)$$

and its energy is

$$E_S=\varepsilon_0+W'+\frac{V^2}{2M\omega_0^2}-2Je^{-F_\theta}+\frac{M_Sv^2}{2}-\frac{eF_\theta}{24J}\left(\frac{V^2}{M\omega_0^2}\right)^2, \quad (117)$$

where

$$M_S=\frac{\hbar^2e^{F_\theta}}{2Jr_0^2},\ W_S=\frac{4r_0JM\omega_0^2}{V^2e^{F_\theta}}.$$

The polaron-soliton in Eq.(116) is localized in the region $\approx W_s$. The polaron effect is defined by the last two terms in Eq.(117). The third term describes the renormalization of the electron level shift due to its translation jumps.

Note that there arises a temperature renormalization of the conductive electron with the mass $m_e=\hbar^2/2r_0^2J$ due to J is replaced by $J\exp[-F_\theta]$, at present. Therefore, the polaron now gets heavier, i.e., $M_s(\theta)>M_s(0)$, and the effective mass becomes temperature dependent in such a case. If $\theta\to 0$, then $M_s(\theta)>M_s(0)(1+3\chi\theta/8\omega_0)$, where χ is a coupling constant between electron and phonon and is given by $\chi=4V^2/3M\hbar^2\omega_0^3$.

The same effect, i.e. increase of mass, was observed for $\theta\neq 0$ in cyclotron resonance experiments by Rodriges and Fedyanin[105]. Such a correspondence is naturally associated with the similar structures of the Pekar-Frolich and Holstein Hamiltonians[98-101]. For a polaron with a small radius, a canonical transformation method has been well developed.

By means of this transformation, Makhankov and Fedyanin [12]examined the main contribution to the electron mass renormalization defined by its strong interaction with the lattice vibration and obtained the Hamiltonian of a "residual" (weak) polaron lattice vibration interaction.

In this case, the lattice Hamiltonian in Eq. (109) is rewritten in terms of the normal coordinates Q_q and ξ_q via the unitary operator $U = \exp(-iL)$, where

$$L = i\sum_q \xi_q^0 \frac{\partial}{\partial \xi_q}, \tag{118}$$

with

$$\xi_q^0 = -\frac{V}{M\omega_q^2}\sqrt{\frac{2}{N}}\sum_n N_n \sin\left(qn + \frac{\pi}{4}\right)$$

The U transformations of the normal coordinates ξ_q, viz, $\overline{\xi}_q = U\xi_q U^{-1} = \xi_q + \xi_q^1 q$ are such that ξ_q^1 may be regarded as the operator of lattice deformation. The overall Hamiltonian, $\overline{H}$ is then divided into three parts: $\overline{H}_{ph}$ which describes harmonic oscillations of lattice sites about their equilibrium positions, the polaron Hamiltonian $\overline{H}_p$

$$\overline{H}_P = (\varepsilon_0 + W')\sum_k N_k - Je^{-F_T}\sum_n \left(b_{n+1}^+ b_n + b_{n+1}^+ b_{n+1}\right) + \\ V\sqrt{\frac{2}{N}}\sum_n N_n \sum_q \xi_q^0 \sin\left(qn + \frac{\pi}{4}\right) + \sum_q \frac{M\omega_q^2}{2}(\xi_q^0)^2, \tag{119}$$

where

$$F_T = \frac{2V^2}{4\hbar MN}\sum_q \frac{1}{\omega_q^3}\left\{\sin\left(qn + \frac{\pi}{4}\right)\sin\left[q(n+1) + \frac{\pi}{4}\right]\right\}\coth\left(\frac{\hbar\omega}{2K_B T}\right)$$

and the Hamiltonian $\overline{H}_{\text{int}}$ of the residual interaction of the polaron with the oscillation of the deformed-lattice. The polaron motion described by Eq. (119) is adiabatic such that the lattice reconstructs completely with oscillations about the new equilibrium positions. $\overline{H}_{\text{int}}$ describes the "friction process", i.e., the polaron motion is accompanied by a change of the phonon number, and the polaron loses its energy and slows down. Such effects decrease the energy gain due to soliton formation.

Using $|\varphi(t)\rangle$ in Equation (113), we now replaced the lattice deformation operator ξ_q^0 by

$$\overline{\xi}_q^0 = \left\langle \varphi(t) \middle| \xi_q^0 \middle| \varphi(t) \right\rangle = -\frac{V}{M\omega_q^2}\sqrt{\frac{2}{N}}\sum_n |\varphi_n(t)|^2 \sin\left(qn + \frac{\pi}{4}\right)$$

Then $\overline{H}_p$ is transformed into $\overline{H}_p'$ which satisfies the following equation

$$i\hbar\frac{\partial}{\partial t}\left|\phi(t)\right\rangle = \overline{H}'_{p}\left|\phi(t)\right\rangle \tag{120}$$

From Equation (120) and the relation $\overline{H}'_{p} = \overline{H}_{p}(\xi_{q}^{0})$, we can obtain the equation satisfied by $\varphi(t)$ in the continuum approximation. Up to $0(\omega_1^2/\omega_0^2)$, the equation is

$$i\hbar\varphi_t(x,t) = (\varepsilon_0 + W + \widetilde{\varepsilon}_0 - 2Je^{-F_T})\varphi(x,t) - Je^{-F_T}\tau_0^2\varphi_{xx} - \frac{V^2}{M\omega_0^2}\left|\varphi(x,t)\right|^2\varphi(x,t) \tag{121}$$

where

$$\widetilde{\varepsilon}_0 = \frac{V^2}{2Mr_0^2\omega_0^2}\int\left|\varphi(x,t)\right|^2 dx$$

Equation (121) is similar to Eq. (115), and therefore, has similar soliton solution as Eqs.(116) – (117). The canonical transformation adopted here is valid for small V so that the renormalization of the band width remains small, i.e. $J\exp(-F_T) >> V^2/M\omega_0^2$ (This is the condition for continuum approximation). If $\exp(-F_T) << 1$, we can neglect the electron translational motion and arrive at the small radius polaron. For the electron- acoustic phonon interaction, Whitfield and Shaw[106] showed that the polaron state becomes a soliton using the deformation potential technique.

7.2. The Features of Soliton-Polarons in Ionic Crystals

Using the adiabatic approximation, Fedyanin et al. [104] studied the properties of polaron in the case where the kinetic energy of the lattice oscillation can be neglected. In this approximation, the electron is assumed to move rapidly in the slowly varying field of the lattice deformation, which may take place if the time between two successive translation leaps of the electron, $r_1 \approx \hbar/J$, is much less than the characteristic relaxation time of the lattice deformation, $r_2 \approx \omega_0^{-1}$, i.e.,. $\hbar\omega_0 << J$. This is reasonable for a sufficiently wide electron band J. In this case, the self- consistent electron amplitude, $\phi(x,t)$ in Eq. (121), satisfies the following equation

$$i\hbar\varphi_t(x,t) = (\varepsilon_0 + W - 2J)\varphi(x,t) - Jr_0^2\varphi_{xx} - \frac{V^2\left|\varphi\right|^2\varphi}{M\omega_0^2}, \tag{122}$$

in the continuum limit.

Equation (122) is similar to Eq. (115), but the coefficients are constant. According to the above approach, Eqs. (115) – (117), the total energy of the electron and the lattice in the adiabatic approximation is given by

$$E_S = \frac{M_S v^2}{2} - \frac{1}{48J}\left(\frac{V^2}{M\omega_0^2}\right)^2$$

where $M_S = \hbar^2 / 2Jr_0^2$ is the soliton mass which is the same as that of the electron in conduction band in the given approximation, i.e., there are no temperature renormalization of the energy shift due to the translation leaps ($-2J$) and the soliton localization area $W_s^0 = (2Jr_0^2 / V^2)M\omega_0^2$. This implies the electron being weakly connected with the lattice, since $V \to 0 (W_s^0 \to \infty)$, so that we have $\hbar\omega = \varepsilon_0 + W - 2J + m_z v^2 / 2$ and $\phi(x,t) \approx \exp[i(\omega t - kx)]$, i.e., the soliton wavefunction degenerates into a plane wave, in other words, the soliton is delocalized as an electron.

In their study, Fedyanin et al. [104] used the normal coordinate ξ_q and its conjugate momentum P_q, and introduced an energy functional $E = \langle \phi(t) | H | \phi(t) \rangle$, with $|\phi(t)\rangle$ given by Eq.(113), which is a function of ξ_q and P_q for a given $\phi_n(t)$. The equation satisfied by ξ_q determines the motion of a conventional oscillator under a constant external force. The solution of the equation in such a case can be simply written as

$$\xi_q(t) = \xi_q(0)\cos(\omega_q t) - \frac{V^2}{M\omega_q^2}\sqrt{\frac{2}{N}}\sum_n |\varphi_n(t)|^2$$

Inserting the above into the equation

$$\frac{\partial}{\partial t}|\varphi(t)\rangle = \hat{H}|\varphi(t)\rangle$$

yields, in the continuum approximating, the following equation for $'\varphi_n(x,t)$

$$i\hbar\varphi_t(x,t) = (\varepsilon_0 + B - 2J)\varphi(x,t) - Jr_0^2\varphi_{xx}(x,t) - \frac{V^2}{M\omega_0^2|\varphi|^2}\varphi(x,t) \tag{123}$$

where

$$B = \sum_q \frac{V^2}{MN\omega_q^2}\left[\sum_n \frac{\partial}{\partial t}|\varphi_n(t)^2|\sin\left(qn + \frac{\pi}{4}\right)\right]^2$$

is the kinetic energy of lattice deformation consistent with the electron motion, and

$$\varepsilon_0 = \sum_q \frac{V^2}{MN\omega_q^2}\left[\sum_n |\varphi_n(t)^2| \sin\left(qn + \frac{\pi}{4}\right)\right]^2$$

is the elastic energy of the lattice. Note that in the above adiabatic approximation, we have neglected the energy of harmonic oscillations about the shifted equilibrium positions

$$E' = T' + V' = \sum_q \frac{1}{2} M\omega_q^2 \xi_q^2$$

which would lead to a constant correction to the excitation dispersion. Equation (123) is similar to Equations (115) and (122), and thus has soliton solution similar to Equation (116). It indicates that polaron in the nonlinear system is a soliton and satisfies the nonlinear Schrödinger equation. It describes a quasiparticle of mass $M_s = m\exp(F_\theta)$, localized in the region of $W_s = W_0 \exp(-F_\theta)$. The energy of the soliton,

$$\varepsilon + W - 2J\exp(-F_\theta) + V^2/(2M\omega_0^2),$$

is less than that of the electron level, $\varepsilon + W + V^2/(2M\omega_q^2)$.

Here

$$m = \frac{\hbar^2}{2Jr_0^2}, W_S^0 = \frac{4JM\omega_0^2}{v^2}, F_\theta = \frac{V^2\theta}{2M\hbar^2\omega_0^4} + O\left(\frac{\omega_1^2}{\omega_0^2}\right)$$

These properties of the polaron are independent of the method of calculation. For soliton satisfying the nonlinear Schrödinger equation (123), its "kinetic energy" is

$$E = \frac{1}{2}(m + \Delta m)v^2,$$

and

$$m = \frac{\hbar^2}{2Jr_0^2}, \quad \Delta m = \frac{4}{15}\frac{V^2}{M\omega_q^2}\left(\frac{r_0}{W_S^0}\right),$$

with the latter being the temperature independent renormalization of the polaron mass. Using $W_s^0 = 4JM\omega_0^2/V^3, \Delta m$ is given by

$$\Delta m = \frac{(V^2/M\omega_0^2)^4}{240J^3 r_0^2 \omega_0^2}.$$

8. The Soliton Excitations of Small Polarons and Its Properties in Semiconductors

We now discuss the nonlinear quantum properties of small polarons. Brown et al. [107-108] proposed a theory of polaron formation based on the Fröhlich Hamiltonian in the transportless limit[99-100]. It was found that two related processes are involved in this formation. In the first process, energy interpretable as the polaron binding energy is released by the exciton system into the normal modes of the lattice.

Another process results in the dispersion of this energy throughout the lattice and results in the development of a persistent deformation around the region occupied by the exciton. This indicates that a polaron cannot be exist until the polaron binding energy is dispersed and the lattice deformation is completed. The amount of energy released into the lattice during polaron formation and the time dependence of this energy transfer are independent of the detail with which the initial bare exciton is distributed. Ivic and Brown[109 -110] further studied the nonlinear localization of the small polarons in this case.

It is well known that a foreign particle or excitation in a deformable solid causes often local distortions of the host which can significantly affect the character and dynamics of the quasiparticle.

For example, interstitial hydrogen isotopes in metal hydrides can cause large volume dilations; mobilities of photon injected charge carriers in organic molecular crystals, such as anthracene and naphthalene exhibit novel temperature dependences; excitonic spectra in organic molecular crystals such as pyrene and α-perylene are significantly affected by local distortions; and certain features in Raman spectra of biological materials such as *l*-alanine have been identified with local vibrational modes. The quasiparticle transport in these deformable media was investigated by Ivic and Brown[109-110] using the fpllowing Fröhlich Hamiltonian:

$$H=\sum_{n}E_n b_n^+ b_n-\sum_{nm}J_{mn}b_m^+ b_n+\sum_{q}\hbar\omega_q a_q^+ a_q+\sum_{qn}\hbar\omega_q\left(x_n^q a_q^+ + x_n^{q*}a_q\right)b_n^+ b_n \quad (124)$$

where J_{nm} is a transfer integral between different sites in the medium. For the translationally invariant acoustic chain model, they choose the phonon dispersion relation ω_q and the dimensionless coupling function χ_n^q to be

$$\omega_q=\omega_B\sin\left(\frac{1}{2}|qr_0|\right),\ x_n^q=x^q e^{-iqR_n}=\frac{2ix\sin(qr_0)}{\sqrt{2NM\hbar\omega_q^3}}e^{-iqR_n},$$

respectively, respectively, where r_0 is the lattice constant, b_n and a_q are the annihilation operators of the small-polaron and phonon with wave vector q and frequency ω_q, respectively. Exact solutions corresponding to the Fröhlich Hamiltonian are known only in the limit of infinite effective mass, wherein the transfer integrals between different sites in the medium are all zero. In general the solutions are found by the most approximate approaches which are either small-polaron theories, usually employing perturbation methods based on a set of polaron basis states, or large-polaron theories, usually employing variational methods to obtain nonlinear evolution equations. Among the latter are theories which describe quasiparticle transport in terms of envelope solitons. Ivic et al. [109-110] used the latter to display the unification of polaron and soliton theories of quasiparticle transport in the small-polaron regime based on Eq.(124).

Following the notation adopted by Ivic et al. [109-110], we denote wave function related to small- polaron by a tilde,

$$|\overline{D}(t)\rangle = \sum_{a} \varphi_n(t)\overline{b}_n^{+}|0\rangle \otimes |\tilde{\beta}(t)\rangle,$$

$$|\tilde{\beta}(t)\rangle = \exp\left\{\sum_{n} \tilde{\beta}_q(t)\tilde{a}_q^{+} - \tilde{\beta}_q^{*}(t)\tilde{a}_q^{+}\right\}|0\rangle, \tag{125}$$

where $\overline{b}_n = Ub_nU^{+} = b_n\exp[\sum_q(\beta_{nq}a_q^{+} - \beta_{nq}^{*}a_q)]$ and $\overline{a}_q = Ua_qU^{+} = a_q + \sum_q \beta_{nq}b_n^{+}b_n$ are annihilation operators of the small-polaron and the dressed phonon, respectively, with

$$U = \exp\left[-\sum_{n}\left(\beta_{nq}a_q^{+} - \beta_{nq}^{*}a_q\right)b_n^{+}b_n\right] \tag{126}$$

The $|\overline{D}(t)\rangle$ states contain the exact small-polaron states in the infinite-mass limit $(\overline{\beta}_q(t) = 0)$.

The Hamiltonian in Eq.(124) is not diagonal in the small-polaron states when $J \neq 0$. However, the interaction between small polarons ($(\overline{b}, \overline{b}^{+})$ and the dressed phonons $(\overline{a}, \overline{a}^{+})$ is weaker than the interaction between the bare quasiparticles (*b, b+*) and bare phonons (*a, a+*). In this case the microscopic dynamics can be well described by a factored polaron state $|\overline{D}(t)\rangle$. Since the phase mixing part of $\beta_{qn}(t)$ is chosen to be time independent, Ivic and Brown [109 -110] considered instead the dressed polarons or solitons. To determine the equations of motion of polarons, Ivic et al. [109-110] applied the variational principle of time-dependent quantum mechanics,

$$\delta \int_{t_1}^{t_2} dt \left\langle \Psi(t) \right| (i\hbar \frac{d}{dt} - H) \left| \Psi(t) \right\rangle = 0,$$

which was also used by others (Skrinjar, Kapor, and Stojanovic [111], Zhang, Romero-Rochin [112]) and obtained the following evolution equations,

$$i\hbar\frac{d\varphi_n(t)}{dt}+i\hbar\sum_a\left[x_n^q\tilde{\beta}_q^*(t)-x_n^{q*}\tilde{\beta}_q(t)\right]\varphi_n(t)=\frac{\partial\langle H\rangle}{\partial\varphi_n^*(t)}, \tag{127}$$

$$i\hbar\frac{d\tilde{\beta}_q(t)}{dt}-i\hbar\sum_n x_n^q\frac{d}{dt}\left|\varphi_n(t)\right|^2=\frac{\partial\langle H\rangle}{\partial\tilde{\beta}_q^*(t)} \tag{128}$$

where $\langle H\rangle$ is the expectation value of the Hamiltonian, (124), in the state $|\overline{D}\rangle$. From Equations (127) and (128), we can eliminate the explicit phonon variables in the latter, to obtain an integro- differential equation for the small-polaron probability amplitudes,

$$\begin{aligned}i\hbar\frac{d\varphi_n(t)}{dt}&=(E-\sum\nolimits_q\left|\chi^q\right|^2\hbar\omega_q)\varphi(t)-\sum\nolimits_m\tilde{J}[\varphi_{n+1}(t)+\varphi_{n-1}(t)]+\\&\int_0^t d\tau\sum\nolimits_j G_{nj}(t-\tau)\frac{d}{dt}\left|\varphi_j(\tau)\right|^2\varphi_n(t)+g_n(t)\varphi_n(t)\end{aligned} \tag{129}$$

where

$$\tilde{J}=J\exp\left[-4\sum_{-q}\left|x^q\right|^2\sin^2\left(\frac{1}{2}qr_0\right)\right]$$

is the renormalized tunneling matrix element, obviously it is reduced from the bare value J by interactions with phonons, and

$$G_{nj}(t)=2\sum_q x_n^q x_j^{q*}\hbar\omega_q\cos(\omega_q t),$$

$$g_n(t)=\sum_a\hbar\omega_q\left[x_n^q\tilde{\beta}_n^*(0)e^{i\omega_q t}+x_n^{q*}\tilde{\beta}_q(0)e^{-i\omega_q t}\right]$$

The kernels $G_{nj}(t)$ are related to the thermal part of the fluctuations $g_n(t)$, as discussed by Wang et al. [113-114]. Ivic et al. [109-110]. expressed a small-polaron Bloch state by

$$|\psi(t)\rangle=\sum_n\tilde{u}(k)\exp\left\{i\left[kR_n-\frac{\tilde{E}(k)}{\hbar}\right]t\right\}\tilde{b}_n^+|0\rangle \tag{130}$$

$$\tilde{E}(k)=\left(E-\sum_q\left|x^q\right|^2\hbar\omega_q\right)-2\tilde{J}\cos(kr_0). \tag{131}$$

$$\varphi_n(t)=\tilde{u}(k)\exp\left\{ikR_n-\frac{i\tilde{E}(k)t}{\hbar}\right\}$$

and $\overline{\beta}_q(0) = 0$. Then Eq. (129) is linearized, indicating that the small-polaron, Bloch states are exact stationary solutions of the nonlinear evolution equation. In the continuum limit, Ivic et al. finally obtained the nonlinear Schrödinger equation given below by searching for solutions with the D'Alembert property $|\varphi(y,\tau)|^2 = |\varphi(y - v\tau|^2$, and choosing $\beta_q(0)$ corresponding to a dressed soliton at zero temperature,

$$i\hbar\frac{d\varphi(x,t)}{dt} = -\frac{\hbar^2}{2\tilde{m}}\frac{\partial^2}{\partial x^2}\varphi(x,t) + \tilde{E}(0)\varphi(x,t) - \frac{\upsilon^2}{\upsilon_0^2}G(\upsilon)|\varphi(x,t)|^2\varphi(x,t), \quad (132)$$

where $\overline{m} = \hbar^2 / 2Jr_0^2$ with $J = J\exp(-\delta^2 F)$.

$$F = 4\sum_q |x^q|^2 \sin(r_0 q/2)$$

is the effective mass of the small-polaron, $\tilde{E}(0) = E - \delta(2-\delta)E_0 - 2J\exp(-\delta^2 F)$ is the bottom of the energy band of the small-polaron, as given by Equation (131), and

$$G(v) = \frac{4\chi^2 r_0}{\omega}\frac{v_0^2}{v_0^2 - v^2}$$

is a velocity-dependent nonlinearity parameter. In the above, δ is defined by

$$\beta_{nq}(t) = -\delta\chi_n^q + \overline{\beta}_q(t) \text{ and } \overline{\beta}_{nq}(t) = \delta\chi_n^q.$$

The above result shows that the small polaron is a soliton which satisfies the nonlinear Schrödinger equation. This shows clearly that the dynamic properties of the small polarons can also be described by nonlinear quantum mechanics. In this case this soliton solution of equation (132) is

$$\varphi(x,t) = \sqrt{\frac{\mu}{2}}\frac{e^{-i[\overline{E}(0)/\hbar]t}e^{i(kx-\omega t)}}{\cosh[\mu(x-ut)]} \quad (133)$$

where

$$\hbar k = \tilde{m}v,\ \hbar\mu = \frac{\tilde{m}G(v)}{2\hbar}\frac{v^2}{v_0^2} \text{ and } \hbar\omega = \frac{1}{2}\tilde{m}v^2 - \frac{v^4}{v_0^2}\frac{\tilde{m}G(v)^2}{8\hbar^2}.$$

The solution in Eq. (133) of this equation are not bare Bloch states, but are small-polaron soliton with a larger effective mass $(\overline{m} > m)$ and a lower energy $[\overline{E}(0) < E(0)]$ compared to a bare Block state, due to the dressing of the quasiparticle. Therefore, the properties of the small polaron-soliton can be described completely by nonlinear quantum mechanics and are similar with those of microscopic particles in nonlinear quantum mechanics [13,93].

References

[1] Peierls R. E., *Quantum theory of solids*, London: Oxford University Press, 1955, P108.

[2] Friedman, L. and Holstein, T., *Ann. Phys.*, 21 (1963) 494.

[3] Shore, H. B. and Sander, L. M., *Phys. Rev.* B, 7 (1973) 4537.

[4] Bardeen, J., Cooper, L. N. and Schreiffer, R., *Phys. Rev.*, 108 (1957) 1175.

[5] Alexandrov, A. and Ranninger, J., *Phys. Rev.* B 45 (1992) 13109.

[6] Nasu, K., *Phys, Rev.* B, 37 (1988) 5075.

[7] Kuper, C. G. and Whitield, G. D., *Polarons and excitons*, New York: Plenum Press, 137 (1963).

[8] Habbard, L., *Proc. Roy. Soc.* A176 (1963) 238.

[9] Lindner, U., V. Fedyanin, *Phys. Stat.* s01. (b) 89 (1978) 123, 95 (1978) k83.

[10] Makhankov, V., Makhaidiani, N. and Pashaev, O., *Phys. Lett.* 81A (198) 167.

[11] Makhankov, V., *Phys. Lett.* A81 (1981) 156.

[12] Makhandov, V. G. and Fedyanin, V. K., *Phys. Rep.* 104 (1984) 1-80.

[13] Niu, Q., Wang, X. D., Kleimman, L., and Liu, W. M, et al. Adiabatic dynamics of local spin moments in Itinerant magnets. *Phys. Rev. Lett.*, 83(1999) 207.

[14] Li, H., Shen, S. Q., Liang, J. Q., Liu, W. M., *Phys. Rev.* B, 72 (2005) 014546.

[15] Zheng, H., *Phys. Rev.* B, 37 (1988) 7419.

[16] Zheng, H., *Phys. Rev.* B, 38 (1988) 11865.

[17] Zheng, H., *Phys. Rev.* B, 41 (1990) 4723.

[18] Feinberg, D., Ciuchi, S., de Pasquale, F., *Inter. J. Mod. Phys.* B, 4 (1990) 1317.

[19] Zheng, H., Feinberg, D. and Avegmon, M., *Phys. Rev.* B, 39 (1989) 9405.

[20] Hirsch, J. E., *Phys. Rev. B*, 31 (1985) 6022.

[21] Fradkin, E. and Hirsch, J. E., *Phys. Rev. B*, 27 (1983): 1680.

[22] Lang, I. C. and Firsov, Y. A., *Zh. Exsp.Ter. Fiz.* 1 43 (962) 1843. (Sov. Phys. JETP 16: (1963) 1301).

[23] Chen, H., Zheng, Y. M., Wu, X., *Phys. Rev. B*, 54 (1996) 3273.

[24] Manka, R., *Phys. Stat. Sol.* (b), 93: (1979) 53.

[25] Lo, C. F., Manousakis, E., Sollie, R., and Wang, Y. L., *Phys. Rev. B*, 50 (1994) 418.

[26] Eremk, A. A., *Z. Phys.* B, 104 (1997) 765.

[27] Kim, M. S. and Sanders, B. C., Kim, M. S., and Sanders, B. C., *Phys. Rev. A*, 1996, 53: 3694.

[28] Sahoo, S., *J. Phys. Condensed Matter* 10 (1998) 1999.

[29] Jeckelman, E., Zhang, C. L., Mite, S. R., *Phys. Rev.* B, 60 (1999): 7950.

[30] Zolotaryuk, Y., Christiansen, P. L., Rasmussen, J. J., *Phys. Rev. B*, 58 (1998) 14305.

[31] Robin, J. M., *Phys. Rev. B,* 58 (1998) 14335.

[32] Wang, W. Z., Bishop, A. R., Gaammel, J. T., and Silver, R. V. *Phys. Rev. Lett.*, 80 (1998) 3284.

[33] Leitner, D. M., *Phys. Rev. B*, 64 (2001) 094201.

[34] Tanner, P. A., *Phys. Rev. B,* 65: (2002) 214305.

[35] Xiao, J. L., Xiao, W., *Phys. Rev. B*, 58 (1998) 1678.

[36] Fan Hong-Yi, and Xiao Min, *Phys. Lett A*, 1996, 220:81.
[37] Pang Xiao-feng, *J. Phys. Chem. Sol.,* 61 (2000) 701.
[38] Pang Xiao-feng, *J. Mat. Sci. Technol.*, 151 (1999) 369.
[39] Pang Xiao-feng, *Chinese Phys. Lett.*, 16 (1998) 129.
[40] Pang Xiao-feng, *Phys. Stat. Sol.* (b), 217 (2000) 887.
[41] Pang Xiao-feng, *Acta. Physica Sinica*, 8 (1999) 598.
[42] Majernikova, E. and Pang Xiao – feng, *Phys. Lett.* 30 (1997) 89.
[43] Pang Xiao-feng, *J. Phys. Chem. Solids*, 62: (2001) 491.
[44] Pang Xiao-feng, *Science in China*, G, 51 (2008) 225-336.
[45] Borghi, G. P., Girlando, A., Painelli, A., and Voit, J., *Europhys. Lett.,* 1996, 24:127.
[46] Cooper, L. N., *Phys. Rev.* 104 (1956) 11-89.
[47] Schrieffer, J. R., *Superconductivity*, Benjamin, New York, 1969.
[48] Schrieffer, J. R., *Theory of Superconductivity*, Benjamin, New York, 1964.
[49] Parks, R. D., *Superconductivity*, Marcel. Dekker, 1969.
[50] Landau, L. D. and E. M. Lifshitz, *Quantum Mechanics*, Pergamon Press, Oxford, 1987.
[51] Davydov, A. S. and V. N. Ermakov, *Phys. Stat. Sol.* B148 (1988) 305.
[52] Ginzberg, V. I. and L. D. Landau, *Zh. Eksp. Theor. Fiz.* 20 (1956) 1064.
[53] Ginzburg, V. L. and L. P. Pitayevsky, *Sov. Phys. JETP* 7 (1968) 85.
[54] Ginzberg, V. L. and D. A. Kirahnits, *Problems in High-Temperature Superconductivity*, Nauka, Moscow, 1977.
[55] Gorkov, L. P., *Sov. Phys. JETP* 9 (1958) 189.
[56] Gorkov, L. P., *Sov. Phys. JETP* 9 (1959) 1364.
[57] Abrikosov, A. A., *Zh. Eksp. Theor. Fiz.* 32 (1957) 1442.
[58] Abrikosov, A. A. and L. P. Gorkov, *Zh. Eksp. Theor. Phys.* 39 (1960) 781.
[59] Perez-Garcia, V. M., M. Michinel and H. Herrero, *Phys. Rev.* A57 (1998) 3837.
[60] Pang, Xiao-feng, J. *Kunming Tech. Sci. Univ.*, No. 3 (1991) 57.
[61] Pang, Xiao-feng, J. *Kunming Tech. Sci. Univ., Sin.* No. 4 (1991) 13.
[62] Rogovin, D. and M. Scully, *Ann. Phys.* 88 (1969) 371.
[63] Rogovin, D. and M. Scully, *Phys. Rep.* 25 (1976) 176.
[64] Rogovin, D., *Ann. Phys.* 90 (1974) 1.
[65] Rogovin, D., *Phys. Rep.* 251 (1976) 1786.
[66] Rogovin, D., *Phys. Rev.* B11 (1975) 1906; ibid. *Phys. Rev.* B12 (1975) 130.
[67] Ronognan, J. P., *Solid State Commun.* 14 (1974) 83.
[68] Leggett, A. J., *Prog. Theor. Phys.* (Supp.) 69 (1980) 80.
[69] Leggett, A. J., In: Percolation, *Localization and Superconductivity*, eds. by A. M. Goldlinan, S. A. Bvilf, Plenum Press, New York, 1984. pp. 1- 41.
[70] Leggett, A. J., In: *Low temperature physics,* Springer, Berlin, 1991, pp. 1-93.
[71] Niek, I. R. and J. C. Fraser, *J. Low Temp. Phys.* 3 (1970) 225.
[72] Pang Xiao-feng, *J. Electronic Science and Technology of China*, 6(2) (2008) 205-211.
[73] Caradoc-Davies, B. M., R. J. Ballagh and K. Bumett, *Phys. Rev. Lett.* 83 (1999) 895.
[74] Matthews, M. R., et al., *Phys. Rev. Lett.* 83 (1999) 249.
[75] Madison, K. W., et al., *Phys. Rev. Lett.* 84 (2000) 806.
[76] Tonomara, A., *Foundation Phys.* 28 (1998) 59.
[77] Pang, Xiao-feng, *J. Southwest Inst. For Nationalities Sin.* 15 (1989) 97.
[78] Blackbunu, J. A., et al., *Phys. Rev.* B11 (1975) 1053.
[79] Jacobson, D. A., *Phys. Rev.* B8 (1965) 1066; *Adv. Phys.* 14 (1965) 419.

[80] Pang, Xiao-feng, *J. Science Exploration Sin.* No. 4 (1986) 70.
[81] Pang, Xiao-feng, *J. Southwest Inst. For Nationalities Sin.* 17 (1991) 1 and 1.
[82] Josephson, B., *Phys. Lett.* 1 (1962) 251; *Adv. Phys.* 14 (1965) 419.
[83] Josephson, B., *Thesis, unpublised*, Cambridge University (1964).
[84] A. Blackburn J., et al., *Phys. Rev.* B 11 (1975) 1053.
[85] P. Romagnan, J., et al., *Solid State Commun.* 14 (1974) 83.
[86] Dewitt, B. S., *Phys. Rev. Lett.* 16 (1966) 1092.
[87] Kusayanage, E., T. Kawashima and K. Yamafuji, *J. Phys. Soc. Japan* 33 (1972) 551.
[88] Pang, Xiao-feng, *J. Low Temp. Phys.* 58 (1985) 333.
[87] Pang, Xiao-feng, The Properties of Soliton Motion for Superconductivity Electrons. *Proc. ICNP*, Shanghai, 1989, p.139.
[90] Cai, S. Y. and A. Bhattacharjee, *Phys. Rev.* A43 (1991) 6934.
[91] Pang Xiao-feng, *Theory of Nonlinear Quantum Mechanics*, Chinese Chonqing Press, Chonqing, 1994, P573-583.
[92] Pang Xiao-feng, *Soliton Physics, Sichuan.* Sci. Tech. press. Chengdu, 2002, p714-771.
[93] Pang Xiao-feng, *Nonlinear Quantum Mechanics*, Beijing, Chinese Electronic Industry Press, Beijing, 2009, p. 20-63.
[94] Pang Xiao-feng, *Chinese. J. Low Temp. Supercond.* No. 4 (1989) 33 and No. 3 (1982) 62.
[95] Pang Xiao-feng, *Nature Sciences*, 2(1) (2007) 42.
[96] Pang, Xiao-feng, *Chinese J. Nature*, 5 (1982) 254.
[97] Landau, L., *Phys. Z. Sowjetunion*, 3 (1933) 664.
[98] Pekar, S., *J. Phys.* (Moscow) 10 (1946) 341 and 10 (1946) 347.
[99] Fröhlich, H., Pelzer, H. and Zienau, S., *Plilos. Mag.* 41 (1950) 221.
[100] Fröhlich, H., *Adv. Phys.* 3 (1954) 325.
[101] Holstein, T., *Ann. Phys.* (N.Y.) and 8(1959)325 and 8 (1959) 343.
[102] Devreese, J. T., P*olarons in ionic crystals and polarsecaicounductors*, Amsterdam. North Holland 1972.
[103] Fedyanin, V. S. and Yushankhai, V., *Teor. Mat. Fiz*, 35 (1978) 240.
[104] Fedyanin, V., Yakushevich, L., *Phys. Lett.* 85A (1981) 100.
[105] Rodriges, K. and Fedyanin, V., JINR P17-81-169, Dubna, 1981; JINR p. 17-81-8, Dubna 1981.
[106] Whitfield, G. and P. Shaw, *Phys. Rev.* B14 (1976) 3346.
[107] Brown, D. W., Lindemberg, K. and West, B. J., *Phys. Rev.* B37 (1988) 2946.
[108] Brown, D. W., Lindemberg, K. and West, B. J., *J. Chem. Phys.* 87 (1987) 6700 and 84 (1986).
[109] Ivic, Z., Brown, D. W., *Phys. Rev. Lett.* 63 (1989) 423.
[110] Brown, D. W. and Ivic, Z., *Phys. Rev.* B40 (1989) 9876.
[111] Skrinjar, M. J., Kapor, D. V. Stojanovic, S. D., *Phys. Rev.* A38 (1988) 6402.
[112] Zhang, Q., Romero-Rochin, V., Silbey, R., *Phys. Rev.* A38 (1988) 6409.
[113] Wang, X. D., Brown, D. W., Lindenberg, K., West, B. J., *Phys. Rev.* A33 (1986) 4101.
[114] Brown, B. W., West, B. J. and Lindenerg, K., *Phys. Rev.* A33 (1986) 4110.

[108] Brown, D. W., Lindemberg, K. and West, B. J., *J. Chem. Phys*. 87 (1987) 6700 and 84 (1986).
[109] Ivic, Z., Brown, D. W., *Phys. Rev. Lett*. 63 (1989) 423.
[110] Brown, D. W. and Ivic, Z., *Phys. Rev*. B40 (1989) 9876.
[111] Skrinjar, M. J., Kapor, D. V. Stojanovic, S. D., *Phys. Rev*. A38 (1988) 6402.
[112] Zhang, Q., Romero-Rochin, V., Silbey, R., *Phys. Rev.* A38 (1988) 6409.
[113] Wang, X. D., Brown, D. W., Lindenberg, K., West, B. J., *Phys. Rev*. A33 (1986) 4101.
[114] Brown, B. W., West, B. J. and Lindenerg, K., *Phys. Rev*. A33 (1986) 4110.

In: Solitons
Editor: Matthew C. Williams

ISBN: 978-1-62618-234-9

Chapter 2

THE SOLITON EXCITATIONS AND ITS EXPERIMENTAL EVIDENCES IN MOLECULAR CRYSTALS

Pang Xiao-Feng*

Institute of Life Science and Technology, University of Electronic Science and Technology of China, Chengdu, P.R. China

Abstract

We here first establish Hamiltonian of molecular crystal and corresponding nonlinear dynamic equation of exciton through determining the mechanism of generation and forms of nonlinear interaction based on its molecular structure. Subsequently we find out the exciton-soliton and phonon-soliton solutions of the dynamic equations of exciton and phonon in acetanilide and study their properties, respectively. We further calculate the eigenenergy spectrum and specific heat of acetanilide and investigate theoretically the features of Raman scattering and infrared absorption of acetanilide arising from these soliton excitations, and obtain the rule of exponential decreases of infrared absorption strength of $1650cm^{-1}$ band, which corresponds to the soliton, with increasing the temperature of acetanilide using the soliton model. Finally we measure the experimental results of Raman scattering and infrared absorption to verify the correctness of theoretical results mentioned above, in which we obtain that the frequencies of peaks in Raman and infrared spectra consist exactly with the theoretical result in eigenenergy levels of acetanilide and the experimental rule of change of strength of the peaks at $1650cm^{-1}$ with varying the temperature concord also the above theoretical result obtained by the soliton theory. Meanwhile, to use Mark free-electron laser captures truly the information of the soliton and obtain that the soliton lifetime is 15ps in this experiment and 35ps in pump-probe spectroscopy measures, respectively. These results obtained from theoretical and experimental studies represent clearly and confirm further that the exciton-soliton is really existent in the molecular crystal.

* E-mail address: pangxf2006@yahoo.com.cn, Tel: 00862883202595, Address: Institute of Life Science and Technology, University of Electronic Science and Technology of China, Chengdu 610054, P.R. China. (Corresponding author)

1. The Soliton Excitation of the Excitons in Molecular Crystal

1.1. The Structure of Molecular Crystals and Soliton Excitation of the Exciton

We first discuss a particular molecular crystal, the acetanilide $(CH_3COHNC_4H_5)_x$. Two close chains of hydrogen-bonded amide-I groups which consists of atoms of carbon, oxygen, nitrogen and hydrogen (CONH), run through the acetanilide crystal [1-6]. This crystal structure has been determined and a unit cell of acetanilide is shown in Figure 1. The space group is $D_{2h}^{15}(P_{bca})$ and the unit cell or factor group is D_{2h} for this crystal. The average lattice constants are a = 1.9640 nm, b = 0.9483 nm, and c = 0.7979 nm. There are eight molecules in an unit cell and at the amide- I frequency, each of these has one degree of freedom (d.f.). Thus, there are three infrared-active modes (B_{1u}, B_{2u}, and B_{3u}), four Raman-active modes (A_g, B_{1g}, B_{2g}, and B_{3g}), and one inactive mode (A_u). However, at low frequency (< $200\,cm^{-1}$), each molecule exhibits 6 d.f. (three translations and three rotations. This gives 48 low-frequency modes: 24 Raman active modes ($6A_g + 6B_{1g} + 6B_{2g} + 6B_{3g}$), 18 infrared-active modes ($6B_{1u} + 6B_{2u} + 6B_{3u}$) and six ($A_u$) modes corresponding to acoustic modes of translation and rotation). All of these active modes are seen in infrared absorption and Raman experiments. Meanwhile, in acetanilide, the nearly planar amide-I groups have bond lengths, which are close to those found in polypeptide (see Figure 2). Since the physical properties of such hydrogen bonded amide-I systems are very sensitive to bond lengths, hence, studies of acetanilide can reveal new phenomena. For example, in experiments of infrared absorption and Raman scattering, a new amide-I band red-shifted from the main peak at 1666 cm^{-1} by about 16 cm^{-1}, when the crystalline acetanilide is cooled from 320 to 10 K, but no other major changes occur from 4000 to 800 cm^{-1}.

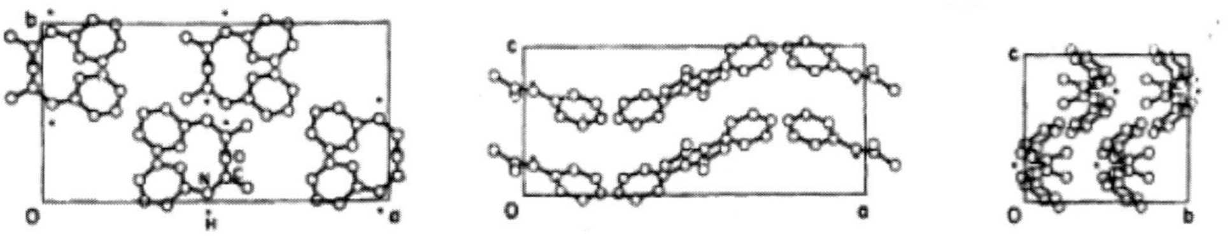

Figure 1. Various views of the cell unit of acetanilide, with cell parameters a = 19.640 °A, b = 9.483 °A, and c = 7.979 °A [Reference 1-6].

The intensity of this new band increases steadily from room temperature to 70 K. The band at $1650\,cm^{-1}$ is not present in amorphous materials or acetanilide methylated at the position where hydrogen-bonded distances occur, but it is recovered after annealing. Similar phenomena can be observed in Raman scattering experiments [7-25].

As it is known, the characteristic feature of amide-I group, CONH, in polypeptides is amide-I mode, which mainly involves the stretching of C=O bond. This mode is observed as an infrared absorption peak at 1666 cm^{-1} in acetanilide and near this value in a wide variety of materials, including the amide-I groups.

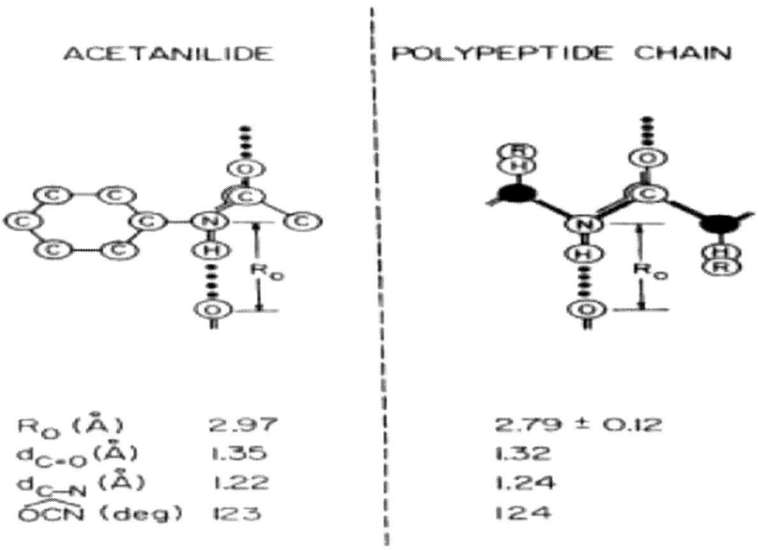

Figure 2. Comparisons between of acetanilide and a protein molecule [Reference 1-6].

The corresponding spectroscopic evidence of the new band at 1650 cm^{-1} has been mentioned earlier, but detailed measurements of the crystal structure and specific heat as a function of temperature peptide groups preclude assignment of the new band to (1) a conventional amide-I mode, (5) crystal defect states, (3) Fermi resonance or (4) frozen kinetics between two different subsystems.

The correct assignment is the self-trapping of the amide-I vibrational energy. These conclusions are based on the following experimental facts: (1) the ^{15}N substitution induces a small shift to the amide-I at 1666 cm^{-1}, and the new band is also shifted by the same amount; (2) deuterium substitution at the NH position strongly affects both the amide-I and the new band in a complicated way; (3) upon cooling a decrease in the integrated absorption of the normal amide-I and a corresponding increase in the integrated absorption of the 1650 cm^{-1} band are observed; (4) the 1650 cm^{-1} band and the amide-I band show the same dichroism over the temperature range integrated; and (5) the measurements of specific heat, the dielectric constant and the volume expansion as a function of temperature rule out the occurrence of rotational isomerism or polymorphic transition which would affect some other infrared and Raman absorption bands, but not the new band. The self-trapping mechanism of amide-I vibrational energy used by Scott and Eilbeck et al. [8-12] comes from the Davydov model of vibration energy transport in alpha-helix protein molecules [16-20] which was described in his book [17]. They have given a good account of the properties of first excited state in this model. Scott and co-workers, Alexander and Krumbansl have also obtained an exponential dependence of the absorption intensity on temperature, exp $(-\beta T^2)$ [14, 15], and explained the experimentally observed intensity change of the new band with decreasing temperature in terms of a complementary polaron or soliton of the self-trapping state of exciton on the basis of the Davydov model (Davydov referred to the intramolecular excitation occurred in the systems as exciton in such a case).

However, the red shift of a few cm^{-1} given by this model is much smaller than the experimental value of 16 cm^{-1}. This indicates that these models need further improvement and development.

1.2. The Theory of Soliton Excitation of the Excitons

Pang and co-workers [26-35] also proposed that the new band of amide-I is caused by the self- trapping of amide-I vibrational quanta (exciton) in their soliton model (the vibrational quantum is referred to as exciton, in accordance with the exciton theory by Davydov). But this model is different from Davydov's soliton model or Alexander's complementary polaron model.

Here it is nonlinear interaction between the localized amide-I vibrational-quantum and low frequency vibrations of the lattice (phonons) that result in self-trapping of the exciton as a soliton.

The mechanism of this soliton model can be described as follows: Through their intrinsic nonlinear interaction, an amide-I vibrational quantum acts as a source of low-frequency phonons and causes shifts in the average positions of the lattice molecules in ground state. These shifts, or lattice distortions, in turn, react through the same nonlinear interaction, as a potential well to trap the amide-I vibrational quantum, resulting in the soliton, and preventing the energy dispersion of the amide-I vibrational quantum via the dipole-dipole interaction that occurs in neighboring peptides with certain electric moments.

Evidently, the soliton is formed by the self-trapping of excitons interacting with low frequency lattice phonons, and is a dynamic self-sustaining entity. This entity propagates, together with the lattice deformation along the molecular chains.

The main property of the soliton is that it can move over macroscopic distances at velocity v and retain its wave form, energy, momentum, and other quasi-particle properties. Using this model, Pang and coworkers also assigned an exponential dependence of the infrared-absorption intensity to the temperature which is consistent with experimental result [26-35].

From these studies, we know that collective excitations can result from a localized fluctuation of the amide-I vibrational quanta and structural deformation of peptide groups due to photon-excitation in molecular crystals. The Hamiltonian of the systems given by Pang [26-35] is as follows:

$$H = H_{ex} + H_{ph} + H_{\text{int}} = \frac{1}{2m}\sum_n p_n^2 + \frac{m\omega_0^2}{2}\sum_n r_n^2 + \frac{m\omega_1^2}{2}\sum_n r_n r_{n+1}$$
$$+\frac{1}{2M}\sum_n P_n^2 + \frac{\beta}{2}\sum_n (R_n - R_{n-1})^2 + m\chi_1 \sum_n (R_{n+1} - R_{n-1}) r_n^2$$
$$+ m\chi_2 \sum_n (R_{n+1} - R_n) r_n r_{n+1} - m\chi_2 \sum_n (R_{n-1} - R_n) r_n r_{n-1} \tag{1}$$

This Hamiltonian includes vibrational excitation of amide-I caused by localized fluctuation and vibration of the peptide groups. This is, in turn, caused by structural deformation of molecular chains, and the interaction between the two modes of motion in the respective crystals. Here m is the mass of amide-I vibrational quantum (exciton), ω_0 and ω_1 are diagonal and off-diagonal elements of the dynamic matrix of vibrational quantum, ω_0 is also a Einstein vibrational frequency of the exciton, and $m\omega_1^2 r_n r_{n+1}/2$ is the interaction between the nearest neighboring excitons caused by the dipole-dipole interaction in molecular chains.

Likewise r_n and $\mathrm{p}_n = m(\dot{r}_n)$ *are* the normal coordinates of the nth excitons and its canonical conjugate momentum, respectively, M the mass of a peptide group in the unit cell, $\chi_1=\partial\omega_0^2/\partial R_n$ and $\chi_2=\partial\omega_1^2/\partial R_n$ are the change of energy of exciton and of coupling interaction between the neighbouring excitons for an unit extension of molecular chain, respectively. R_n and $\mathrm{P}_n = M\dot{\mathrm{R}}_n$ are the canonically conjugate operators of displacement and the momentum of peptide group and β the elastic constant of molecular chains. H_{ex} in H is the Hamiltonian of harmonic vibration of amide-I including the off-diagonal factor, H_{ph} is the Hamiltonian of harmonic vibration of the peptide group and H_{int} is the interaction Hamiltonian between the two modes of motion.

This Hamiltonian is significantly different from that of the Davydov mode 1 [16-20] used by Scott and co-workers [7-8, 11-12]. As far as the vibration of amide-I is concerned, we adopt a harmonic oscillator model with optical vibration that includes an off-diagonal factor, which comes from the interaction between excitons, and interaction with displacement of peptide groups. Thus, vibrational frequencies of amide-I are related to displacements of peptide groups, which shows the occurrence of the interaction between amide-I vibrations and displacement of peptide groups. Their relationship is described by:

$$\omega_0^2(R_n) \approx \omega_0^2 - \frac{\partial\omega_0^2}{\partial R_n}(R_n - R_{n-1}) = \omega_0^2 + \chi_1(R_n - R_{n-1}),$$

$$\omega_1^2(R_n) \approx \omega_1^2 + \chi_2(R_n - R_{n-1}).$$

Inserting the above into the vibrational Hamiltonian of amide-I, and taking into account the effect of the neighboring peptide groups on the excitons in both sides, the result seen in equation (1) can be obtained. Therefore, the Hamiltonian given above has a one-to-one correspondence for the interactions.

However, the Hamiltonian in the Davydov mode does not have such a one-to-one correspondence.

In other words, there is no interaction between the amide-I vibrational quantum and displacement of the peptide groups resulting from the interaction between the neighboring excitons in the Davydov model [16-20].

Thus, the Davydov model has encountered many difficulties in the study of the dynamics of energy transport in the systems.

The Hamiltonian given above does include not only the optical vibration of amide-I, but also the resonant interaction caused by the dipole-dipole interaction between neighboring excitons, and it also takes into account both the change of the relative displacement of neighboring peptide groups resulting from the vibration of amide-I *and* the correlation interaction between neighboring excitons.

Therefore, the Hamiltonian given above has a one-to-one correspondence in H_{ex} and H_{int}. Thus it can give a more complete description on the dynamics of systems compared to Davydov model and other models.

Since amide-I vibration and vibration of peptide groups are all quantized, we here introduce the following canonical second quantization transformation,

$$r_n = \sqrt{\frac{\hbar}{2m\omega_0}}\left(b_n^+ + b_n\right), p_n = -i\sqrt{\frac{2m\omega_0}{\hbar}}\left(b_n - b_n^+\right) \tag{2}$$

$$R_n = \sum_q \left[\frac{\hbar}{2NM\omega_q}\right]^{1/2} e^{iqnr_0}\left(a_{-q}^+ + a_q\right), P_n = i\sum_q \left[\frac{M\hbar\omega_q}{2N}\right]^{1/2} e^{iqnr_0}\left(a_{-q}^+ - a_q\right) \tag{3}$$

where $i=\sqrt{-1}$, $\omega_{\mathrm{q}} = 2(\beta/M)^{1/2}\sin(r_0 q/2)$ is the frequency of the phonon with wave vector q, N the number of unit cells in the molecular chain, r_0 the distance between the molecules, $\mathrm{b}_n^+(b_n)$ and $a_q^+(a_q)$ are the creation (annihilation) operator of the exciton and phonon, respectively.

Using Equations (2) and (3), equation (1) becomes

$$H = \sum_n \varepsilon_0\left(b_n^+ b_n + \frac{1}{2}\right) - J\sum_n\left(b_n^+ b_{n+1} + b_n b_{n+1}^+\right) + \sum_q \hbar\omega_q\left(a_q^+ a_q + \frac{1}{2}\right) +$$

$$\sum_{n,q}\left[g(q)\left(b_n^+ b_n + b_n b_n^+\right) + g_1(q)\left(b_n^+ b_{n+1} + b_n b_{n+1}^+\right) - g_2(q)\left(b_n^+ b_{n-1} + b_n b_{n-1}^+\right)\right]\left(a_q + a_{-q}^+\right)e^{inr_0 q}, \tag{4}$$

where

$$\varepsilon_0 = \hbar\omega_0, J = \hbar\omega_1^2/4\omega_0, g(q) = \left(\frac{\hbar}{2NM\omega_q}\right)^{1/2}\left(\frac{\hbar\chi_1}{2\omega_0}\right)\left[e^{ir_0 q} - e^{-ir_0 q}\right],$$

$$g_1(q) = \left(\frac{\hbar}{2NM\omega_q}\right)^{1/2}\left(\frac{\hbar\chi_2}{2\omega_0}\right)\left[e^{ir_0 q} - 1\right], g_2(q) = \left(\frac{\hbar}{2NM\omega_q}\right)^{1/2}\left(\frac{\hbar\chi_2}{2\omega_0}\right)\left[e^{-ir_0 q} - 1\right].$$

Because the collective excitations generated by the localized fluctuation of the excitons and structural deformation of the peptide groups have a coherent feature , thus Pang [25-30] represented the excitation states of the excitons and vibrations of peptide group (phonons) by the following wave function:

$$|\Phi\rangle = |\varphi(t)\rangle U(t)|0\rangle_{ph} = \frac{1}{\rho}\left(1+\sum_n \varphi_n(t) b_n^+\right)|0\rangle_{ex} \cdot U(t)|0\rangle_{ph} \tag{5}$$

with

$$U(t) = \exp\left\{\sum_n \frac{1}{i\hbar}\left[u_n(t)P_n - \pi_n(t)R_n\right]\right\}$$

or

$$U(t) = \exp\left\{\sum_q \left[\alpha(t)a_q^+ - \alpha_q^*(t)a_q\right]\right\} \tag{6}$$

where $|0\rangle_{ex}$ and $|0\rangle_{ph}$ are the excitonic and phononic ground-states, respectively. The ρ is the normalization factor. We assume hereafter that $\rho = 1$ for convenience of calculation unless it is explicitly noted otherwise. The $\phi_n(t)$, $u_n(t) = \langle\Phi|R_n|\Phi\rangle$ and $\pi_n(t) = \langle\Phi|P_n|\Phi\rangle$ are three unknown functions.

In Equation (5) the wave function of the excitons is

$$|\varphi\rangle = \frac{1}{\rho}\left(1+\sum{}_n \varphi_n(t) b_n^+|0\rangle_{ex}\right),$$

which is different from $|\varphi_D\rangle = \sum{}_n \varphi_n(t) B_n^+|0\rangle_{ex}$ in the Davydov wave function $|D_2\rangle$, seen in his book [17], and it is not an excitation state of a single particle, but rather a coherent state, or more accurately, a quasi-coherent state. In the case of small $\varphi_n(t)$ (i.e., $|\phi_n(t)| << 1$) and it can be approximately written as

$$|\varphi\rangle \approx \frac{1}{\rho}\exp\left[-\frac{1}{2}\sum_n |\varphi_n(t)|^2\right]\exp\left[\sum_n \varphi_n(t) b_n^+\right]|0\rangle_{ex} = \frac{1}{\rho}\exp\left[\sum_n \varphi_n(t) b_n^+ - \varphi_n^*(t) b_n\right]|0\rangle_{ex}$$

in which the final representation is a standard coherent state, which is normalized. Since the condition $\sum{}_n |\varphi_n(t)|^2 = 1$, is required in this calculation, then the condition, $|\phi_n(t)| << 1$, is naturally satisfied for the acetanilide consisting of several hundreds peptide groups.

Therefore, the above representation is correct. Thus, the wave function in Equation (5) is also normalized at $\rho = 1$. In fact, the wave function of the exciton state in Equation (5) should be viewed as an effective truncation of a standard coherent state, in which the aforementioned two terms are retained.

Thus it is referred to as a quasi-coherent state. However, it is not an eigenstate of the number operator $\hat{N}=\sum_n b_n^+ b_n$ since

$$\hat{N}|\varphi\rangle=\sum_n b_n^+ b_n|\varphi\rangle=|\varphi\rangle-|0\rangle_{ex}\neq|\varphi\rangle.$$

Hence, $|\varphi\rangle$ represents a coherent superposition of the one-quantum amide-I vibrational state and the ground state of the exciton.

But, the number of quantum are determinate in this state and it contains only one exciton because:

$$N=\langle\varphi|\hat{N}|\varphi\rangle=\langle\varphi|\sum_n b_n^+ b_n|\varphi\rangle=\sum_n|\varphi_n(t)|^2=1$$

Therefore, $|\varphi\rangle$ given in Equation (5) not only exhibits coherent feature of collective excitations of the excitons caused by the nonlinear interaction generated by means of the exciton-phonon interaction, which makes the wave functions of the states of the exciton and phonon symmetrical (because the wave function of the phonon was already a standard coherent sate), but it can also make the number of the excitons conserved for the Hamiltonian in Equation (4) as well.

Meanwhile, the wave function given above has another advantage, i.e., the equation of motion of the exciton can be obtained from the following Heisenberg equations for the creation and annihilation operators in Equations (4) – (6)

$$i\hbar\frac{\partial}{\partial t}\langle\Phi|b_n|\Phi\rangle=|\Phi\rangle[b_n,H]|\Phi\rangle, i\hbar\frac{\partial}{\partial t}\langle\Phi|a_q|\Phi\rangle=|\Phi\rangle[a_q,H]|\Phi\rangle \tag{7}$$

From these equations we can obtain the following equations of $\varphi_n(t), \alpha_q(t)$ and $\alpha_{-q}^*(t)$

$$i\hbar\dot{\varphi}_n=\hbar\omega_0\varphi_n-J(\varphi_{n+1}+\varphi_{n-1})+\sum_q\{2g(q)(\alpha_q+\alpha_{-q}^*)\varphi_n+ \\ (g_1(q)-g_2(q))[(\alpha_q+\alpha_{-q}^*)\varphi_{n+1}+(\alpha_q+\alpha_{-q}^*)\varphi_{n-1}]\}e^{ir_0qn}, \tag{8}$$

$$i\hbar\dot{\alpha}_q=\hbar\omega_q\alpha_q+\sum_n\left\{2g(q)|\varphi_n(t)|^2+[g_1(q)-g_2(q)](\varphi_n^*\varphi_{n+1}+\varphi_n^*\varphi_{n-1})\right\}e^{-ir_0qn} \tag{9}$$

$$i\hbar\dot{\alpha}_{-q}^*=-\hbar\omega_q\alpha_{-q}^*-\sum_n\left\{2g(q)|\varphi_n(t)|^2+[g_1(q)-g_2(q)](\varphi_n^*\varphi_{n+1}+\varphi_n^*\varphi_{n-1})\right\}e^{-ir_0qn},$$

respectively. From Equations (9) and (10), we can get:

$$\ddot{\alpha}^*_{-q} + \ddot{\alpha}_q = -\omega_q^2\left(\alpha_q + \alpha^*_{-q}\right) - \sum_n \left\{ \frac{4g(q)\omega_q}{\hbar}|\varphi_n|^2 + \frac{2g_1(q) - 2g_2(q)}{\hbar}\omega_q\left(\varphi^*_n\varphi_{n+1} + \varphi^*_n\varphi_{n-1}\right)\right\} e^{-ir_0qn}.$$

Also, from Equation (3) we get the Fourier transformation of the variable, $u_n = \langle\Phi|R_n|\Phi\rangle$, as:

$$u_n = \frac{1}{\sqrt{N}}\sum_q u_q(t)e^{iqx},$$

where $x = nr_0$ and $u_q(t) = \sqrt{\frac{\hbar}{2M\omega_q}}\left(\alpha_q + \alpha^*_q\right)$.

Thus we get in the long-wave length approximation

$$\ddot{u}_q + \upsilon_0^2 q^2 u_q = \frac{1}{\sqrt{N}}\sum_n \left\{ \frac{2iq\hbar r_0\chi_1}{M\omega_0}|\varphi_n|^2 + \frac{iq\hbar r_0\chi_2}{M\omega_0}\left(\varphi^*_n\varphi_{n+1} + \varphi^*_n\varphi_{n-1}\right)\right\} e^{-ir_0qn}$$

where $\omega_q = v_0 q$, $v_0 = r_0\left(\beta/M\right)^{1/2}$.

Now, multiplying the above equation by $\exp(ir_0qn)/\sqrt{N}$, summing over the wave number, q, and making the continuum approximation, we [26-32] get:

$$\frac{\partial^2 u(x,t)}{\partial t^2} - \upsilon_0^2\frac{\partial^2 u(x,t)}{\partial x^2} = \frac{2\hbar r_0\left(\chi_1 + \chi_2\right)}{M\omega_0}\frac{\partial|\varphi(x,t)|^2}{\partial x}. \qquad (11)$$

At the same time, equation (7) becomes:

$$i\hbar\frac{\partial\varphi(x,t)}{\partial t} = \left(\hbar\omega_0 - 2J\right)\varphi(x,t) - Jr_0^2\frac{\partial^2\varphi(x,t)}{\partial x^2} + \frac{2\hbar r_0\left(\chi_1 + \chi_2\right)}{\omega_0}\varphi(x,t)\frac{\partial u(x,t)}{\partial x} \qquad (12)$$

Inserting Equation (11) into Equation (12), we can get the following nonlinear Schrödinger equation for motion of the exciton:

$$i\hbar\frac{\partial\varphi(x,t)}{\partial t} = \left(\hbar\omega_0 - 2J\right)\varphi(x,t) - Jr_0^2\frac{\partial^2\varphi(x,t)}{\partial x^2} + \frac{4\hbar^2 r_0^2\left(\chi_1 + \chi_2\right)^2}{M\omega_0\left(\upsilon^2 - \upsilon_0^2\right)}|\varphi(x,t)|^2\varphi(x,t) \qquad (13)$$

where $\hbar^2/2m = Jr_0^2$. Here m is the effective mass of the exciton. Equation (13) is a nonlinear Schrödinger equation, its solution is a soliton and can be represented by:

$$\varphi(x,t)=\sqrt{\frac{\mu}{2}}\operatorname{sec}h\left[\mu'\left(x-x_0-\upsilon t\right)\right]\exp\left\{\frac{i}{\hbar}\left[\frac{\hbar^2\upsilon\left(x-x_0\right)}{2Jr_0^2}-E_{sol}t\right]\right\}, \tag{14}$$

$$u(x,t)=-\frac{\hbar\left(x_1+x_2\right)}{2\omega_0\beta\left(1-s^2\right)}\tanh\left[\mu'\left(x-x_0-\upsilon t\right)\right] \tag{15}$$

where

$$\mu=\frac{G}{4J},G=\frac{4\hbar^2r_0^2\left(\chi_1+\chi_2\right)^2}{M\upsilon_0^2\omega_0^2\left(1-s^2\right)},\mu'=\frac{\mu}{r_0},s=\frac{\upsilon}{\upsilon_0},\varepsilon=\hbar\omega_0-2J,\varepsilon_0=\hbar\omega_0, \tag{16}$$

Using

$$\alpha_{\mathrm{q}}=\sqrt{\frac{M\omega_{\mathrm{q}}}{2\hbar}}\mathrm{u}_{\mathrm{q}}+\mathrm{i}\sqrt{\frac{1}{2M\hbar\omega_{\mathrm{q}}}}\pi_{\mathrm{q}},$$

and the results given above, we found that $\alpha_q(t)$ in Equation (3) is given by:

$$\alpha_q(t)=\frac{1}{\sqrt{N}}\sum_n\alpha_n(t)e^{-iqx}=\frac{i\hbar\left(\chi_1+\chi_2\right)\pi M\left(\omega_q+\upsilon q\right)}{2M\upsilon_0^2\omega_0\left(1-s^2\right)\hbar N\omega_q}\sinh\left(\frac{\pi r_0 q}{2\mu}\right)e^{iq\upsilon t}=\alpha_q e^{iq\upsilon t}. \tag{17}$$

These results clearly show that the excitation, and motion of the excitons in the acetanilide, are a nonlinear quantum mechanical problem, and can satisfy or obey the laws of nonlinear quantum mechanics. Thus the exciton-soliton has a wave-corpuscle duality and all properties of microscopic particlesm [36-37].

Obviously, in this model the motions of the exciton and phonon and their interaction were all considered, the nonlinear interactions for the exciton and phonon occur simultaneously. In this case the nonlinear interaction energy (*G*), is provided by the coupling between the excitons and phonons. Just so, the exciton is a self- trapped as a soliton in this case [36-37].

1.3. The Features of Soliton Excitation in Acetanilide

The energy of the soliton can be found by using Equations (14) and (15) and is given by:

$$E_{sol}=\int_{-\infty}^{\infty}Hd\xi=\varepsilon_0-2J+\frac{\hbar^2\upsilon^2}{4Jr_0^2}-\frac{\mu^2J}{3}=E_0+\frac{M_{sol}\upsilon^2}{2},$$

where

$$E_0 = \varepsilon_0 - 2J - \frac{\hbar^4(\chi_1+\chi_2)^4}{3\beta^2 J\omega_0^4}$$

is the rest energy of the soliton. The mass of the soliton is given by:

$$M_{sol} = m + \frac{4\hbar^4(\chi_1+\chi_2)^4(1-3s^2/2-s^4/2)}{3\beta^2 J\omega_0^4\upsilon_0^2(1-s^2)^2} > m$$

Therefore, the energy and the rest energy of the soliton are $\hbar^4(\chi_1+\chi_2)^4/3\beta^2 J\omega_0^4(1-s^2)$ and $\hbar^4(\chi_1+\chi_2)^4/3\beta^2 J\omega_0^4$, respectively, both of which are lower than those of the exciton, $E' = \varepsilon_0 - 2J + \frac{1}{2}\text{mv}^2$ and $E_0' = \varepsilon_0 - 2J$, respectively.

Therefore, the exciton-soliton is stable. But the mass of the soliton is greater than that of the exciton, m.

1.4. The Behaviors of Soliton Excitations of Phonons in Molecular Crystal

From the above studies we know that exciton-soliton or localization of exciton occurs due to nonlinear exciton-phonon interaction. In such a case, the phonon is also localized, at the same time, and moves in a soliton in the systems. This can be verified from Equations (11) – (12). In such a case, from Equations (14) – (15) and

$$\mu = (\chi_1+\chi_2)^2\hbar^2 / JM\upsilon_0^2\omega_0^2(1-\text{s}^2)^2,$$

where $s = v/v_0$, we can get

$$\frac{\partial}{\partial x}|\varphi(\chi,\ \text{t})|^2 = \frac{2(\chi_1+\chi_2)^3\hbar^3}{r_0^3\beta J^2(1-\text{s}^2)\omega_0^3}\left[u - \frac{4\omega_0^2(1-\text{s}^2)^2\beta^2}{\hbar^2(\chi_1+\chi_2)^2}u^3\right] \tag{18}$$

Substituting Equation (19) into Equation (11), we obtain

$$u_{\text{tt}} - \upsilon_0^2 u_{\text{xx}} = \frac{4(\chi_1+\chi_2)^4\hbar^4}{Mr_0^2\beta J^2(1-\text{s}^2)\omega_0^4}\left[u - \frac{4\omega_0^2(1-\text{s}^2)^2\beta^2}{\hbar^2(\chi_1+\chi_2)^2}u^3\right] \tag{19}$$

This is a ϕ^4-equation for *u(x, t)*. It has a soliton solution given in Equation (15). Thus phonon has been localized in this case. Obviously, the localization of phonon is due to the nonlinear exciton-phonon interaction in acetanilide. If $\chi_1 = \chi_2 = 0$, then Equation (20) reduces to a linear wave equation, $u_{tt} - v_0^2 u_{xx} = 0$, which has a solution of plane wave, $u = A\sin(\omega t - \theta_0)$, with dispersive effect.

This manifests that the peptide groups performs harmonic vibration in this case. This denotes that the localization of phonons is carried out by virtue of the exciton- phonon interaction, which results in nonlinear interaction of $B'u^3$ in Equation (19), where $B' = 16(\chi_1 + \chi_2)^2 \hbar^2 \beta(1 - s^2)/M v_0^2 \omega_0^2 J$.

Hence, the motion of the phonon should be described by nonlinear quantum mechanics, instead of linear quantum mechanics, in the system.

These results show clearly that both microscopic particles (or excitons) and background field (or phonons) are acted by the nonlinear interactions.

In other words, the microscopic particles (excitons) and vibration of background field (or phonons) simultaneously are localized and possess the wave-corpuscle duality under action of the nonlinear interaction. This consists with the conclusion in nonlinear quantum mechanics [36-37], i.e., the action and counteraction of the nonlinear interaction simultaneously exist for interacting two particles in nonlinear quantum mechanics, although their sizes are different.

2. The Features of the Solitons in Anharmonic Vibration of Lattices in Acetanilide

Pang et al. [26-31] also studied the influence of anharmonic vibration of the peptide groups on the soliton. Then the Hamiltonian of peptide group in Equation (1) is now replaced by

$$H'_{ph} = \frac{1}{2M}\sum_n P_n^2 + \sum_n [\frac{\beta}{2}(R_n - R_{n-1})^2 + \frac{1}{3}\lambda_1 (R_n - R_{n-1})^3] \tag{20}$$

It was found that term of $(\hbar\omega_0 - 2J)\varphi(x,t)$ in Equation (13) is only changed in this case, thus the phase, velocity, energy, amplitude and width of the soliton are only varied.

Inserting Equation (3) into Equation (20) we obtain

$$H_{ph} = \sum_q \hbar\omega_q \left(a_q^+ a_q + \frac{1}{2}\right) + \sum_{q_1 q_2} F(q_1, q_2)\left(a_{-q_1}^+ + a_{q_1}\right)\left(a_{-q_2}^+ + a_{q_2}\right)\left(a_{-(q_1+q_2)}^+ + a_{(q_1+q_2)}\right) \tag{21}$$

where

$$F(q_1, q_2) = \frac{8}{3} i\lambda_1 \left(\frac{\hbar}{2M}\omega_{q_1}\omega_{q_2}\omega_{(q_1+q_2)}\right)^{1/2} \sin(r_0 q_1/2)\sin(r_0 q_2/2)\sin[r_0(q_1+q_2)/2]$$

From Equation (21) and the phonon part, $U|0\rangle_{ph}$, in Equation (5) and the Heisenberg equation (7), we can get

$$i\hbar\dot{\alpha}_q = \hbar\omega_q\alpha_q - \sum_k \left\{F(k-q)\left(\alpha_k + \alpha^*_{-k}\right)\left(\alpha_{k-q} + \alpha^*_{k-q}\right)\right\} \tag{22}$$

$$i\hbar\dot{\alpha}^*_{-q} = -\hbar\omega_q\alpha^*_{-q} + \sum_k \left\{F(k-q)\left(\alpha_k + \alpha^*_{-k}\right)\left(\alpha_{k-q} + \alpha^*_{k-q}\right)\right\} \tag{23}$$

From Equations (22) and (23), we get:

$$\ddot{\alpha}^*_{-q} + \ddot{\alpha}_q = -\omega_q^2\left(\alpha_q + \alpha^*_{-q}\right) - \sum_n \left\{\frac{2F(k-q)\omega_q}{\hbar}\left(\alpha_k + \alpha^*_{-k}\right)\left(\alpha_{k-q} + \alpha^*_{k-q}\right)\right\} \tag{24}$$

Because

$$u_n = {}_{ph}\langle 0|U|R_n|U|0\rangle_{ph} = \frac{1}{\sqrt{N}}\sum_q u_q(t)e^{iqx},$$

where $x=n\,r_0$ and $u_q(t) = \sqrt{\frac{\hbar}{2M\omega_q}}\left(\alpha_q + \alpha^*_q\right)$,

then, equation(24) becomes

$$\ddot{u}_q + \omega_q^2 u_q = \frac{-8}{3M} i\lambda_1 \sin(r_0 q_1/2)\sum_k \sin(r_0 k/2)\sin[r_0(q-k)/2]u_k u_{q-k}e^{-ir_0(k-q)} \tag{25}$$

Taking into account the dispersion relation,

$$\omega_q \approx \frac{1}{2}\sqrt{\frac{\beta}{M}}[\frac{1}{2}v_0 q - \frac{1}{3!}(\frac{1}{2}r_0 q)^3] = v_0 q\left(1 - \frac{1}{24}(r_0 q)^2\right),$$

Equation (25) in long wavelength and the second order approximation can be written as

$$\ddot{u}_q + v_0^2 q^2 u_q - \frac{1}{12}v_0^2 r_0^2 q^4 u_q = \frac{-i}{3M}\lambda_1 r_0^3 q\sum_k (q-k)k u_k u_{q-k}$$

In the continuum approximation, applying the following relation,

$$\frac{\partial u(x,t)}{\partial x}=\frac{1}{\sqrt{N}}\sum_q (iq)u_q(t)e^{iqx}$$

The above equation can be represented by

$$\frac{\partial^2 u(x,t)}{\partial t^2}-\upsilon_0^2\frac{\partial^2 u(x,t)}{\partial x^2}-\frac{v_0^2 r_0^2}{12}\frac{\partial^4 u(x,t)}{\partial x^4}=\frac{\lambda_1 r_0^3}{3M}\left|\frac{\partial u(x,t)}{\partial x}\right|^2 \tag{26}$$

Letting $Q(x,t)=-\partial u(x,t)/\partial x$, then the above equation can be rewritten as

$$\frac{\partial^2 Q(x,t)}{\partial t^2}-\upsilon_0^2\frac{\partial^2 Q(x,t)}{\partial x^2}-\frac{v_0^2 r_0^2}{12}\frac{\partial^4 Q(x,t)}{\partial x^4}=\frac{\lambda_1 r_0^3}{3M}|Q(x,t)|^2 \tag{27}$$

This is a nonlinear equation of the phonon. It shows again that the phonon is a nonlinear microscopic particle in such a case. Applying the boundary conditions, $Q(\pm\infty)=u(\pm\infty)=0$, it has the following soliton solution

$$Q(x,t)=\frac{3M}{4\sqrt{2}\lambda_1 r_0^3}(v^2-v_0^2)\mathrm{sec}h^2\left[\sqrt{2}(x-x_0-\upsilon t)/(r_0\sqrt{(v^2/v_0^2)-1})\right],(v>v_0) \tag{28}$$

Thus

$$u(x,t)=\frac{-3M}{4\sqrt{2}\lambda_1 r_0^2}(v^2-v_0^2)^{3/2}\tanh\left[\sqrt{2}(x-x_0-\upsilon t)/(r_0\sqrt{(v^2/v_0^2)-1})\right],(v>v_0) \tag{29}$$

where v is the velocity of the phonon-soliton. The shape of the soliton is similar to the soliton solutionin Equation (15). But Equation (29) is a supersonic soliton ($v>v_0$), which satisfies the nonlinear equation (26), while Equation (15) is a subsonic soliton ($v<v_0$), which satisfies the ϕ^4 − equation (19). Therefore, the mechanism and properties of localization of the phonons in Equations (19) and (27) are obviously different in the two cases. These results show clearly that both exciton and phonon become as solitons with wave-corpuscle duality under action of the nonlinear interaction arising from their motion and interaction between them. Thus, their rules and properties of motion can be described using nonlinear quantum mechanics [36-37].

3. Raman Scattering Effect Arising from the Motions of the Exciton-Solitons

3.1. The Studied Method

At present, we study the physical phenomena and effects arising from the nonlinear excitations and motions of the excitons, such as, Raman scattering and infrared absorption, in the following three sections.

In order to study the Raman scattering effect we first should diagonalize the soliton Hamiltonian equation in nonlinear quantum mechanics [36-37]. Then we should transform it to the intrinsic reference frame moving with the soliton at velocity v [26-31]. In this case the (4) must be replaced by

$$\overline{H} = H - \sum_{\kappa} \hbar k v (b_{\kappa}^{+} b_{\kappa} + a_{\kappa}^{+} a_{\kappa}) \tag{30}$$

where

$$b_k^+ = N^{-1/2} \sum_n \exp(iknr_0) b_k^+, b_k = (b_k^+)^+$$

Since the soliton excitation is always connected with the deformation of molecular chain or intramolecular distances, it is very necessary to pass in Equation (30) to new phonons taking this deformation into account. In the light of Eremko, et al.'s method [25], such a transition is realized by the replacement $(R_i, P_i) \rightarrow (a_q^+, a_q) \rightarrow (A_q^+, A_q)$ according to the relations: $A_q^+ = a_q^+ - N^{-1/2}\alpha_q^*, A_q = a_q - N^{-1/2}\alpha_q$, where A_q^+ (A_q) is the creation (annihilation) operator of the new phonon. The coherent phonon state (lattice distortion) in Equation (6), thus, becomes as the vacuum state of new phonons

$$\left|\hat{0}\right\rangle_{ph} = \exp[(\alpha_q(t) a_q^+ - \alpha_q^*(t) a_q^+]\left|0\right\rangle_{ph}$$

where $A_q\left|\tilde{0}\right\rangle_{ph} = 0$. We now carry out the canonical transformation for the partial diagonalization of the soliton Hamiltonian by the following transformation:

$$B_\lambda^+ = \sum_n \varphi_\lambda(t) b_n^+, B_\lambda = (B_\lambda^+)^+,$$

with

$$\sum_\lambda \varphi_\lambda^*(n)\varphi_\lambda(n') = \delta_{nn'}, \sum_n \varphi_\lambda^*(n)\varphi_{\lambda'}(n) = \delta_{\lambda\lambda'}$$

The so-called partial diagonalization of the Hamiltonian implies the diagonalization of that part of the Hamiltonian which does not contain the creation and annihilation operators of the new phonons. The conditions imposed onto the function φ_λ to realize such a diagonalization are equivalent in the continuum approximation to the following problem on eigenfunction φ (n/λ) and the eigenvalues E_λ:

$$\left[-J\frac{\partial^2}{\partial n^2}-(\frac{i\hbar v}{r_0})\frac{\partial}{\partial n}-(2\mu^2 Jr_0^2)\sec h^2(\mu n)+\varepsilon_0-2J\right]\varphi\begin{pmatrix}n\\ \lambda\end{pmatrix}=E_{\lambda\varphi}\varphi\begin{pmatrix}n\\ \lambda\end{pmatrix}.$$

It has the unique bound state solution:

$$\varphi\begin{pmatrix}n\\ s\end{pmatrix}=(\frac{\mu}{2})^{1/2}\sec h(\mu n)\exp\left[\frac{i\hbar vn}{2Jr_0^2}\right],$$

$$E_s=E_0-2J-[(\hbar^2v^2)/(4Jr_0^2)]-\mu^2 J \tag{31}$$

and an unbound state solution given by

$$\varphi\begin{pmatrix}n\\ k\end{pmatrix}=\left\{\frac{\mu\cdot th(\mu n)-ikr_0}{\sqrt{2}\pi(\mu-ikr_0)}\right\}\exp\left[i(\frac{\hbar v}{2Jr_0^2})+(kr_0)n\right],$$

$$E_\kappa=E_0-2J-[(\hbar^2v^2)/(4Jr_0^2)]+(kr_0)^2 J \tag{32}$$

E_s is less than that of lowest unbound state $E_\lambda^0=\varepsilon_0-2J$ by μ^2 J in the statistical state. Thus, the partially diagonalized Hamiltonian in Equation (31) becomes

$$\begin{aligned}\overline{H}=&E_sB_s^+B_s+\sum_k E_kB_k^+B_k+\sum_q\hbar(\omega_q-vq)A_q^+A_q+\frac{1}{\sqrt{N}}\sum_q\hbar(\omega_q-vq)(A_q^+\alpha_q+A_q^+\alpha_q^*)(1-B_s^+B_s)+\\&\frac{1}{N}\sum_{qk}\overline{F}(k,q)(B_s^+B_{-k}+B_k^+B_s)(A_{-q}^++A_q)+\frac{1}{\sqrt{N}}\sum_{qk}F(k,q)(B_{k+q}^+B_k)(A_q+A_{-q}^+)+\frac{G^2}{48J}\end{aligned} \tag{33}$$

where

$$\tilde{F}(k,q)=\left[\frac{2ir_0qJ\pi\hbar}{\omega_0(G+4Jir_0k)}\right](\frac{\hbar J}{GM\omega_q})^{1/2}[x_1(e^{iqr_0}-e^{-iqr_0})+x_2(e^{iqr_0}-e^{-iqr_0})]\sec h\left[\frac{2\pi r_0J(q-k)}{G}\right],$$

and

$$F(k,q)=(\frac{\hbar^3}{2M\omega_0^2\omega_q})^{1/2}[x_1(e^{iqr_0}-e^{-iqr_0})+x_2(e^{iqr_0}-e^{-iqr_0})]\left\{1-\frac{4iGqr_0}{(G-4ikr_0J)[G-4o(l+q)r_0J]}\right\}$$

are the coupled constants of the bound state and the unbound state with the new phonons, respectively. Obviously, the distribution amplitude of the amide-I vibrational excitations, Equations (31) and (32), can be found if an exactly specified deformation potential is assumed to be given a priori. Thus this treatment mentioned above makes us see that the Hamiltonian in Equations (32) or (31) describing the molecular chain in the reference frame moving with velocity v possesses both the soliton state, equation (31), with the energy

$$E_{sol} = E_s + W = \varepsilon_0 - 2J - (\frac{\hbar^2 v^2}{4Jr_0^2}) - \frac{G^2}{48J},$$

and also the delocalized state corresponding to the exciton in the undeformed chain with the energy E_k in Equation (32). The former is a localized coherent structure with a size of the order of $2\pi\ r_0\ /\ \mu$ that propagates with velocity v and can transfer energy $E_{sol} < \varepsilon_0$. Unlike a bare exciton that can be scattered by interactions with the phonons, the soliton state describes a quasi-particle consisting of the exciton plus the lattice deformation (for this point we can see from E_{sol}) and hence it already includes the interaction with the acoustic phonons. Therefore, it is not scattered and speeded by this interaction, and can maintain its form, energy, momentum and other quasi-particle properties over a macroscopic distance in the absence of external fields and /or impurities.

The above representation shows sufficiently that the soliton energy is lower than that of the lowest exciton state (k =0) by $G^2/48J$. Thus an energy gap between the soliton state and the exciton state occurs which corresponds to the binding energy of the soliton. This is due to the fact that the soliton, unlike the exciton, is localized, and is a dynamically self-sustaining entity resulting from a self-trapping of exciton through interaction with the deformation of the molecular chain. Because of the self-trapping, the energy of the exciton drops about $G^2/24J$ which is the same as the deformation energy of the lattice,

$$\frac{1}{N}\sum_q \hbar(\omega_q - vq)\left|\alpha_q\right|^2 = \frac{G^2}{24J}$$

The energy of the soliton formed by this mechanism is about $G^2/48J$ less than that of the exciton. Thus, the energy gap between them is $G^2/48J$. The energy gap can result in the red shift from the main peak at 1666 cm^{-1} in the infrared absorption and Raman spectra. The red shift of about 16 cm^{-1} from main peak 1666 cm^{-1} was observed experimentally in infrared and Raman spectra in the acetanilide as mentioned earlier.

Utilizing the following values of the physical parameters of acetanilide [1-6, 26-31],

β=(9.6~26)N/m; M=(3.42~4.50)$\times10^{-25}$kg , $\chi = \hbar\chi_1/2\omega_0 = (56-62)PN$, J=3.4-4$cm^{-1}$, $\chi' = \hbar\chi_2/2\omega_0 = (6-9)PN$, r_0=4.0 $\overset{0}{A}$, $\omega_0 = (2-4)\times10^{14}s^{-1}, \omega_1 = (3-6)\times10^{13}s^{-1}$ ε_0 =(0.205~0.210)eV, (34)

Pang obtained the binding energy of the soliton or the energy gap between the soliton and exciton to be 18.1–33 cm^{-1} which is basically consistent with the experimental value of the red shift, 16 cm^{-1}. However, the binding energy of the soliton is about 3–7 cm^{-1} for the Davydov model, which is too small. The soliton or the localization of the exciton is formed due to the fact that the nonlinear interaction energy G=(1.5-3.2)×10^{-21}J balances and suppresses the dispersive energy J=0.795×10^{-22}J. This indicates clearly that the observed red shift is caused by the soliton. Thus it is experimentally confirmed that the soliton is actually localized in the acetanilide.

Identification of the corresponding Stokes component in the Raman scattering spectrum is one of the ways of experimentally determining the energy gap between the soliton and exciton and, hence, the solitons themselves.

Therefore, it is necessary to calculate the differential cross-section of the Raman scattering due to the solitons. We assume here that the Raman scattering process is activated by some intermediate states of molecular chains associated with the soliton, e.g., electronic excitation whose Hamiltonian is

$$H_e = \sum_{kn} \varepsilon_n(k) D^+_{kn} D_{kn}$$

where $D^+_{kn}(D_{kn})$ is the creation (annihilation) operator of the electronic excitation of nth band with wave vector k and the energy $\varepsilon_n(k)$. At the same time, let the incident light wave be quantized in the volume $V'=Nr_0S'$ and be denoted by the Hamiltonian

$$H_Q = \sum_{Q\sigma} \hbar\omega_Q C^+_{Q\sigma} C_{Q\sigma}$$

where $C^+_{Q\sigma}(C_{Q\sigma})$ is the creation (annihilation) operator of the photon with the wave vector Q , the energy $\hbar\omega_Q$ and unit polarization vector $\vec{e}_\sigma(Q)$. In such a case, the interaction Hamiltonian among the soliton, electronic excitation states and light wave leading to the Raman scattering can be denoted by

$$H_{\text{int}} = \frac{1}{N}\sum_{kk'}\sum_{q} X_{nn}(q) D^+_{k+qn} D_{kn} B^+_{k'-q} B_{k'} + \frac{1}{N}\sum_{kk'}\sum_{q} \overline{X}_{nn}(q) D^+_{kn} D_{kn'} B^+_{k'-q} B_{k'} (B^+_k B_{k'} - B^+_k B_k) + \sqrt{N}\sum_{Q\sigma n} U_{n\sigma}(Q)(C^+_{Q\sigma} + B_{-Q\sigma})(D^+_{Qn} - D_{-Qn}) \quad (35)$$

where $X_{nn'}$ and $\bar{X}_{nn'}$ are the interaction coefficients among the excitons and solitons and electronic excitations with different wave vectors, respectively, $U_{n\sigma}(Q)$ is the coupled constant between the electronic excitations and the light wave, i.e.,

$$U_{n\sigma}(Q) = -i\varepsilon_n(0)(\vec{e}_\sigma(Q).\vec{d}_n$$

where $\vec{d}_n$ is the dipole moment of the transition from the ground state of electron into the state of nth band, the Z –axis is directed along the molecular chain.

Pang calculated the transition probability per unit time of the electron-light-soliton (exciton) system from the initial state:

$$|m\rangle = C^{+}_{Q_0\sigma_0}|0\rangle_Q B^{+}_{s}|0\rangle_{ex}\prod_q \frac{1}{\sqrt{\bar{n}_q}}(A^{+}_q)^{\bar{n}_q}|\bar{0}\rangle_{ph}|0\rangle_e,$$

to the final state:

$$|fk\rangle = C^{+}_{Q\sigma}|0\rangle_Q B^{+}_{k}|0\rangle_{ex}\prod_q \frac{1}{\sqrt{n_q!}}(a^{+}_q)^{n_q}|0\rangle_{ph}|0\rangle_e$$

caused by the perturbation potential, Equation (34), where the initial state consists of the photons with the wave vector $\vec{Q}_0$ and polarization $\vec{e}_{\sigma_0}(\vec{Q}_0)$ and the soliton moving with the velocity v and the new phonons and the electronic excitation, while the final state consists of the photons with the wave vector $\vec{Q}$ and polarization $\vec{e}_\sigma(\vec{Q})$ and the exciton and a certain number of ordinary phonons and electronic excitations, where $|0\rangle_Q$ and $|0\rangle_e$ are the vacuum states of the photons and the electronic excitation states, respectively. According to the perturbation theory in quantum theory, in the case of interest the probability of transition from the initial state to the final state can be determined by

$$\frac{d}{dt}\overline{W}[(Q_0,\sigma_0)\to(Q,\sigma)] = \frac{d}{dt}\Big[\frac{1}{\hbar^6}\sum_{k,j}P^{(ph)}\Big|\int_{-\infty}^{t} dt_1\int_{-\infty}^{t_1} dt_2\int_{-\infty}^{t_2} dt_3\langle fk|\overline{H}_{\text{int}}(t_1)\overline{H}_{\text{int}}(t_2)\overline{H}_{\text{int}}(t_2)|m\rangle\Big|^2\Big]$$

where

$$\overline{H}_{\text{int}}(t_1) = \exp[iH_0t]H_{\text{int}}(t)\exp[iH_0t]$$
$$H_0 = H_0 + H_Q + H_e - \hbar\sum_{kn} kvD^{+}_{kn}D_{kn} - \hbar\sum_{Q\sigma}(\vec{Q}.\vec{v})C^{+}_{Q\sigma}C_{Q\sigma},$$

However, we are more interested in the long-time behavior of $d\overline{W}/dt$ for estimating this probability. By straightforward calculation the final transition probability can approximately be represented by

$$\lim_{t\to\infty}\frac{d\overline{W}}{dt}\approx\frac{\pi^2 r_0^2}{4V'^2}\frac{\gamma\sqrt{J}}{\mu^2 J+\Delta_{ks}}(\frac{\gamma}{2.42\omega_0})^{g_0}\{[(\Delta_{ks}(0)-W)^2+(\hbar\gamma)^2][\sqrt{(\Delta_{ks}(0)-W)^2+(\hbar\gamma)^2}-$$

$$\sqrt{(\Delta_{ks}(0)-W)^2}]\}^{-1/2}\left|\sum_{nn_o}\frac{(Q_{0z}-Q_z)U_{\sigma_0 n_0}(Q_o)[X_{nn_0}(Q_{0z}-Q_z)+\overline{X}_{nn_0}(Q_{0z}-Q_z)]U^*_{\sigma n}(Q)}{[(\varepsilon_{n_0}(Q_{0z})-\hbar\omega_{Q_{0z}})(\varepsilon_n(Q_z)-\hbar\omega_{Q_z})]}+\right. \tag{36}$$

$$\left.\frac{(Q_z-Q_{0z})U_{\sigma_0 n_0}(Q_o)[X_{nn_0}(Q_z-Q_{0z})+\overline{X}_{nn_0}(Q_z-Q_{0z})]U^*_{\sigma n}(Q)}{[(\varepsilon_{n_0}(Q_{0z})-\hbar\omega_{Q_{0z}})(\varepsilon_n(Q_z)-\hbar\omega_{Q_z})]}\right|^2\times$$

$$\{2+\tanh^2[\frac{\pi}{2\mu}(\sqrt{\frac{\Delta_{ks}(0)}{J}}-(Q_{0z}-Q_z)r_0)]+\tanh^2[\frac{\pi}{2\mu}(\sqrt{\frac{\Delta_{ks}}{J}}-(Q_z-Q_{0z})r_0)]$$

where

$$\gamma=\frac{2J\mu r_0 K_B T}{\hbar^2 v_o},\Delta_{ks}=E_k-E_s-\mu^2 J/3+\hbar(\omega_Q-\omega_{Q_0})$$

3.2. Characteristics of Raman Scattering of Acetanilide

According to quantum theory we find out the cross-section of Raman scattering is closely associated with the transition probability, Equation (36).

Due to the fact that this transition probability has a sharp peak, when the frequency of scattering light is in vicinity of the frequency of the incident light, ω_{Q_0}, therefore we only are of interest to the cross-section of Raman scattering, which is

$$\frac{dN}{NdS}=\frac{d^2\delta}{d\omega_Q d\Omega}=\frac{V'^2\omega_{Q_0}^2}{4\pi^2}\frac{d\overline{W}}{dt}\approx\frac{V'^2\omega_{Q_0}^2 r_0^2}{16V'^2\hbar^2}\frac{\hbar\gamma\sqrt{J}}{\mu^2 J+\Delta_{ks}}(\frac{\gamma}{2.42\omega_0})^{g_0}\{[(\Delta_{ks}(0)-W)^2+(\hbar\gamma)^2]\times$$

$$[\sqrt{(\Delta_{ks}(0)-W)^2+(\hbar\gamma)^2}-\sqrt{(\Delta_{ks}(0)-W)^2}]\}^{-1/2}\times$$

$$\left|\sum_{nn_o}\frac{(Q_{0z}-Q_z)U_{\sigma_0 n_0}(Q_o)[X_{nn_0}(Q_{0z}-Q_z)+\overline{X}_{nn_0}(Q_{0z}-Q_z)]U^*_{\sigma n}(Q)}{[(\varepsilon_{n_0}(Q_{0z})-\hbar\omega_{Q_{0z}})(\varepsilon_n(Q_z)-\hbar\omega_{Q_z})]}+\right. \tag{37}$$

$$\left.\frac{(Q_z-Q_{0z})U_{\sigma_0 n_0}(Q_o)[X_{nn_0}(Q_z-Q_{0z})+\overline{X}_{nn_0}(Q_z-Q_{0z})]U^*_{\sigma n}(Q)}{[(\varepsilon_{n_0}(Q_{0z})-\hbar\omega_{Q_{0z}})(\varepsilon_n(Q_z)-\hbar\omega_{Q_z})]}\right|^2\times$$

$$\{2+\tanh^2[\frac{\pi}{2\mu}(\sqrt{\frac{\Delta_{ks}(0)}{J}}-(Q_{0z}-Q_z)r_0)]+\tanh^2[\frac{\pi}{2\mu}(\sqrt{\frac{\Delta_{ks}}{J}}-(Q_z-Q_{0z})r_0)]$$

where dΩ=dθ dφ sinθ is the element of solid angle in direction of the scattering light propagation, θ_0 is an angle between the incident light $\vec{Q}_0$ and the z-axis, θ is an angle between the $\vec{Q}_0$ and the scattering light $\vec{Q}$, φ is a angle between the projection of the vector $\vec{Q}$ on the xy-plane and x-axis as is shown in Figure 3.

Obviously, the size of the Raman scattering depends closely on the size of the energy gap and the direction and intensity of incident light and the size and orientation of the dipole moment of the transition from the ground state to the resonance electronic level. This shows that the properties of the scattering light are different for different incident lights in the systems. If the frequency of the incident light is in the following region,

$$0 < \hbar\omega_{Q_0} < \mu^2 J - \varepsilon_n(0)$$

or

$$\mu^2 J + \varepsilon_n(0) < \hbar\omega_{Q_0} < 2\varepsilon_n(0),$$

then Equation (37) represents a resonant Raman scattering. The inequalities ensure not only the smoothness of the cross-section as a function of ω_{Q_0} but also the applicability of the above perturbation theory to the calculation of the cross-section.

However, when the frequency of the incident light satisfies $\mu^2 J << \hbar\omega_{Q_0} << \varepsilon_n(0)$ for all n, then Equation (37) represents a nonresonant Raman scattering. In such a case, a part of energy of incident light will be expanded and converted into the excitation energy of the soliton.

Thus, the angular-dependent part in Equation (37) may approximately be treated by a constant.

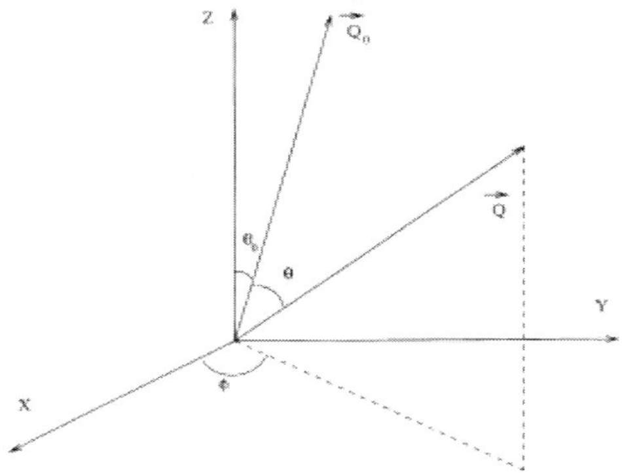

Figure 3. The schematic figure of directions for the incident light $\vec{Q}$ and scattering light $\vec{Q}$, here the molecular chain is in the z axis.

If studying in detail the significances of Equation (37), the Raman scattering has the following properties: (1)When the frequencies of the scattering light is close to that of the incident light, then the scattering cross-section is largest. (2) For each fixed direction of the scattered light, the cross-section and the shapes of the spectral line are mainly determined by the factor,

$$\frac{\hbar\gamma\sqrt{J}}{\mu^2 J+\Delta_{ks}}\frac{\omega_{Q_0}^2 r_0^2(16\hbar^2)^{-1}(\gamma/2.42\omega_0)^{g_0}}{\{[(\Delta_{ks}(0)-W)^2+(\hbar\gamma)^2][\sqrt{(\Delta_{ks}(0)-W)^2+(\hbar\gamma)^2}-\sqrt{(\Delta_{ks}(0)-W)^2}]\}^{1/2}}$$

whether the resonant Raman scattering or nonresonant Raman scattering. (3) In general, the scattering indicatrices (angular dependence of the scattering cross-section) are different for the resonant scattering and nonresonant scattering.

For the former the scattering indicatrix depends strongly on the orientation of the dipole moment of the transition from the ground state to the resonance electronic levels. For the latter when the frequencies of the scattered light, ω_Q , are close to that of the Raman peak, its scattering indicatrix is not related with the coupling strength, and has a simple form. (4) The intensity of the scattered light reaches a maximum when the frequency ω_Q approaches $\omega_{Q_0}+2J-G^2/48J-\varepsilon_0$, This maximum scattering intensity is associated with the transition of the system from the soliton state, which is accompanied by lattice deformation, to the delocalized exciton state without lattice deformation.

Therefore, the soliton becomes unstable as the incident light is scattered. These characteristics of Raman scattering and the remarkable asymmetry exhibited in the spectral line shape (sharper on the short wave side and smoother on the long wave side), which is connected to the corresponding energy dependence of the density of the delocalized states, allow us to confirm the existence of the soliton state of excitons in the organic molecular crystals through the Raman experimental results.

4. Changes of Properties of Infrared Absorption of Acetanilide Arising from Exciton-Soliton

4.1. Calculation of Coefficient of Infrared Absorption in Acetanilide

In the light of Hamilton theory in the nonlinear quantum mechanics [36-37], Pang expressed the effective Hamiltonian corresponding to Equation (13) as

$$H_{eff}=\frac{1}{r_0}\int(\frac{1}{2}Jr_0^2|\varphi_x|^2+\frac{1}{2}\varepsilon|\varphi|^2-\frac{1}{4}G|\varphi|^4)dx. \quad (38)$$

The corresponding discrete form is

$$H_{eff}=\frac{J}{2}\sum_n(\varphi_n\varphi_{n+1}^*+\varphi_{n-1}\varphi_n^*)+\sum_n(\frac{1}{2}\varepsilon|\varphi_n|^2-\frac{1}{4}G|\varphi_n|^4) \quad (39)$$

In earlier discussion, we considered only properties of exciton-soliton which is formed through coupling of amide-I oscillators (excitons) with longitudinal acoustic phonons. To explain the temperature dependence of intensity of the infrared absorption of the exciton-soliton, one has to take the interaction between the soliton and optical phonons in acetanilide into account. It is believed that this phenomenon is related to the zero-phonon line spectrum arising from random thermal modulations of the exciton-solitons by the optical phonons. In this case, the Hamiltonian of this system can be written as

$$H = H_{eff} + H_{op} + H^{'} \tag{40}$$

where H_{eff} is the Hamiltonian of the effective amide-I oscillator system given by Equation (39), which satisfies $H_{eff}\left|m\right\rangle = \hbar\omega(m)\left|m\right\rangle$. Here the eigenstate and the energy eigenvalue are collectively denoted by $\left|m\right\rangle$ and $\hbar\omega(m)$, respectively, in terms of a quantum number m. H_{op} is the Hamiltonian for the optical-phonon system which is taken to be a set of harmonic oscillators,

$$H_{op} = \sum_n h_n(Q_n) = \sum_n [-\frac{\hbar^2}{2M^*}(\frac{\partial^2}{\partial Q_n^2}) + \frac{1}{2}M^*\omega_n^2 Q_n^2] \tag{41}$$

with

$$h_n(Q_n)\Phi_{\kappa n}(Q_n) = \hbar\omega_n(\kappa n + \frac{1}{2})\Phi_{\kappa n}(Q_n) = \hbar\omega_n(\kappa n)\Phi_{\kappa n}(Q_n),$$

$$\kappa n = 0, 1, 2, \ldots,\ \omega\ (N) = \sum_n \omega_n(\kappa n),$$

where h_n, Q_n and ω_n are the Hamiltonian, the normal coordinate and the eigenfrequency of the nth harmonic oscillator, respectively, in which M^* is its effective mass, and $\Phi_{k_n}(Q_n)$ is an eigenfunction of h_n characterized by the integer k_n. $H^{'}$ in Equation (40) describes the interaction between the amide-I oscillator system and the optical phonon system which we take to be linear in the $Q_n^{'}s$, *i.e.*,

$$H^{'} = \sum_{n^{'}} V_{n^{'}}(\varphi_{n^{'}})Q_{n^{'}} \tag{42}$$

We can find that

$$V_{n^{'}}(\varphi_{n^{'}}) = \frac{1}{\sqrt{n}}\sum_{n^{''}} \lambda_{n^{''}}\varphi_{n^{'}}^2 \vec{e}(\vec{k},\eta)e^{i\vec{k}\bullet\overrightarrow{u_{n^{'}}}} \sin(\vec{k}\bullet\overrightarrow{u_{n^{'}}}) \tag{43}$$

where $\vec{e}(\vec{k},\eta)$ is a polarization vector associated with the wave vector $\vec{k}$ and branch $\eta(\eta=1,2,3$, two transverse and one longitudinal), $\lambda_{n''}$ is an interaction constant in which the index n'' indicates the range of the oscillator-phonon interaction.

In order to study the properties of the entire system given by Equation (40), we use an adiabatic approximation, *i.e.*, the phonon state depends adiabatically on the amide-I oscillator state. This can be justified because the characteristic frequency of about 1650 cm^{-1} of the amide-I oscillator system is much higher than the maximum frequency of the phonon system which is about 200 cm^{-1}.

In order to obtain the adiabatic potential U_m associated with the state $|m\rangle$ of the amide-I oscillator system, we neglect the change in wave function of the amide-I oscillator system caused by the perturbation H'. Thus the effective Hamiltonian H_{eff} of this system for the state $|m\rangle$ can be written in terms of a set of the Hamiltonians of the displaced harmonic oscillators of the phonon system

$$H_{eff}(m)=\hbar\omega(m)+\sum_{n'}\frac{1}{2}M^{*}\omega_{n'}^{2}Q_{n'}^{2}-\sum_{n'}\frac{\hbar^{2}}{2M^{*}}\frac{\partial^{2}}{\partial Q_{n'}^{2}}+\sum_{n'}q_{n'}(m)Q_{n'}$$

$$=U_{m}-\sum_{n'}\frac{\hbar^{2}}{2M^{*}}\frac{\partial^{2}}{\partial Q_{n'}^{2}}=\sum_{n'}h_{n'}[Q_{n'}-Q_{0n'}(m)]+const., \tag{44}$$

where

$$q_{n'}(m)=\langle m|V_{n'}(m)|(m)\rangle, Q_{0n'}(m)=-\frac{q_{n'}(m)}{M\omega_{n'}^{2}}.$$

Using this Hamiltonian, we can find the infrared absorption coefficient arising from the motion of the soliton. In accordance with the first order Born approximation in quantum radiation and theory of absorption of light, the absorption coefficient per unit volume of the amide-I oscillator system described by Equation (44) can be expressed as

$$B_{fn'(\omega)}=\frac{4\pi\omega}{\varepsilon' chV_{o}}\int dte^{-i\omega t}\langle\sum_{m_f}\langle m_{n'}|P(t)|m_f\rangle\langle m_f|P|m_{n'}\rangle\times$$

$$\prod_{n'}\sum_{k_n'}\Gamma(\kappa_{n'},\kappa_n',m_n,m_f)^{2}\,e^{i[\omega(\kappa_n')-\omega(\kappa_{n'})]t}\rangle_T \tag{45}$$

where

$$P_{(t)} = e^{iH_{eff}t/\hbar} P e^{-iH_{eff}t/\hbar}$$

$$\Gamma(\kappa_{n'}, \kappa'_{n'}, m_n, m_f) = \int_{-\infty}^{\infty} d\zeta_{n'} \Phi_{\kappa_{n'}} [\zeta_{n'} - \zeta_{0n'}(m_n)] \Phi_{\kappa'_{n'}} [\zeta_{n'} - \zeta_{0n'}(m_f)],$$

in which

$$\zeta_{n'} = Q_{n'} \sqrt{\frac{M^* \omega_{n'}}{\hbar}},$$

$$Q_{0n'}(m) \sqrt{\frac{M^* \omega_{n'}}{\hbar}} = -q_{n'}(m) \sqrt{\frac{\hbar}{M^* \omega_{n'}}} \frac{1}{\hbar \omega_{n'}} << 1$$

In the above equations, $m_{n'}(k_n)$ and $m_f(k'_n)$ with $\omega(N_n) = \sum_{n'} \omega_{n'}(k_n)$ and $\omega(N_f) = \sum_{n'} \omega_{n'}(k'_n)$ are the quantum numbers denoting the initial and the final states of the amide-I oscillator system (the nth harmonic oscillator), respectively.

The quantities, c, P, ε' are the speed of light, the effective dipole moment of the amide-I oscillator and the refractive index, respectively, V_0 is the volume of the total system, $\langle \ldots \rangle_T$ denotes the average over all initial states in the canonical equilibrium.

Following procedure of Takeno [38-39], we get

$$\Gamma(\kappa_{n'}, \kappa'_{n'}, m_n, m_f) = \begin{cases} 1 - \frac{1}{2}(\kappa_{n'} + \frac{1}{2}) \zeta^2_{0n'}(m_n, m_f), \; for \; \kappa'_{n'} = \kappa_{n'} \\ \sqrt{\frac{\kappa_{n'} + 1}{2}} \zeta_{0n'}(m_n, m_f), \qquad for \; \kappa'_{n'} = \kappa_{n'} + 1 \\ -\sqrt{\frac{\kappa_{n'}}{2}} \zeta_{0n'}(m_n, m_f), \qquad for \; \kappa'_{n'} = \kappa_{n'} - 1 \\ 0, \qquad for \; \kappa'_{n'} \neq \kappa_{n'}, \kappa_{n'} \pm 1 \end{cases}$$

and

$$\prod_{n'} \sum_{k'_{n'}} \Gamma(\kappa_{n'}, \kappa'_{n'}, m_n, m_f) e^{i[\omega(k'_{n'}) - \omega(k_{n'})]t} \cong \prod_{n'} [1 + Z_{n'}(t, m_n, m_f)]$$

where

$$Z_{n'}(t, m_n, m_f) = \frac{1}{2} \zeta^2_{0n'}(m_n, m_f) [(\kappa_{n'} + 1) e^{-i\omega_{n'} t} + \kappa_{n'} e^{-i\omega_{n'} t} - 2\kappa_{n'} + 1],$$

$$\zeta_{0n'}(m_n, m_f) = \zeta_{0n'}(m_n) - \zeta_{0n'}(m_f) << 1$$

In accordance with the above approximation procedure, the thermal average in the integrand of Equation (45) can be expressed separately for the amide-I oscillator state and the phonon state, which are denoted by the angular bracket with subscript *sT* and *aT*, respectively. Taking first the average over the phonon state, we get

$$\left\langle \prod_{n'}[1+Z_{n'}(t,m_n,m_f)]\right\rangle_{aT} = \left\langle \exp_A[\sum_{n'} Z_{n'}(t,m_n,m_f)]\right\rangle_{aT}$$
$$= \exp\{\left\langle \exp_A[\sum_{n'} Z_{n'}(t,m_n,m_f)]-1\right\rangle_{aT}^{D}\} \approx \exp[\left\langle \sum_{n'} Z_{n'}(t,m_n,m_f)\right\rangle_{aT}] \qquad (46)$$

where the symbol $\exp_A(\sum_{n'} Z_{n'})$ is a leveled exponential function that levels off the product $Z_{n'}$ s by neglecting terms which contain higher orders than unity in any of the $Z_{n'}$ s, and the superscript (*D*) attached to the bracket symbol indicates a cumulative average, as proposed by Kabo [40]. Using Equation (45), equation (46) can be written as

$$B_{fn'(\omega)} = \frac{4\pi\omega}{\varepsilon' chV_o}\int dte^{-i\omega t}\langle\sum_{m_f}\exp[-z_0(T,m_n,m_f]\exp[-z_1(T,t,m_n,m_f]>\times$$
$$\langle m_{n'}|P(t)|m_f\rangle\langle m_f|P|m_{n'}\rangle_{aT}$$

$$z_0(T,m_n,m_f) = \sum_{n'}\left\langle \varsigma_{on'}^2(m_n,m_f)(k_{n'}+1/2)\right\rangle$$
$$z_1(T,t,m_n,m_f) = \frac{1}{2}\sum_{n'}\left\langle \varsigma_{on'}^2(m_{n'},m_f)[(k_{n'}+1)\exp(i\omega_{n'}t)+(k_{n'}\exp(-i\omega_{n'}t)\right\rangle_{aT}$$

The function $\exp[z_1(T,t,m_n,m_f]$ can be expanded as

$$\exp[z_1(T,t,m_n,m_f)] = 1+z_1(T,t,m_n,m_f)+\frac{1}{2!}z_1^2(T,t,m_n,m_f)+......,$$

where the first, the second and the third terms on the right-hand side are contributions to the zero, one and two-phonon processes. In the general case, one and multi-phonon processes give weak side-band effects [38-40]. Therefore, we may pay attention exclusively to the zero-phonon process and the main absorption line. Then we get

$$B_{fn'(\omega)} = \frac{4\pi\omega}{\varepsilon' chV_o}\int dte^{-i\omega t}\langle\sum_{m_f}\exp[-z_0(T,m_n,m_f]\langle m_{n'}|P(t)|m_f\rangle\langle m_f|P|m_{n'}>\rangle_{aT} \qquad (47)$$

In order to determine $Z_0(T, m_n, m_f)$, we have to find out $\varsigma_{0n}(m_n, m_f)$. Evaluating the matrix element of H_i in Equation (42) with respect to $|m\rangle$ for low frequency phonons and in the adiabatic approximation, $\varsigma_{0n'}(m_n, m_f)$ is separated into two parts, phonon and soliton,

$$\zeta_{0n'}(m_n, m_f) \cong \zeta_{0n'}^{(1)} \zeta_s^{(2)}(m_n, m_f), \tag{48}$$

where

$$\zeta_{0n'}^{(1)} \cong \sqrt{2} \sum_l \lambda_l \vec{e}(n') \frac{\mu_{n'}}{\upsilon_{0n'}} \sqrt{\frac{1}{\hbar M^* \omega_{n'}}} = \frac{F_{n'}}{\sqrt{\omega_{n'}}},$$

$$\zeta_s^{(2)}(m_n, m_f) \cong \frac{1}{\sqrt{N}} \sum_i (\langle m_n | Q_i Q_i^* + Q_i^* Q_i | m_n \rangle - \langle m_f | Q_i Q_i^* + Q_i^* Q_i | m_f \rangle)$$

here

$$F_{n'} = \sum_l \sqrt{2} \lambda_l \vec{e}(n') (\frac{\mu_{n'}}{\upsilon_{on'}}) \frac{1}{\sqrt{\hbar M^*}}, \quad \omega_{n'} \cong \upsilon_{on'} k,$$

v_{0n} is the speed of the nth phonon mode. Since the n' -dependence of $\vec{e}(n')$ and $v_{0n'}$ comes mainly from $\omega_{n'}$, we can neglect the n_i-dependence of $F_{n'}$ and rewrite it as F. Inserting the above formula into Equation (47), and using the Debye's approximation for the frequency spectrum of phonons, Pang finally obtained

$$Z_o(T, m_n, m_f) = Z_o(T) = [\zeta_s^{(2)}]^2 \sum_{n'} \frac{1}{\omega_{n'}} \left\langle F_{n'}^2 \left(k_{n'} + \frac{1}{2} \right) \right\rangle_{aT}$$

$$\approx \left[\zeta_s^{(2)} \right]^2 \mathrm{F}^2 \sum_{n'} \frac{1}{\omega_{n'}} \left(\left\langle k_{n'} \right\rangle_{aT} + \frac{1}{2} \right) = C + \lambda T^2, \text{ for } T << \theta \tag{49}$$

where C and, λ of which the explicit expressions are omitted here, are constants, and θ is the Debye temperature of the acetanilide. If $T << \theta$, we have $z_0(T) = C + \lambda T^2$.

4.2. The Properties of Infrared Absorption of Acetanilide

Until now, no specification has been given for the quantum number m. It may be divided into two parts, one being associated with the bound states or the soliton states of the microscopic particle, the other with continuum or scattering states of excitons, denoted by m_s and m_e, respectively.

Correspondingly, the absorption coefficient can be divided into the soliton absorption line $B_{sf(\omega)}$ and the exciton absorption line $B_{ex(\omega)}$, appearing at about 1650 cm^{-1} and 1665 cm^{-1}, respectively, i.e.,

$$B_{n'f(\omega)} = B_{sf(\omega)} + B_{ex(\omega)}$$

From Equations (47) and (48), the following relation can be derived

$$\langle m_s | Q_{n'} Q^*_{n'} + Q^*_{n'} Q_{n'} | m_s \rangle >> \langle m_e | Q_{n'} Q^*_{n'} + Q^*_{n'} Q_{n'} | m_e \rangle$$

Thus we have,

$$\zeta_{0n}(m_{sn}, m_{sf}) >> \zeta_{0n}(m_{en}, m_{ef})$$

or

$$Z_o(T, m_{sn}, m_{sf}) >> Z_o(T, m_{en}, m_{ef})$$

Since the mean square displacement of the exciton system associated with the state $|m_e\rangle$ is a sum of a large number of oscillatory functions with respect to the space variable n', the temperature dependence of $B_{ex}(\omega)$ is very weak as compared to that of $B_{sf}(\omega)$. When the soliton density is so small that their mutual interactions are negligible, we can consider only a single excited state corresponding to the first excited state of single-soliton states. Pang then approximately obtained

$$B_{ex}(\omega) = \frac{4\pi\omega}{\varepsilon' chV_o} \int_{-\infty}^{\infty} dte^{-i\omega t} \langle P_1(t)P_1 \rangle_{eT} \tag{50}$$

$$B_{sf}(\omega) = \frac{4\pi N_s\omega}{\varepsilon' chV_o} e^{-Z_o(T)} \int_{-\infty}^{\infty} dte^{-i\omega t} \langle P_1(t)P_1 \rangle_{m_sT} \approx e^{-(C+\lambda T^2)} B_s^{(0)}(\omega)$$

where

$$B_s^{(0)}(\omega) = \frac{AP_1^2}{\pi} \frac{W_s\omega}{(\omega^2 - \omega_s^2)^2 + W_s^2\omega^2}$$

and N_s and P_1 are the soliton number and an effective dipole moment associated with a single soliton state, respectively, ω_s is the eigenfrequency of the soliton, W_s and A are, respectively, a natural width of the single-soliton absorption line and a constant. Since the soliton energy is much higher than that of phonons, $B_s^{(0)}(\omega)$ does not depend on temperature in the above approximation.

From Equation (49), we see that the soliton absorption line spectrum $B_{sf}(\omega)$ is a product of the phonon- free soliton absorption line spectrum $B_s^{(0)}(\omega)$ and the Debye-Waller factor $\exp[-(C+\alpha T^2)]$. This indicates that the intensity of infrared absorption due to the nonlinear excitation of the excitons in acetanilide depends on the temperature exponentially, which is consistent with results obtained by Alexander and Krumbansl, *i.e.*, $\exp(-\beta T^2)$ [15], using a polaron model. This temperature dependence of the infrared absorption line at about 1650 cm^{-1} is analogous to that of localized or resonant harmonic phonon modes due to impurities in alkali halide crystals, where the impurity-induced infrared absorption spectra have a characteristic temperature dependence in the form of $\exp(-\alpha T^2)$ (α = constant) at low temperatures.

Although the physical origins of the localized modes are different in these two cases, the Debye-Waller factor results from the same origin, *i.e.*, random thermal modulations of low frequency optical phonons. Furthermore, the temperature dependence of the exciton absorption line, if any, is quite different from that of the soliton in the absence of the Debye-Waller factor in $B_{ex}(\omega)$. If we identify $B_{sf}(\omega)$ and $B_{ex}(\omega)$ with the absorption spectra of the acetanilide at about 1650 cm^{-1} and 1665 cm^{-1}, respectively, which were observed experimentally, the experimental results in acetanilide can be explained. Therefore, it is firmly believed that soliton can exist in acetanilide and the above theory gives a correct description of it.

Pang et al. [30-35] used again the exciton-soliton model and other methods for the interaction of the amide-I vibrational quanta (exciton) with optical phonons to study infrared absorption at finite temperature. They used an improved theory for color centers in solid state and the nonlinear quantum perturbation method [36,37]. Their numerical result for the exponential temperature dependence of the intensity of the anomalous band at 1650 cm^{-1} is basically consistent with the experimental data [41-42].

5. Properties of Eginenergy- Spectrum of Soliton Excitation in Molecular Crystals

5.1. The Method Calculating Eginenergy-Spectrum

We now calculate the properties of energy spectra of the nonlinear excitations of the excitons, or quantum energy-levels of vibrational amide-I in the acetanilide with two channels in virtue of the Hamiltonian and wave function given in Equations (4) -(5) in Pang's model by using quantization method [36,37,49]. and study its optical features [26-35]. Thus we have to discretize Equations (11) and (13), which can be denoted as

$$i\hbar\dot{\varphi}_n(t)=\varepsilon_0'\varphi_n(t)-J[\varphi_{n+1}(t)+\varphi_{n-1}(t)]+\lambda'[u_{n+1}(t)-u_n(t)]\varphi_n(t) \tag{51}$$

and

$$\ddot{u}_n(t)=(\beta/M)[u_{n+1}(t)-2u_n(t)+u_{n-1}(t)]+\lambda'/M[|\varphi_{n+1}(t)|^2-|\varphi_n(t)|^2] \tag{52}$$

where $\varepsilon_0=\hbar\omega_0,\beta=Mv_0^2/r_0^2,\lambda'=2\hbar(x_1+x_2)/\omega_0$ and the following transformation relation between continuous and discrete functions are used

$$\varphi(x,t)\to\varphi_n(t)$$

and

$$u(x,t)\to u_n(t),$$

$$\varphi_{n\pm1}(t)=\varphi_n(t)\pm r_0\frac{\partial\varphi_n(t)}{\partial x}+\frac{1}{2!}r_0^2\frac{\partial^2\varphi_n(t)}{\partial x^2}\pm....$$

$$u_{n\pm1}(t)=u_n(t)\pm r_0\frac{\partial u_n(t)}{\partial x}+\frac{1}{2!}r_0^2\frac{\partial^2 u_n(t)}{\partial x^2}\pm....$$

From Equations (51)-(52) we can get the equation of motion for $\varphi_n(t)$ as follows

$$(i\frac{d}{dt}-\omega_0)\varphi_n+J(\varphi_{n+1}+\varphi_{n-1})+\frac{1}{2}G(|\varphi_n|^2)\varphi_n=0 \tag{53}$$

where G is nonlinear interaction constant, $G=4\hbar^2r_0^2(x_1+x_2)^2/M\omega_0(v^2-v_0^2)$ for the new model , where $\hbar=1$, s=v/v$_0$, v is velocity of exciton, v$_0$ is the sound velocity in peptide lattice. The effective Hamiltonian of excitons corresponding to Equation (53) is

$$H_{eff}=\sum_n[\omega_0|\phi_n|^2-J(\phi_{n+1}^*\phi_n+\phi_n^*\phi_{n+1})-\frac{1}{2}G|\phi_n|^4] \tag{54}$$

In addition to the energy H$_{eff}$, Equation (53) has another constant of motion, i.e. the particle number:

$$N=\sum_n|\varphi_n|^2\,. \tag{55}$$

Equation (53) is just a discretely nonlinear Schrödinger or self-trapping dynamic (DSTD) equation [10-13,43-48] for the acetanilide, in which each chain has three modes of motion, $\varphi_n(t)$, $\varphi_{n-1}(t)$, $\varphi_{n+1}(t)$.

Thus Equation (53) denotes the equation of motion of six modes for the acetanilide with two channels. Then we can write it as the following matrix form

$$(i\frac{d}{dt}-\omega_0)\varphi+\overline{M}\varphi+GD(|\varphi|^2)\varphi=0 \tag{56}$$

where φ=Col.($\varphi_1, \varphi_2, \varphi_3, \varphi_4, \varphi_5, \varphi_6,$) is a complex 6-vectors for acetalinide with two chains, D($|\varphi|^2$) denotes the diagonal matrix diag.($|\varphi_1|^2$, $|\varphi_2|^2$, $|\varphi_3|^2$,$|\varphi_4|^2$, $|\varphi_5|^2$, $|\varphi_6|^2$), and $\overline{\mathrm{M}}$ is a real symmetric matrix with six orders, which represents various neighboring interactions between amide-Is, i.e.,

$$\overline{M}=\begin{bmatrix} 0 & J & \varepsilon & \varepsilon & \varepsilon & J \\ J & 0 & \mathrm{J} & \varepsilon & \varepsilon & \varepsilon \\ \varepsilon & J & 0 & \mathrm{J} & \varepsilon & \varepsilon \\ \varepsilon & \varepsilon & \mathrm{J} & 0 & J & \varepsilon \\ \varepsilon & \varepsilon & \varepsilon & \mathrm{J} & 0 & J \\ J & \varepsilon & \varepsilon & \varepsilon & J & 0 \end{bmatrix} \tag{57}$$

Upon quantization [49], the complex mode amplitudes are replaced by harmonic oscillator creation and annihilation operators $\hat{B}_j^+$ and $\hat{B}_j$ with the following properties:

$$\hat{B}_j^+|m_j>=\sqrt{(m_j+1)}\,|m_j+1>, \hat{B}_j\,|m_j>=\sqrt{m_j}\,|m_j-1>,$$

and $[B_j, \hat{B}_l^+]=\delta_{jl}$. The particle number and the energy of the systems corresponding to Equation (57) then become operators

$$\hat{N}=\sum_j^6(\hat{B}_j^+\hat{B}_j+\frac{1}{2}), \tag{58}$$

$$\hat{H}=(\omega_0-\frac{1}{2}\gamma)\hat{N}-J\sum_j^6\hat{B}_j^+\hat{B}_j-\frac{1}{2}\gamma\sum_j^6\hat{B}_j^+\hat{B}_j\hat{B}_j^+\hat{B}_j \tag{59}$$

where we take $\hbar$=1 and assume that both frequencies and energies are measured in the same unit (cm^{-1}).

Choosing basic function set, $|m,k\rangle=|m_1\rangle|m_2\rangle|m_3\rangle|m_4\rangle|m_5\rangle|m_6\rangle$, where $m(=m_1+m_2+m_3$ $m_4+m_5+m_6)$ is the total quantum number, k=1,2,..., d(m), and d(m) is equal to the number of ways that m particles are distributed in six states. We now consider the common eigenfunction, $|\psi_m\rangle$, of the above both operators. If we take

$$|\psi_m\rangle=C_1|m,1\rangle+C_2|m,2\rangle+\ldots+C_d(m)|m, d(m)\rangle, \qquad (60)$$

then the Hamiltonian equation $\hat{H}\,|\,\psi_m\rangle = E_m\,|\,\psi_m\rangle$ can reduce to an algebra equation on column vector $\overline{C}_m = col\,(C_1, C_2, \ldots, C_{d(m)})$

$$H_m\overline{C}_m = E_m\overline{C}_m, \qquad (61)$$

where H_m is a d(m)xd(m), real, symmetrical matrix with the diagonal elements $H_{mr}=\langle m, r\,|\,\hat{H}\,|m, r\rangle$ and the off-diagonal elements $H_{rs}=\langle m, r\,|\,\hat{H}\,|\,m, s\rangle$, $1\le r\le d(m), 1\le s\le d(m)$, and $r\neq s$.

5.2. The Features of Energy Spectrum of Acetanilide

Applying the above formulae and the values of parameters in Equation (34) for acetanilide, we can find the parameter values in Equation (61) in this calculations, which are ω_0^1=1687 cm^{-1}, J^1=3.8 cm^{-1}, γ^1=32.48 cm^{-1}, for the *Davydov* 's model [18-20] and ω_0^2 =1687.4 cm^{-1}, J^2=3.8 cm^{-1}, γ^2 =37.29 cm^{-1} for the *Pang* 's model [26-35], respectively.

Thus we easily calculate numerically the quantum vibrational energy levels of the acetanilide. The calculated results of the energy spectrum of the nonlinear excitations or quantum vibrational energy levels for the Davydov's and *Pang* 's models together with the experimental data [2-11, 41-42, 49-51] are presented in Table 1. This is the first time to get vibrational energy spectra of the acetanilide which are never calculated in detail before.

From table 1 we see that the energy spectra have the following features: (1) They consist of a series of manifolds, i.e., there are several energy levels corresponding to each vibrational quantum-number m.

For example, the first excited state (from 1649.71-1670.73 cm^{-1}) contains 4 levels, there are 8, 16 and 24 levels at m=2 (from 3141.29-3298.61 cm^{-1}), m=3 (from 4763.41-4992.27 cm^{-1}) and m=4 (from 6231.24-6568.21 cm^{-1}), respectively.

The vibrational spectra have a strongly local-mode pattern, or otherwise speaking, the distance between energy levels depends strongly on the nonlinear interaction, γ as we see from Equations (58) and (61). This is due to the fact that γ is greater than J. (3) The Local-mode degeneracies of energy levels appear for higher-lying vibrational states starting from m>2. Precisely speaking, there are degeneracies at 4763cm^{-1} for m=3, and at 6231 cm^{-1},6257cm^{-1},6287cm^{-1} and 6532cm^{-1}for m=4, respectively. This shows that degeneracy of energy levels increases with increasing m.

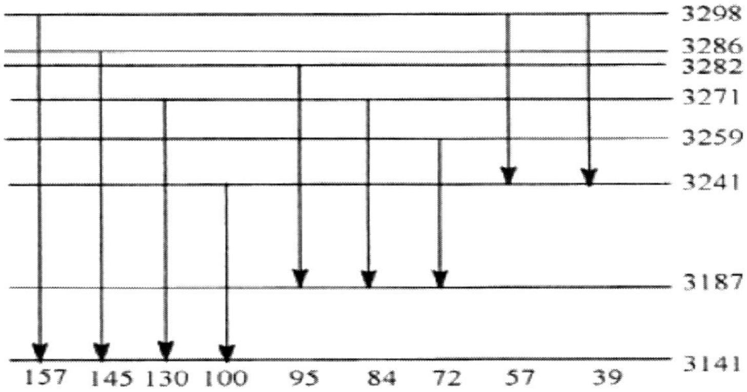

Figure 4. Distribution of energy levels at m=2 and corresponding Raman lines of acetanilide.

Table 1. Quantum vibrational energy levels (in cm^{-1}) of acetanilide

m	Species	Exp.[a]	Cal.[b]	Cal (c)	m	Species	Exp.[a]	Cal[b]	Cal[c]
	A_1	1666	1670.73	1670.86		A1	6225	6231.24	6243.17
	E	1662	1662.23	1662.23		A1		6231.49	6243.98
1	A_1	1659	1659.43	1602.11	4	A2	6245	6257.19	6278.12
	A_1	1650	1649.71	1654.52		A2		6257.87	6278.12
						E	6271	6287.17	6322.17
2	A_1	3150.3	3141.29	3182.14		E		6287.96	6320.17
	E	3205.1	3187.59	3242.57		A_2	6291	6296.92	6361.22
	A_1	3216	3241.96	3245.25		A_1	6301	6316.21	6367.45
	E	3250	3259.13	3254.21		E		6347.92	6370.05
	A_1	3267	3271.92	3274.67		A_2	6354	6371.67	6375.92
	A_1	3279	3282.61	3286.76		A1		6409.21	6384.24
	E	3281	3287.67	3290.13		E	6409	6414.97	6425.71
	A_1		3298.61	3300.59		A2		6425.23	6436.64
3	A_1	4752	4763.41	4774.21		A1		6430.16	6447.09
	A_1	4762	4763.66	4788.94		A2		6433.29	6452.91
	E	4793	4793.19	4801.37		E		6468.31	6472.80
	A_2	4803	4803.43	4822.61		A2		6469.96	6483.95
	E	4813	4839.87	4845.52		E		6472.77	6486.26
	A_2	4841	4856.15	4861.37		A1		6474.87	6527.74
	A_2	4864	4863.64	4874.92		A1		6513.54	6541.05
	A_1		4872.71	4879.26		E	6515	6532.46	6564.13
	A_1		4876.41	4886.27		E		6532.86	6564.24
	E		4880.13	4893.49		A2		6546.21	6571.37
	A_2		4896.23	4897.67		A1		6568.21	6684.49
	A_2		4909.76	4913.47					
	E		4936.89	4941.15					
	A_1		4956.87	4961.28					
	A_1		4981.19	4990.20					
	A_2		4992.27	5601.75					

[a]From the compilation of Reference [1-6, 21-23].
[b]the results for our model.
[c]the results for the Davydov model.

It is a major advantage of the above method that it allows to treat more than one amide-I quantum; the validity (in zero dispersion limit) is directly testable through Equation (61). It is clear from this table that theoretically calculated values are in good agreement with the experimental data of infrared absorption and Raman scattering in the acetanilide, by Careri et al. [6-9].

Obviously, we see from Table 1 that the theoretical results obtained by Pang′s mode approach better the experimental data than these values by Davydov′s model. This shows that the model introduced here is suitable for studying the vibrational spectra of the acetanilide with fewer parameters and simpler computational program.

Utilizing the energy spectra shown in Table 1 we can get the Raman lines in the acetanilide, which are shown in Figure 4 for m=2. The nine lines occurring in this figure, 157 cm^{-1},145 cm^{-1},130 cm^{-1},100 cm^{-1},95 cm^{-1},84 cm^{-1},72 cm^{-1},57 cm^{-1},39 cm^{-1} are quite consistent with the experimental data of nine peaks in the Raman spectrum of acetanilide at T=50K by G. Careri et al. [1-7] as 154 cm^{-1},148 cm^{-1},127 cm^{-1},104 cm^{-1},92 cm^{-1},84 cm^{-1},75 cm^{-1},60 cm^{-1},38 cm^{-1}, respectively, which is shown in Figure 8.

6. The Property of Specific Heat of Acetanilide

Careri et al. [3-6] measured the specific heat of powdered crystalline acetanilide from liquid nitrogen to room temperature. The measured data were then fitted to the formula

$$C(T) = 4.59\times 10^{-3}T + 1.505 \tag{62}$$

in units of $Jg^{-1}K^{-1}$, where T is temperature in °C

The temperature-dependence of specific heat can be understood from the thermal motion of exciton-soliton in acetanilide with finite temperature. In this case we must find the energy of the system at finite temperature. Thus we should obtain the dynamic equation of the exciton-soliton at temperature T≠0. Then we need utilize the model referring to the soliton coupled with the thermal phonons to study this phenomenon [33-35].

For the system with finite temperature, the collective excitations and motions of the excitons and phonons will be changed due to the thermal effect. A straightforward result is that thermal excitation of the phonons occurs. Thus the part, U(t)l0>ph, in Equation (5) is replaced by the following representation:

$$|\alpha_v> = U_n^+|v> = \exp\left\{\sum_{qn}\left[\alpha_{nq}(t)a_q^+ - a_{nq}^*(t)a_q\right]\right\}\prod_q (v_q'!)^{-1/2}(a_q^+)^{v_q'}|0>_{ph} \tag{63}$$

Where

$$|v> = \prod_q (v_q'!)^{-1/2}(a_q^+)^{v_q'}|0>_{ph}$$

is the phononic complete set, which represents the elementary excitation state of the single phonon due to the finite temperature T≠0K.

Utilizing the Equations (4)-(6) and Equation (63), we can obtain the equations of motion for the exciton and phonon in the system. However, because the molecular crystal we study here is in contact with a thermostat at finite temperature T≠0K, therefore ,after finishing this calculation of the expectation value in quantum theory for Equation (6), we should calculate the thermal mean values of these quantities by means of the formula:

$$\overline{Y} = < Y > = Tr(\rho_{vv})_{ph} Y = \sum_{v} < v | \rho | v >_{ph} < \Phi_v | Y | \Phi_v > \quad (64)$$

where the density matrix,

$$\sum_{v} < v | \rho | v >$$

is given in the form:

$$< v | \rho | v >_{ph} = (\rho_{vv})_{ph} = < v | \exp[-\frac{H_{ph}}{K_B T}] | v > / \sum_{v} < v | \exp[-\frac{H_{ph}}{K_B T}] | v >$$
$$= \frac{< v | \exp[-\sum_{q} \hbar\omega_q (a_q^+ a_q) / K_B T] | v >}{\sum_{v} < v | \exp[-\sum_{q} \hbar\omega_q a_q^+ a_q / K_B T] | v >} \quad (65)$$

where the diagonal matrix elements of the Hamiltonian are:

$$< \Phi \; | H | \Phi = < \Phi | H_{ex} + H_{\text{int}} | \Phi > + < \alpha_v | H_{ph} | \alpha_v >$$

Making use of previous method and the following relationships

$$< \alpha_v | (a_q + a_{-q}^+) | \alpha_v > = -[\alpha_{nq}(t) + \alpha_{n-q}^*(t)]; \quad < \alpha_v | a_q^+ a_q | \alpha_v > = (v_q + | \alpha_{n-q} |^2);$$
$$\sum_{v} \exp[-\hbar\omega_q v_q' / K_B T] < v | \exp(\alpha_{nq}^* a_q) \exp(-\alpha_{nq} a_q^+) | v > \; = (v_q + 1) e^{-|\alpha_{nq}|^2 (v_q + 1)};$$
$$v_q = [\exp(\frac{\hbar\omega_q}{K_B T}) - 1]^{-1}; \quad \exp[-\overline{W}_{n,n\pm1}] = < \alpha_{nq} | \alpha_{n\pm1q} >;$$
$$\overline{W}_{nn\pm1} = \exp\left\{ \sum_{q} \left[\alpha_{n\pm q}^* \alpha_{nq} - \frac{1}{2}\left(|a_{n+1q}|^2 + |\alpha_{nq}|^2 \right) \right] \right\} \prod_{q} \left\{ \sum_{m=1}^{n} \frac{(-1)^m | \alpha_{nq} + \alpha_{n\pm1q} |^2 \, v_q' !}{(m\,!)(v_q' - m)!} \right\} \quad (66)$$

we get, after some tedious calculations, the following equations of motion for the exciton and phonon in the molecular crystals at finite temperature.

$$\frac{\partial^2 u}{\partial t^2} - v_0^2 \frac{\partial^2 u}{\partial x^2} = \frac{\hbar r_0 (\chi_1 + \chi_2)}{2M\omega_0} \frac{\partial}{\partial x} | \phi |^2 \quad (67)$$

$$i\hbar \frac{\partial}{\partial t} \phi(x,t) = (\varepsilon_0 - 2JB(T,q)) \phi(x,t) - J r_0^2 B(T,q) \frac{\partial^2}{\partial x^2} \phi(x,t) \; + \frac{\hbar(\chi_1 + \chi_2) r_0}{\omega_0} \frac{\partial u(x,t)}{\partial x} \phi(x,t) \quad (68)$$

From Equations (67)-(68) we can get that the motion of the exciton in the nonlinear quantum systems with finite temperature satisfies now the following nonlinear Schrödinger equation

$$i\hbar\frac{\partial\varphi(x,t)}{\partial t}=\left[\varepsilon_0-2JB(T,q)\right]\varphi(x,t)-Jr_0^2B(T,q)\frac{\partial^2\varphi(x,t)}{\partial x^2}+G|\varphi|^2\varphi \qquad (69)$$

in the first order approximation, where $JB(T,q)=(\hbar^2/2mr_0^2)\exp[-\overline{W}_n]$ with

$$\overline{W}_n\approx\overline{W}_{nn+1}=\sum_q\left(1+v_q\right)\beta_{qn\pm1}\beta^*_{qn}-v_q\beta^*_{qn\pm1}\beta_{qn}+\left(v_q+\tfrac{1}{2}\right)\left(\left|\beta_{qn}\right|^2+\left|\beta_{qn\pm1}\right|^2\right)\approx\tfrac{1}{2}r_0\delta_0\left|\varphi_n\right|^2B'F(q,r)$$

$$\delta_0=\frac{2(x_1+x_2)^2}{8r_0\beta\left(1-s^2\right)J\omega_0^2},\ B'=\frac{\pi r_0(x_1+x_2)^2}{4\omega_0^2\beta v_0^2\hbar\left(1-s^2\right)},\ F(q,r)=\frac{r_0}{\pi N}\sum_q\left(1+r_q\right)|q|$$

The soliton solution of Equation (69) is of the form

$$\varphi(x,t)=\sqrt{\frac{G'(T)}{8J}}\operatorname{sec}h\left[\frac{G'(T)}{2Jr_0}(x-x_0-vt)\right]\times\exp\left[\frac{i\hbar v}{2Jr_0^2}(x-x_0)-\frac{iE_{sol}t}{\hbar}\right]$$

where

$$G'(T)=\frac{\hbar^2(\chi_1+\chi_2)^2r_0^2}{M\omega_0^2v_0^2\left(1-s^2\right)}\left[1-\tfrac{1}{4}B'F(q,T)\right]$$

The corresponding soliton energy is given by

$$E_{sol}=(\varepsilon_0-2J)+\tfrac{1}{2}mv^2-\frac{\hbar^4(x_1+x_2)^4\left(1-5s^2\right)}{24\omega_0^4\beta^2\left(1-s^2\right)^3J}\left[1-\tfrac{1}{4}B'F(q,T)\right]^2 \qquad (70)$$

Thus the specific heat arising from the thermal motion of the exciton-soliton is

$$C_v=\frac{dE}{dT}=\frac{\hbar^4(x_1+x_2)^4B'\left(1-5s^2\right)}{48\omega_0^4\beta^2\left(1-s^2\right)^3J}\times\left[1-\tfrac{1}{4}B'F(q,T)\right]\frac{dF(q,T)}{dT}$$

At high temperature, i.e., $K_BT>>\hbar\omega_q$ we have

$$F(q,T) = \frac{2K_B T}{\hbar v_0 \pi}$$

Thus

$$C_v = K_B(\alpha + bT). \tag{71}$$

where T is the absolute temperature. Using the accepted values of parameters for acetanilide mentioned above in Equation (34), we can obtain $\alpha = 5.15, b = 0.0199K^{-1}$. Thus equation (71) is consistent with experimental data given in Equation (62). Meanwhile, Mrevlishvil [52] and Goldanskii et al. [53] measured the linear specific heat of various biopolymers including proteins. They obtained $C_v = K_B(\alpha' + b'T)$, which resembles Equation (71).

In short, from the above result of specific heat in the acetanilide, which is agreement with experimental datum, we can conclude that the exciton-soliton exist truly and can be described by the nonlinear Schrödinger equation in nonlinear quantum mechanics [36,37,49].

7. Experimental Verifications of Property of Soliton Excitation in Acetanilide

In previous part of this chapter we studied the nonlinear quantum mechanical properties of excitons and phonons and some phenomena arising from their motions in acetanilide, and gave the mechanism of generation of nonlinear interaction in the system [36-37,49]. Because of the form of the exciton-soliton associated with the vibrational features of the C=O stretching and deformation of peptide groups, thus, existence and motion of the exciton-soliton can be observed and determined by infrared absorption and Raman spectra, and so on, except for the specific heat mentioned above, in acetanilide.

Therefore, it is quite necessary to collect the experimental results of infrared absorption and Raman scattering and of lifetime of thermal pulse or ground-state recovery time in acetanilide. In this section we focus these problems.

7.1. Experimental Results of Infrared Absorption and Raman Scattering in Acetanilide

In the early 1970s, Careri et al. [3-14] carried out a novel experimental investigation on protein dynamics. In the measurement of Careri et al. [3-14], the infrared spectra were obtained using three different infrared spectro-photometers. For the study of temperature dependence of the amide-I region, a Nicolet model 7000 Fourier-transform infrared spectrophotometer was used. Spectra were collected for 100 scans using a band width of 0.5 cm^{-1}. The sample was thermostated using a closed-cycle helium refrigerator. Far-infrared absorption spectra were measured using a Michelson interferometer (model 720) equipped with a Golay cell.

Samples consist of pellets obtained from a mixture of grounded acetanilide and polyethylene power. Pure polyethylene pellets were used to measure background transmission. Raman spectra were excited by a coherent radiation model 52 argon ion laser operating at 488nm or 514.5 nm, with stabilized output power of 20-200 mW. Incident light was filtered by proper choice of interference filter and its intensity was monitored using a beam-splitter and a silicon photocell. Scattered light was analyzed by a Jarrel-Ash model 25-300 Raman spectrometer and detected by an ITT model Fw-130 cooled photomultiplier using photon counting electronics.

These measured the infrared spectra of amide-I (1600- 1700 cm^{-1}), amide-II (1500-1600 cm^{-1}), amide-III (1300-1500 cm^{-1}), amide-IV or VI (500 -700 cm^{-1}) and amide-V (700-800 cm^{-1}). The infrared spectra of amide-I in 1600-1700 cm^{-1} and 2500-3500 cm^{-1}, 4750-4850 cm^{-1} and 6200-6400 cm^{-1} are shown in Figure 5. The absorption intensity vs. temperature is shown in Figure6. The Raman spectra of amide-I in 1630- 1700 cm^{-1} and low-frequency modes at 300K and 50 K are shown in Figure 7 and Figure 8, respectively.

According to analysis in Section 1, the eigenenergy spectra of amide-I are 1665 cm^{-1}, 1662 cm^{-1} and 1659 cm^{-1}, which are assigned to the B_{2u}, B_{1u}, B_{3u} modes, respectively. The 1650 cm^{-1} mode in Figure 5(a) and Figure 7 should be assigned to the exciton-soliton. Thus, the soliton excitation, predicted by the above theory indeed exists in these systems.

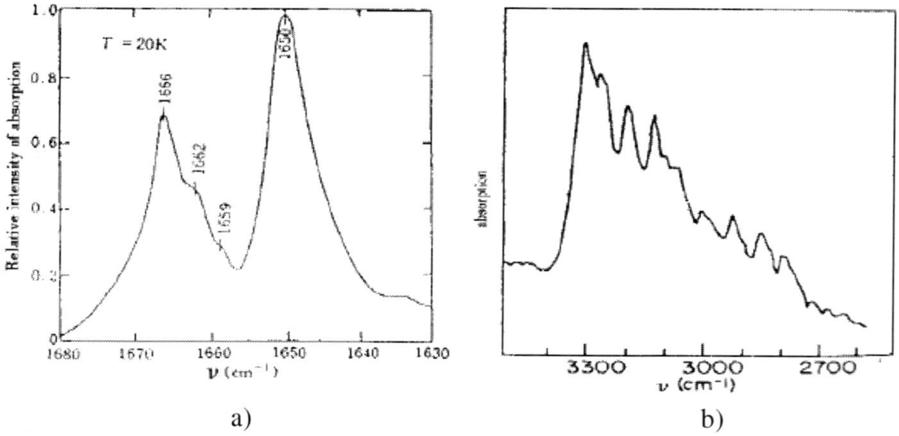

Figure 5. Infrared absorption spectra of acetanilide in (a) 1600-1700 cm^{-1} and (b) 2500-3500 cm^{-1}. (see Reference [3-14]).

From Figure 6, we know that the absorption intensity decreases in the form of $\exp[-\beta T^2]$ with increasing temperature. This is consistent with the theoretical result of $\exp[-(c+\gamma' T^2)]$ given by Alexander and Krumbansl [15] and Pang [49-51] as shown in Section 4. The comparison between experimental (denoted by "*") and the theoretical values (solid curve) is shown in Figure 6(b).

From Figures 5– 7, we see that the peak values of acetanilide infrared absorption agree with nonlinear eigenenergy spectra of the exciton-soliton given in Table 1, were obtained from the above theory proposed by Pang. These interesting results provide experimental evidence to the existence of exciton-soliton in acetanilide and confirm the validity of the above theory.

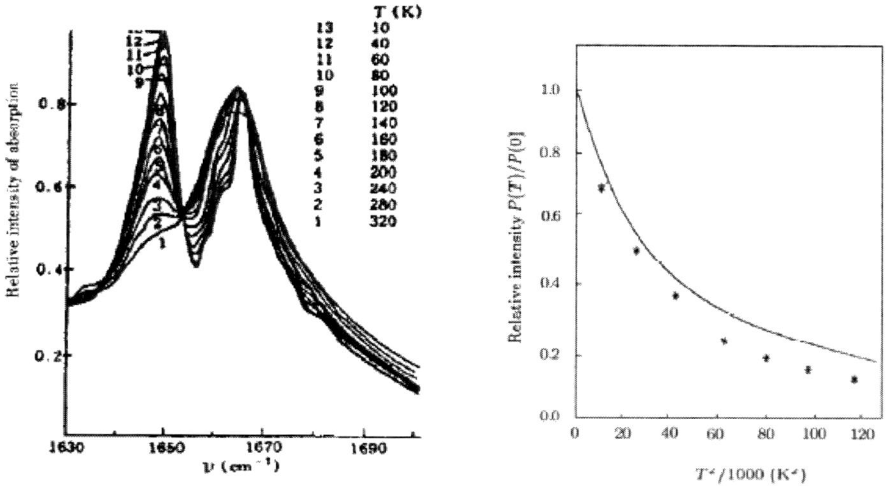

a) b)

Figure 6. (a). The absorption intensity vs. temperature for acetanilide, and (b) comparison of experimental and theoretical values (see Reference [3-6]).

From Figure 8 we know that there are nine peaks at 154 cm^{-1},148 cm^{-1},127 cm^{-1},104 cm^{-1},92 cm^{-1};84 cm^{-1},75 cm^{-1},60 cm^{-1},38 cm^{-1}, in the Raman spectrum of acetanilide at T=50K by G. Careri et al. [3-7], which are quite consistent with the nine lines occurring in this figure 4, 157 cm^{-1},145 cm^{-1},130 cm^{-1},100 cm^{-1},95 cm^{-1},84 cm^{-1},72 cm^{-1},57 cm^{-1},39 cm^{-1} , respectively, which was obtained by the above theory. This result is very interesting because the Raman spectrum shown in Figure 8 have not been clearly explained theoretically up to now. The above agreement is not accidental, it show clearly that the energy spectra in Table 1 are relevant and reliable for acetanilide, and the above method and theory we used are correct.

7.2. Dynamic Test of Soliton Excitation in Acetanilide

The experimental measurement of the exciton-soliton dynamics in acetanilide is of interest. It is useful to resolve the lifetime of the soliton since lifetime and mobility of soliton are important criteria for its' usefulness as a means of energy transport.

Fann et al. [54] measured the relaxation time of the vibrational excitation of the 1650 cm^{-1} band by transient-infrared-bleaching experiments which can set limits on the excitation lifetime. The source required for such measurements must be tunable around 6 μm (1650 cm^{-1}) with adequate spectral resolution to selectively excite the band (10 cm^{-1}).

In addition, pulses with picosecond duration and sufficient intensity to bleach the transition are necessary to observe the bleaching recovery with adequate temporal resolution and signal-to-noise ratio. The source Fann et al. [54] chose to meet these criteria are the Mark free-electron laser (FEL) which produces pulses in bursts (macropulses) of about 1.3 μs in duration.

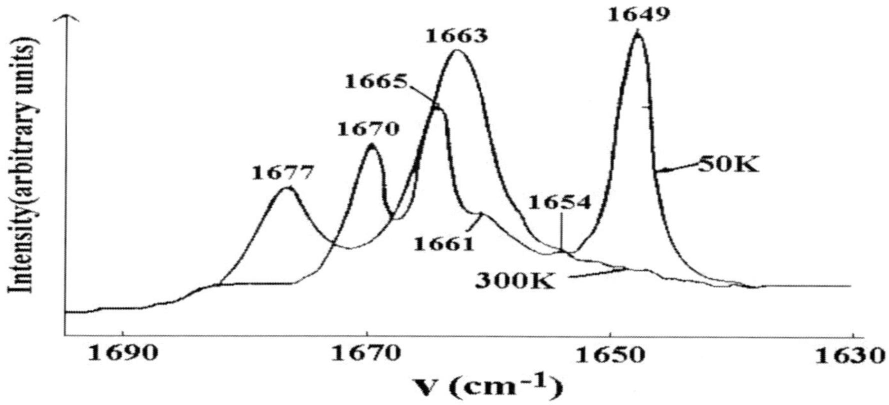

Figure 7. Raman spectra of acetanilide in 1630-1700 cm^{-1} at 300 K and 50 K, respectively [Reference 3-9].

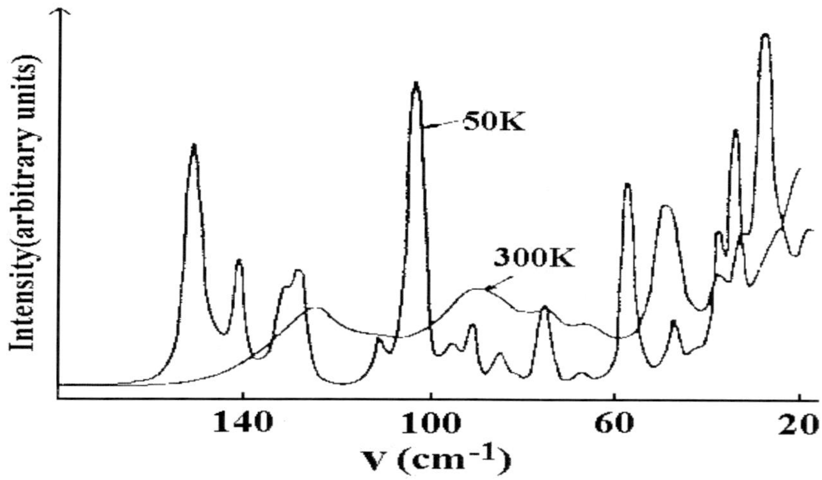

Figure 8. Low frequency Raman spectra of acetanilide at 300K and 50K, respectively [Reference 3-9].

The Mark III laser output is divided into pump (95%) and probe (5%) beams by a CaF_2 wedge plate. Both pulses are focused by a single lens to a spot of 300-μm in diameter on the sample which is held in a closed-cycle refrigerator with CaF_2 windows. The acetanilide was purified by repeated sublimation, crystalline domains were grown by slow cooling from an acetanilide melt between CaF_2 disks of 2 mm thickness. A substantial increase in the probe-pulse transmission at the 1650 cm^{-1} band was observed when the pump pulse was simultaneous with the probe pulse. The transmission increases due to the pump roughly follows the 1650 cm^{-1} band shape at 0 ps delay (pump and probe temporally overlapped), indicating that the effect is associated with the 1650 cm^{-1} band. This show clearly that the exciton-soliton excitation occurs in such a case in the acetanilide.

Figure 9 shows the bleaching recovery dynamics at the band center as assembled from many scans. The fast component in Figure 9 is limited by pulse-duration, while the slow component has a decay time of 15 ± 5 ps which can be associated with saturation recovery due to repopulation of the ground state of the 1650 cm^{-1} mode. The relaxation time or lifetime of 15 ± 5 ps indicates that the 1650 cm^{-1} band is strongly coupled to the lattice, and the 15 ± 5 ps is a typical relaxation value for a vibration line in the acetanilide. These experimental data in acetanilide support the theoretical results obtained by Pang's theory [26-35] mentioned above. Therefore, Pang's theory is appropriate to the acetanilide.

Fann et al. [54] also measured further the transmission recovery of the sample at 100 ps in which the probe decays to the "cold" base line. This is an important fact since it was thought that the 1650 cm^{-1} band should disappear on deposition of energy into the 130 cm^{-1} optical phonons.

Loss of the 1650 cm^{-1} band due to such a heating should result in an increase in transmission approximately by a factor of 2 at 100 ps, which, however, was not observed. Therefore, the time scale of such vibrational relaxations observed is inconsistent with the lifetime of 10^{-10} s predicted by the theory of Pang et al. [49] in protein molecules.

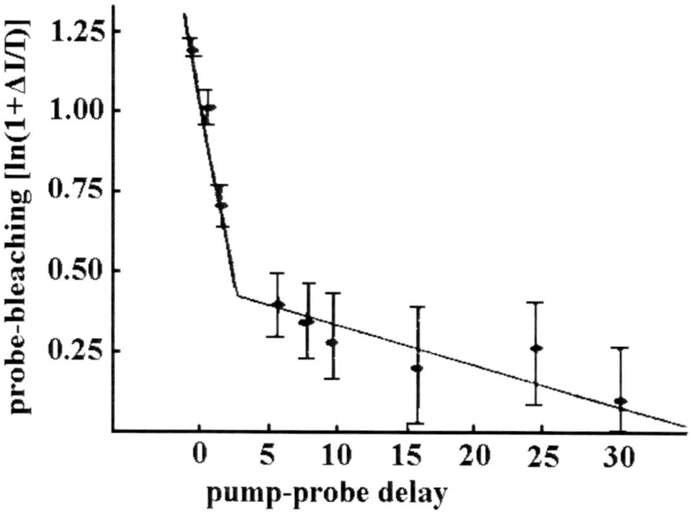

Figure 9. Probe-bleaching vs. pump-probe delay at 1650 cm^{-1} and 80 K [Reference 54].

This manifests that Fann et al's measurement of lifetime of the soliton in acetanilide cannot serve as the experimental evidences of existence of the soliton in protein molecules because the molecular structure and functions of acetanilide are completely different from those of alpha-helix protein molecules. In fact, the alpha-helix proteins is a biomacromolecules of helical structure of three channels, which have certain biological functions.

7.3. The Lifetime of the Soliton in the Acetanilide by Pump- Probe Spectroscopy

Hamm et al. [55-61] measured the properties of infrared spectra of absorption in acetanilide by pump-probe spectroscopy. In the pump-probe experiment the various states of the system are first excited with an intense short IR laser pump pulse, and the response upon that excitation is then probed by a second short, but much weaker probe pulse.

Most of the spectroscopy performed so far on the ACN problem was stationary IR absorption spectroscopy and its temperature dependence as shown in Figure 10. The information gained from the pump-probe experiments, as compared to stationary IR absorption spectroscopy, is manifold.

At the same time, it can measures the anharmonicities of the individual states, in which its first excited state is populated when pumping a state, and the probe pulse is probing both the upward transition to the second excited state and the downward transition back to the ground state through stimulated emission. If the state is harmonic, both upward and downward transitions appear at the same frequency, but with opposite signs, and cancel each other (if taking into account the effect of bleaching the ground state).

On the contrary, when the state is anharmonic, they lead to two bands with opposite signs separated by the anharmonicity of the transition. Therefore, the pump-probe spectroscopy is particularly sensitive to the anharmonic (nonlinear) part of the molecular potentials, which is impossible for the stationary IR absorption spectroscopy.

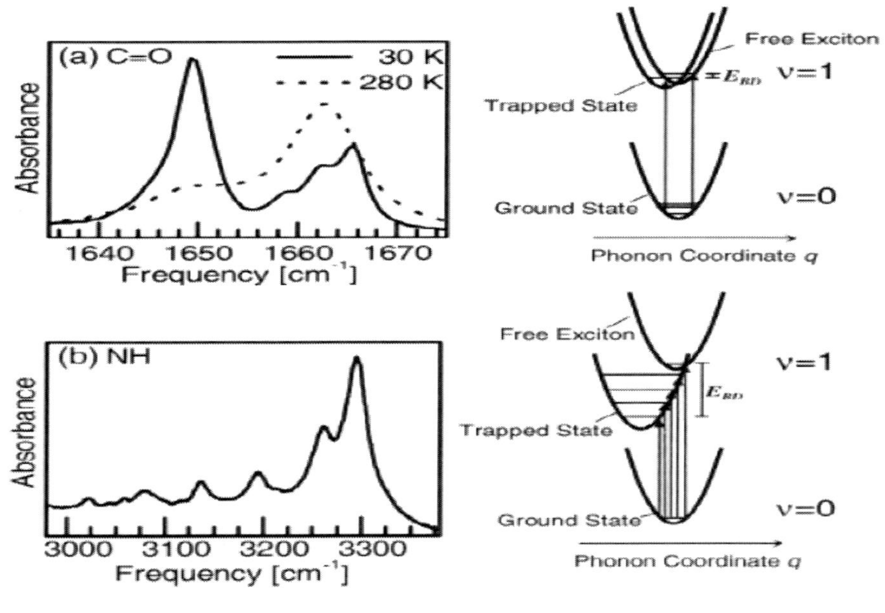

Figure 10. Infrared spectra of absorption of the C=O (a)and the N–H (b) band by pump- probe experiment in ACN and corresponding potential energy curves causing these spectra of the pumped states or soliton excited state by the femtosecond time resolution (Reference [55]).

If two states are coupled, excitation of one state will result in a response, such as a frequency shift, of the other. This means that there are the entangled transitions between them. If pumping two or more coupled states with an ultrashort pulse, whose bandwidth covers these states simultaneously, then quantum beats can be observed. Hence, we can used it to measure the lifetimes of particles.

Figure 11 shows two pump-probe spectra obtained by excitation of either of the two absorption lines of ACN at 90 K, where the *arrows* mark the center frequency of the pump pulse. The pump pulses are spectrally narrow in this experiment (spectral width ≈ 14 cm^{-1}) and excited only one band in them. When resonantly pumping the 1650 cm^{-1} band, it is bleached, but 1644 cm^{-1} band emerges.

When resonantly pumping the 1666 cm^{-1} band, hardly any bleach of the band is observed. This indicates that the 1650 cm^{-1} band responds only with a signal.

In the meanwhile, they discussed also the different response of both bands of acetanilide [56]. They started with the statement that the involved C=O vibrators are slightly anharmonic.

In this case the "anomalous" band with a 0–1 frequency of 1650 cm^{-1} and a 1–2 frequency of 1644 cm^{-1} in Figure 11B displays the expected anharmonic response, but the 1666 cm^{-1} band shows not pump-probe response at all. These results manifest that the anomalous 1650cm^{-1} band shows such an anharmonic response, and hence, is a localized state, but the normal 1666 cm^{-1} band shows no nonlinear response, which can only be the case when it is effectively a harmonic state. (it has been verified that the lifetime of the normal band is long enough to potentially observe a bleach [58]). Hence, the normal band is largely delocalized, i.e., a free exciton, in agreement with the original assignments.

On the other hand, they utilized 2D-IR spectroscopy to clarify the origin of the anomalous amide I band [59]. 2D-IR spectroscopy can distinguish between various kinds of nonlinearities of the molecular Hamiltonian.

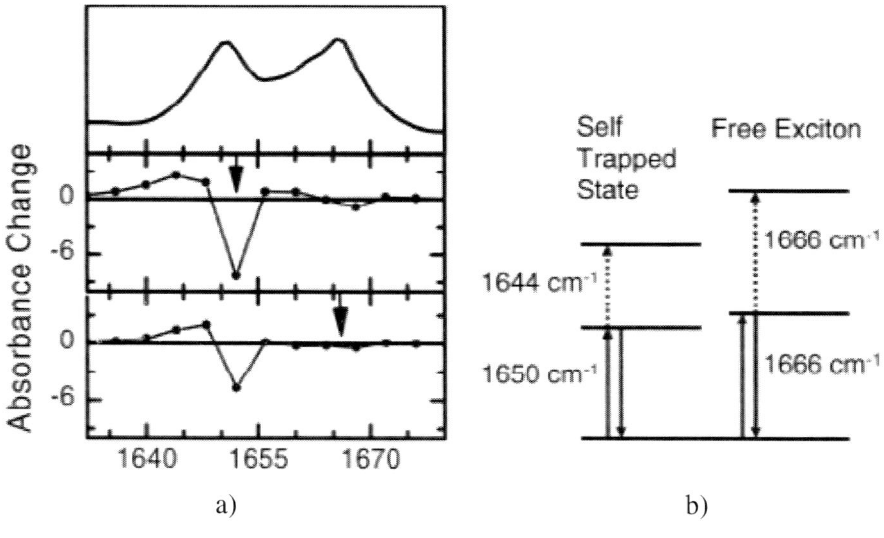

Figure 11. A is linear absorption spectrum (*top*) and pump probe spectra of the C=O mode of ACN at 93 K for two different narrow band pump pulses chosen to be resonant with each of the absorption bands. B is level scheme of the system, explaining the distinctively different response of both modes (see Reference [58]).

Figure 10 shows the comparison of the 2D-IR spectrum of crystalline ACN with those of benzoylchloride and N-methylacetamide(NMA) dissolved in methanol, respectively, where the arrows mark the position of the pump pulse.

These infrared spectra of absorption display clearly the same doublet structure in all three cases, but the 2D-IR spectra differ obviously. Benzoylchloride has a prototype example of a molecule with a strong symmetric Fermi resonance [62], whereas N-methylacetamide dissolved in methanol exists in two spectroscopically distinguishable conformations, one fraction hydrogen bonded to the solvent and the other fraction not [63]. Thus their molecular structures are different from that of ACN. If the two lines in ACN come from molecules in two different conformations or surroundings (topological defects) [64], pumping of one state would not cause any response of the other state (in contrast to what is seen in the experiment). The 2D-IR spectrum should be diagonal, as it is for N-methylacetamide dissolved in methanol (Figure 12). Otherwise, in the case of a Fermi resonance, original explanation of the doublet in the ACN spectrum [66], a very characteristic coupling pattern, is obtained (Figure 12). Thus, investigating these coupling patterns in detail, the compelling evidence against this possibility can be provided as well, and the self- trapping state or soliton state should be substantiated [59].

Figure 13 shows the pump-probe response after exciting the N–H band in ACN obtained by Edler, et al. [59].

In contrast to the C=O work, here they used a pump pulse so short that its bandwidth covers many of the side peaks of the N–H band. They were all excited at the same time, leading to pronounced quantum beats in the response. The Fourier transform of the signal in Figure 13 gave two peaks: a dominant peak at 48 cm^{-1} and a weaker peak at 76 cm^{-1}. The temperature dependence of the two peaks was obtained [59], which are completely consistent with the two lines in the conventional Raman spectrum (Figure 13) [66].

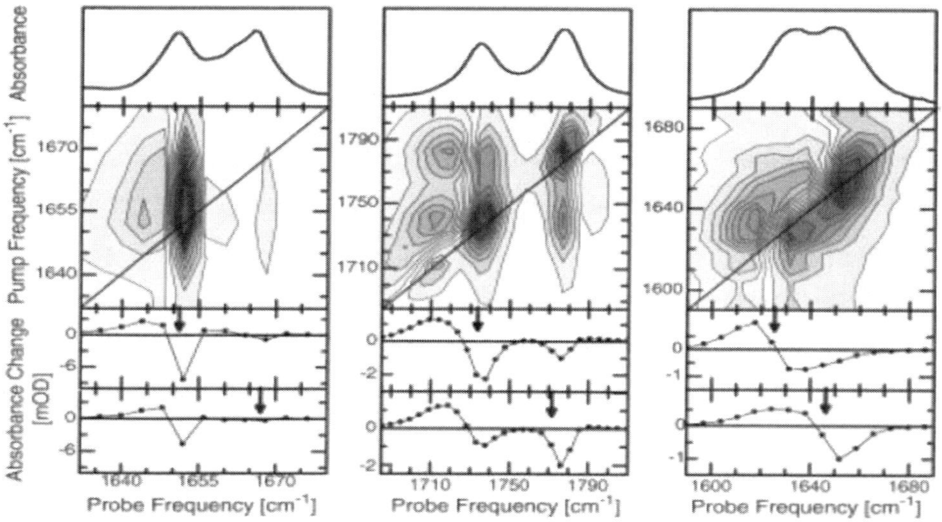

a) b) c)

[a]ACN.
[b]Benzoylchloride.
[c]NMA.

Figure 12. The linear absorption spectra and 2D-IR pump probe spectra of the C=O mode of ACN (at 93 K)(a), benzoylchloride(b), and NMA dissolved in methanol(c) (both at room temperature). *Blue colors* denote the negative absorption change (bleach and stimulated emission), *red colors* positive absorption change (excited state absorption). The *two lower panels* show horizontal cuts through the 2D-IR spectrum for pump frequencies resonant with either of the two bands (see Reference [58]).

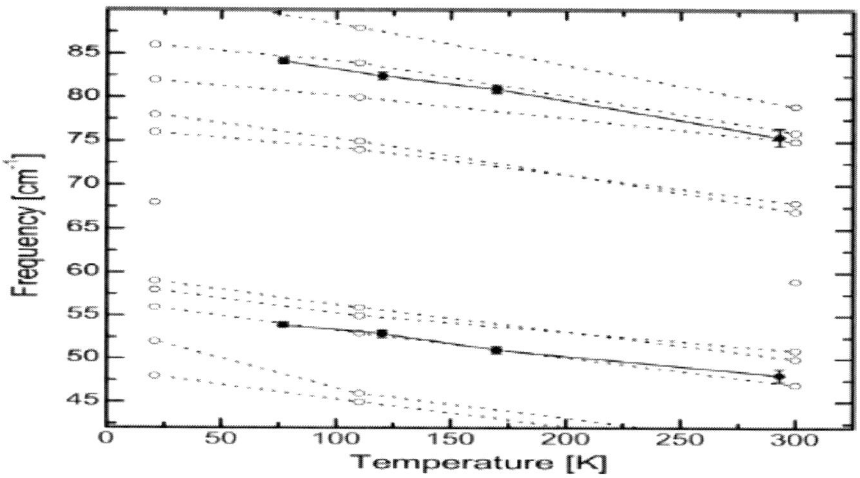

Figure 13. The temperature dependence of the low frequency Raman modes (*open circles*) in ACN (see Reference [66]) and of the beating frequencies observed in the pump-probe experiment (filled circles). The temperature dependence of the beating frequencies perfectly matches that of the Raman modes(see Reference 60]).

This demonstrated that the coherences indeed represent lattice phonons; in fact, this experiment identified the two phonons that mediate vibrational self-trapping in ACN.

Therefore, Hamm and co-workers have identified two phonon modes that mediate the self-trapping and find further evidence for self-trapped states in ACN and N-methylacetamide (NMA) and even in a small polypeptide [61] and speculate that such states may arise in all hydrogen-bonded systems. From these experiments they obtained the lifetime of this soliton in acetanilide is about 36ps [55-56]. These experimental results are very important and interesting, which provide experimental evidence to the existence of soliton in ACN and confirm the validity of above theory in Pang;s model. Thus, the soliton excitation, predicted by the above theory indeed exists in the acetanilide crystal.

Conclusion

In this chapter we established the theory of motion of soliton excitation occurred in molecular crystal and its interaction properties,and provided plenty of experimental evidences of existence of the soliton in this systems. In these investigations we first establish Hamiltonian of molecular crystal and corresponding nonlinear dynamic equation of exciton and phonon through determining the forms of nonlinear interactions and its mechanism of generation based on molecular structure of acetanilide. The soliton excitation of the excitons in molecular crystal and its plenty of properties were obtained from the nonlinear dynamic equations in molecular crystal. Subsequently we calculated the eigenenergy spectrum and specific heat of acetanilide and investigated theoretically the features of Raman scattering and infrared absorption of acetanilide arising from these soliton excitations, in which we obtained the rule of exponential decreases of infrared absorption strength of 1650cm^{-1} band, which corresponds to the soliton, with increasing the temperature of acetanilide using this soliton theory. The theoretical result of specific heat of the acetanilide is consistent with the experimental data. Finally we experimentally measured in detial the results of Raman scattering and infrared absorption of acetanilide to verify the correctness of theoretical results mentioned above. The experimental results show clearly that the frequencies of peaks in Raman and infrared spectra consist exactly with the theoretical results in eigenenergy levels of acetanilide and the experimental rule of change of strength of the peaks at $1650cm^{-1}$ with varying the temperature concord also the above theoretical result obtained by the soliton theory. Meanwhile, the true information of existence of the soliton was captured using the technique of Mark free-electron laser, its lifetime was also obtianed, which is 15ps in this experiment. When the pump-probe spectroscopy was used to measure the lifetime of the soliton, the lifetime of 35ps was obtained. These results obtained from theoretical and experimental studies are very interesting and useful in soliton physics, they represent clearly and can confirm further that the exciton-soliton is really existent in the molecular crystal.

Acknowledgment

The authors would like to acknowledge the National “973”project of China for financial support **(grate No:** 212011CB503701**).**

References

[1] M. Pope, C. E. Swenberg, *Electronic Processes in Organic Crystals*, Clarendon Press, Oxford University Press, Oxford/New York, 1982.

[2] E. A. Silinsh, V. Capek, *Organic Molecular Crystals*, AIP Press, New York, 1994.

[3] Careri, G., A. Gransanti and J. A. Ruple, *Phys. Rev.* A 37 (1998) 2703.

[4] Careri, G., E. Gratton and E. Shyamsunder, *Phys. Rev.* A 37 (1998) 4048.

[5] Careri, G., U. Buontempo, F. Galluzzi, A. C. Scott, E. Gratton and E. Shyamsunder, *Phys. Rev.* B 30 (1984) 4689.

[6] Careri, G., U. Buonttempo, F. Caeta and A. C. Scott, *Phys. Rev. Lett.* 51 (1983) 304.

[7] Scott, A. C., *Physica* D 51 (1990) 333.

[8] Scott, A. C., *Phys. Lett.* A86 (1998) 603.

[9] Scott, A. C., *Phys. Rep.* 217 (1992) 67. [9] Eilbeck. J. C., P. S. Lomdahl, A. C. Scott, *Phys. Rev.* B 30 (1984) 4703.

[10] Scott, A. C., E. Gratton, E. Shyamsunder and G. Careri, *Phys. Rev.* 1332 (1985) 5551.

[11] Scott, A. C., I. J. Bigio, C. T., Johnston, *Phys. Rev.* B39 (1989) 12883.

[12] Careri, G., J. C. Eilbeck, *Phys. lett.* A109 (1985)201.

[13] Careri, G., U. Buontempo, F. Galluzzi, A.C. Scott, E. Gratton, E. Shyamsunder, *Phys. Rev.* B 30 (1984) 4689.

[14] Alexander, D. M. and J. A. Krumbansl, *Phys. Rev.* B 33 (1986) 7172; Alexander, D. M., *Phys. Rev. Lett.* 60 (1985) 138.

[15] Davydov, A. S., *J. Theor. Biol.* 38 (1973) 559 and 66 (1977) 379.

[16] Davydov, A. S., *Phys. Scr.* 20 (1979) 387.

[17] Davydov, A. S., *The Solitons in Molecular Systems*, Dordrecht, Reidel, 1985 p17-87.

[18] Davydov, A. S., Theory of Molecular Excitons, l^{st} *Nauka*, Moscow, 1968; New York, plenum, 1975.
[19] Davydov, A. S., *Biology and Quantum Mechanics*, Pergamon, New York, 1982.
[20] Benkui, T., P. B. John, *Phys. Lett.* A 240 (1998) 282.
[21] Xiao, Y., *Phys. Lett.* A 243 (1998) 174.
[22] Tekec, J., Z. Ivic, Z. Przulj, *J. Phys.: Condens. Matter* 10 (1998) 1487.
[23] Kenkre, V. M., S. Raghvan, L. Cruzeiro-Hansson, *Phys. Rev.* B 49 (1994) 9511.
[24] Eremko, A. A., Yu. B. Gaididel, A. A. Vaknenko, *Phys. Stat. Sol.* b 127 (1985) 703.
[25] Pang Xiao-feng, *J. Phys. Condens. Matter* 2 (1990) 9541.
[26] Pang Xiao-feng, *Eur. Phys. J.* B 10 (1999) 415.
[27] Pang Xiao-feng and Zhang Huai-wu, *J. Phys. Chem. Solids*, 66 (2005) 963.
[28] Pang Xiao-feng, *Chinese J. Atom. Mol. Phys.* 7 (1989) 1235.
[29] Pang Xiao-feng, *Physics, Sin.* 22 (1993) 612.
[30] Pang. Xiao-feng, *Physica* D154 (2001) 138.
[31] Pang Xiao-feng, *J. Phys. Chem. Solid.* 62 (2001) 793.
[32] Pang Xiao-feng, *J. Nature Sin.* 15 (1992) 915.
[33] Pang Xiao-feng, *Chinese. Phys. Lett.*10 (6) (1993) 381.
[34] Pang Xiao-feng, *Acta Phys. Sin.* 42 (1993) 1856.
[35] Pang Xiao-feng, Zhang Huai-wu, *Model Phys. Lett.* 20 (30) (2006) 1923.
[36] Pang Xiao-feng, Feng Yaun ping, *Quantum mechanics in nonlinear system*, World Scientific Publishing Co. Singapore, 2005.
[37] Pang Xiao-feng, *The theory of nonlinear quantum mechanics, Chongqing,* Chinese Chongqing Press, 1994.
[38] Takeno, S., *J. Phys. Soc. Japan* 59 (1991) 3127.
[39] Takeno, S., *Prog. Theor. Phys.* 73 (1985) 853.
[40] Kabo, R., *J. Phys. Soc. Japan* 17(1962)206.
[41] Pang, Xiao-feng, *Chinese J. Infrared Millimeter Wave* 12 (1993) 377.
[42] Pang, Xiao-feng, *Chinese J. Infrared Millimeter Wave* 16 (1997) 288.
[43] Pang Xiao-feng, Chen Xianron, *Chin. Phys. Le*tt. 19 (2000) 1091.
[44] Pang, Xiao-feng, *Acta. Phys. Sin.* 43 (1994) 1987.
[45] Pang Xiao-feng, Chen Xianron, *Int. J. Model Phys.* 20(18) (2006) 2505.
[46] Pang Xiao-feng, Chen Xianron, *Commun. Theor. Phys.* 32 (2000) 437.
[47] Pang Xiao-feng, Chen Xianron, *Phys. Stat. Sol.* (b) 229 (2002) 1397.
[48] Pang Xiao-feng, Chen Xianron, *Chin. J. Atom. Mol.* 19 (2002) 319.
[49] Pang Xiao-feng, *Nonlinear Quantum Mechanics*, Beijing, Chinese Electronic Industry Press, 2009.
[50] Xiao, He-Lan, Cai Guo-ping, Sun Su-qin and Pang Xiao-feng, Chinese. *Atom. Mol. Phys.* 20 (2003) 211.
[51] Pang Xiao-feng, Xiao He-lain, Cue Guo-ping Zhang Huai-wu, Dong Bo, *Int. J. Infrared and Millimeter Waves*, 31 (2010).
[52] Mrevlishvil, G. M., *Usp. Fiz. Nauk* 128 (1979) 273 [Sov. Phys-Usp. 22 (1979) 433].
[53] Goldanskii, V. I., Yu. F. Krupyanskii and V. N. Flerov, *Dokl. Akad. Nauk (SSS*R) 272 (1983) 978.

[54] Fann, W., L. Rothberg, M. Roberso, S. Benson, J. Madey, S. Etemad, R. Austin, *Phys. Rev. Lett*. 64 (1990) 607.
[55] Hamm, P., *J. Biol. Phys*, doi:10.1007/s10867-009-9126-3.
[56] J.Edler, P. Hamm, *J. Chem. Phys*. 117, 2415 (2002).
[57] J. Edler, P. Hamm, *J. Chem. Phys*. 119, 2709 (2003).
[58] J. Edler, P. Hamm, *Phys. Rev. Lett*. 88, 067405 (2002).
[59] J. Edler, P. Hamm, *Phys. Rev*. B 69, 214301 (2004).
[60] J. Edler, R. Pfister, V. Pouthier, C. Falvo, P. Hamm, *Phys. Rev. Lett*. 93, 106405 (2004).
[61] J. F. Bertran, L. Ballester, L. Dobrihalova, N. Sanchez, R. Arrieta, *Spectrochim. Acta, Part* A 24A, 1765 (1968).
[62] S. Woutersen, Y. Mu, G. Stock, P. Hamm, *Chem. Phys*. 266, 137 (2001).
[63] G. B. Blanchet, C. R. Fincher, *Phys. Rev. Lett*. 54, 1310(1985).
[64] C. T. Johnston, B. I. Swanson, *Chem. Phys. Lett*. 114, 547 (1985).
[65] C. T. Johnston, S. F. Agnew, J. Eckert, L. H. Jones, B. I. Swanson, C. J. Unkefer, *J. Chem. Phys*. 95, 5281 (1991).
[66] Pang Xiao-feng, Xiao He-lain, Cue Guo-ping Zhang Huai-wu, Dong Bo, *Int. J. Infrared and Millimeter Waves*, 31, 521 (2010).
[67] Pang Xiao Feng, *Physics of life review*, 8 (2011) 302.
[68] Pang Xiao-feng, *J. Appl. Phys*. 111 (2012) 093519.
[69] H. Ohlich, *Adv. Electron. Electron Phys*. 53 (1980) 86.

In: Solitons
Editor: Matthew C. Williams
ISBN: 978-1-62618-234-9

Chapter 3

SHAPES AND DYNAMICS OF SEMI-DISCRETE SOLITONS IN ARRAYED AND STACKED WAVEGUIDING SYSTEMS

Roy Blit and Boris A. Malomed
Department of Physical Electronics, School of Electrical Engineering,
Tel Aviv University, Tel Aviv, Israel

Abstract

We study the shapes and dynamics of semi-discrete solitons (SDSs) in the known model of the set of linearly-coupled waveguides with the self-focusing cubic nonlinearity. The model describes a planar array of optical fibers, as well as a stack of parallel planar waveguides, in the temporal and spatial domains, respectively. It also applies to the self-attractive Bose-Einstein condensate (BEC) loaded into an array of parallel tunnel-coupled "cigar-shaped" traps. It was known that the interplay of the group-velocity dispersion, discrete diffraction and nonlinearity gives rise to SDSs in this system. We here develop a variational approximation (VA) and additional analytical methods for the description of the SDSs, and study their mobility and collisions by means of direct simulations. The VA and another analytical method (an exact solution of the linearized equation in the core adjacent to the central one) produce an accurate description for the family of fundamental onsite-centered SDS solutions. The VA is developed too for transversely unstable intersite-centered and twisted solitons. In simulations, the solitons are not mobile in the discrete direction (while broader non-soliton modes may be mobile across the array). Collisions between the SDSs traveling in the longitudinal direction reveal a threshold separating the passage and merger, or mutual destruction of the colliding solitons. While the exact shape of the threshold, considered as a function of the solitons' energy, features irregularities, its average form is explained in an analytical form. Collisions between SDSs centered at two adjacent cores are studied too.

Introduction and the Model

Waveguiding arrays offer a basis for the design of a great variety of nonlinear optical [1-9], nano-optic [10-12], and plasmonic [13-16] media, in which many types of discrete and quasi-discrete solitons can be created. Tremendous progress made in experimental and theoretical studies of these areas in the course of the last decade has been surveyed in a number of reviews [1-16]. Among other media created and investigated in this context, arrays of tunnel-coupled optical fibers [6] and nano-wires [11] enable many fundamental and applicative effects.

The most fundamental patterns predicted [17] and realized experimentally [18] in planar waveguiding arrays with the Kerr nonlinearity are discrete solitons, a well-established theoretical model of which is based on the discrete nonlinear Schrödinger equation (NLSE) [19]. More recently, many theoretical results have been published for discrete solitons in similarly organized arrays of plasmonic waveguides [20].

Fiber-optic arrays may give rise to more complex modes, in the form of two-dimensional (2D) *semi-discrete solitons* (SDSs), which are built as continuous temporal solitons in the longitudinal direction (along the array), and, simultaneously, as discrete solitons in the transverse direction (across the array). Solitons of this type in planar fiber arrays had been studied theoretically some time ago, starting from Ref. [21], with expected applications to the collapse-effect compression of the solitons [22] (in the fully continuous limit, the SDSs go over into the *Townes solitons* [23] of the 2D NLSE, which are subject to the *critical collapse* in 2D) and all-optical switching and steering in the arrays [24].

Recently, settings similar to 3D fiber bundles have been created in the form of a set of parallel waveguides written by means of femtosecond pulses in bulk silica, which has made it possible to observe 3D semi-discrete spatiotemporal solitons, which are discrete in the two transverse directions [25] (see also Ref. [8] for a review). Two-component solitons of a different type, which may also be considered as semi-discrete modes, were predicted in systems with the quadratic (second-harmonic-generating) [26] and cubic [27] nonlinearities, with the continuous component propagating in a slab waveguide, and the other one carried by a discrete array attached to the slab.

The subject of the present work is the investigation of the shape, mobility, and interactions of SDSs in the model of the arrayed waveguides. The model is based on the well-known system of linearly-coupled NLSEs for envelopes $u_n(\tau)$ of the electromagnetic waves in this system [21,22,24]:

$$i\frac{\partial u_n}{\partial z}+\frac{1}{2}\frac{\partial^2 u_n}{\partial \tau^2}+\frac{1}{2}\left(u_{n+1}+u_{n-1}-2u_n\right)+\left|u_n\right|^2 u_n=0, \tag{1}$$

where n is the number of the individual fiber (guiding core), τ the reduced temporal variable [28], and z the propagation distance. In the present notation, the group-velocity dispersion (GVD) and nonlinearity coefficient, as well as the nonlinear-coupling constant accounting for the tunneling of light between adjacent waveguiding cores, are all scaled to be 1, assuming the anomalous sign of the GVD and self-focusing sign of the Kerr nonlinearity (alternatively, the normal GVD may be combined with the self-defocusing nonlinearity). The total energy of the wave field, which is a dynamical invariant of Eqs. (1), is

$$E\equiv\sum_{n=-\infty}^{+\infty}\int_{-\infty}^{+\infty}d\tau\left|u_n(\tau)\right|^2 . \tag{2}$$

In addition to E, the system conserves the total momentum in the τ-direction, $i\sum_{n=-\infty}^{+\infty}\int_{-\infty}^{+\infty}d\tau\, u_n\left(\partial u_\tau^* / \partial\tau\right)$ (where the asterisk stands for the complex conjugate), and the Hamiltonian [see Eq. (35) below].

The same system of equations (1) admits an alternative realization in terms of the spatial-domain propagation, for a set of stacked planar waveguides with the self-focusing intrinsic Kerr nonlinearity. In that case, n is the number of the core in the stack, τ denotes the transverse coordinate, the term $(1/2)\partial^2 u_n / \partial\tau^2$ represents the paraxial diffraction in the transverse direction, and Eq. (2) defines the total power of the beam.

Further, Eq. (1) may also be realized as a system of coupled Gross-Pitaevskii equations for a set of parallel cigar-shaped traps confining a self-attractive Bose-Einstein condensate (BEC), which are coupled by the tunneling of atoms across potential barriers separating the individual traps. This setting can be implemented by means of a combination of optical lattices [29], namely, a very deep one isolating the planar layer, and a moderately strong perpendicular lattice splitting the layer into the quasi-discrete set of trapping cores [30]. In terms of the BEC model, Eq. (2) defines the total number of atoms in the condensate.

In previous works [21,22,24], the SDSs in system (1) were constructed in a numerical form. Our first objective is to develop a variational approximation (VA) for them. This is a natural approach, which was elaborated in detail for fully discrete solitons in various models based on the 1D discrete NLSE [31-37], as well as for the symmetry-breaking and switching dynamics of temporal solitons in dual- [38,39] and triple- [40] core fibers. However, this method was not previously applied to SDSs. Actually, it may be quite relevant not only for the model based on Eq. (1), but also for SDSs in other systems. Results produced by the VA are reported in Section 2, and their comparison with numerical solutions is presented in Section 3, demonstrating a reasonably good agreement. The VA is applied to the onsite-centered fundamental solitons and surface solitons in the semi-infinite version of the system [41] (in terms of static state, the surface solitons are tantamount to twisted, i.e., antisymmetric, ones). For the completeness of the description, we also apply the VA to intersite-centered SDSs, which are unstable in the transverse direction (twisted solitons turn out to be unstable too, while their surface counterparts are stable in the semi-infinite system). In the limit case when almost all the energy of the SDS is trapped in the central core, the form of the components in the adjacent cores is directly found in an analytical form.

An issue of fundamental significance is the mobility of discrete solitons [19] and, accordingly, of their semi-discrete counterparts. Furthermore, in case the solitons are mobile, it is relevant to study collisions between them [32,42]. It is well known that sufficiently broad discrete solitons, if kicked, feature efficient mobility in the framework of the one-dimensional (1D) discrete NLSE with the onsite cubic nonlinearity [42], and this property may be studied by means of the VA [32,36]. Moreover, collisions between moving discrete solitons may also be analyzed with the help of this method, although in a rather cumbersome form [32].

The second major objective of the present work is to study the mobility and collisions of SDSs in the framework of Eq. (1), chiefly by means of systematic simulations. Numerical results concerning the mobility are reported in Section 3. The first conclusion is that, unlike the fully discrete solitons in the 1D discrete NLSE, the SDSs are not mobile in the discrete direction (across the array). This is explained by the fact that discrete solitons tend to be mobile when they are broad enough, which corresponds to the quasi-continuum limit [32,43], while, as mentioned above, the 2D continual counterpart of the SDS is the *Townes soliton*, that, in turn, is subject to the collapse, i.e., the catastrophic self-compression [23]. Thus, the onset of the collapse converts broad quasi-continual solitons back into the essentially discrete (narrow) ones [44]. The self-compression eventually arrests the collapse, making the SDSs strongly pinned to the underlying lattice structure, i.e., immobile. For the same reason, fully discrete 2D solitons in lattices with the cubic onsite nonlinearity demonstrate no mobility either. 2D discrete solitons are effectively mobile in settings with the saturable [45] or quadratic (second-harmonic-generating) [46] nonlinearity, where the collapse does not occur (in the continuum limit), allowing for the existence of stable broad solitons.

On the other hand, the SDSs are obviously mobile in the continual direction, which suggests the possibility to consider collisions between them. This possibility is pursued in Section 4, for both head-on collisions and those with a transverse shift between the solitons being investigated. In the former case, the collisions are quasi-elastic if the initial kick, which sets each soliton in motion, exceeds a certain critical value; below the critical value, which decreases inversely proportional to the soliton's energy, the collisions lead to *merger* of the two solitons into a single mode, or their destruction, if the solitons' energy is too small. In Section 4, the dependence of the critical kick on the energy is explained by means of an analytical estimate. A minimum value of the kick necessary to make the collision elastic is identified too for collisions between solitons with a transverse shift, whose centers are placed at adjacent cores of the array. However, in this case the collision does not cause the merger of the solitons below the critical value. Rather, the collision features inelasticity in the form of a conspicuous transfer of the energy from one soliton to the other. The latter regime is very sensitive to small changes in the initial conditions.

The Variational Approximation

Fundamental Onsite-Centered Solitons

Stationary solutions to Eq. (1) with propagation constant k are looked for as

$$u_n(\tau,z)=\exp(ikz)U_n(\tau),$$

with real functions $U_n(\tau)$ obeying the following coupled equations:

$$-kU_n+\frac{1}{2}\frac{d^2U_n}{d\tau^2}+\frac{1}{2}\left(U_{n+1}+U_{n-1}-2U_n\right)+U_n^3=0, \tag{3}$$

which can be derived from the corresponding Lagrangian,

$$L=\frac{1}{2}\sum_{n=-\infty}^{+\infty}\int_{-\infty}^{+\infty}d\tau\left[-(k+1)U_n^2-\frac{1}{2}\left(\frac{dU_n}{d\tau}\right)^2+U_nU_{n+1}+\frac{1}{2}U_n^4\right]. \tag{4}$$

Approximate solutions for stationary onsite-centered SDSs are sought for in the form of the following variational *ansatz*, which is a product of the usual ones adopted for discrete [31] and continual [39] solitons:

$$U_n = A\exp(-\alpha|n|)\operatorname{sech}(\eta\tau). \tag{5}$$

Here α^{-1}, η^{-1} and A are, respectively, widths of the soliton in the discrete and continual directions, and its amplitude. The total energy of the ansatz, calculated as per Eq. (2), is $E=(2A^2/\eta)\coth\alpha$, which suggest us to eliminate the amplitude in favor of the energy:

$$A^2 = \frac{1}{2}\eta\, E\tanh\alpha. \tag{6}$$

The substitution of ansatz (5) into Eq. (4) and straightforward calculations lead to the following effective Lagrangian, as a function of E, η, and α:

$$L_{\text{eff}} = -(k+1)E - \frac{1}{6}\eta^2 E + E\operatorname{sech}\alpha + \frac{1}{12}\eta E^2\left(1+\tanh^2\alpha\right)\tanh\alpha. \tag{7}$$

Next, the corresponding Euler-Lagrange equations, $\partial L_{\text{eff}}/\partial\alpha = 0$ and $\partial L_{\text{eff}}/\partial\eta = 0$, take the following form:

$$\sinh\alpha = \frac{1}{12}\eta E\left(4-3\operatorname{sech}^2\alpha\right), \tag{8}$$

$$\eta = \frac{1}{4}E\tanh\alpha\left(1+\tanh^2\alpha\right). \tag{9}$$

The third variational equation, $\partial L_{\text{eff}}/\partial E = 0$, yields an expression for the propagation constant,

$$k = -1 - \frac{\eta^2}{6} + \operatorname{sech}\alpha + \frac{\eta E}{6}\tanh\alpha\left(1+\tanh^2\alpha\right). \tag{10}$$

Using Eqs. (8) and (9), one can eliminate η to derive an expression for E in terms of α:

$$E^2 = \frac{48\cosh^5\alpha}{\left(2\cosh^2\alpha-1\right)\left(4\cosh^2\alpha-3\right)}. \tag{11}$$

A straightforward analysis of Eq. (11) demonstrates that the soliton's energy takes values above a finite *threshold* (minimum),

$$E_{\min}^{(\mathrm{VA})} = 4.076 \tag{12}$$

This minimum energy may be compared to the constant energy of the Townes solitons in the 2D NLSE with two continuous coordinates [23],

$$E_{\mathrm{Townes}} \approx 5.75 . \tag{13}$$

Note that the variational prediction for the Townes-soliton energy, produced by the isotropic Gaussian ansatz [47], is

$$E_{\mathrm{Townes}}^{(\mathrm{VA})} = 2\pi . \tag{14}$$

The global characteristic of the soliton family predicted by the VA is dependence $E(k)$, as obtained from Eqs. (9), (10) and (11), which is displayed in Figure 1. This dependence allows one to predict the stability of the solitons by means of the well-known Vakhitov-Kolokolov (VK) criterion [48,23]: Stable may be the branch of the plot with the positive slope, $dP/dk > 0$, i.e., at $k > k_{\min} \approx 0.56$, as seen in Figure 1. Strictly speaking, the VK criterion yields only a necessary stability criterion, but, in the present relatively simple system, it is a sufficient one too, as confirmed by numerical simulations, see below.

An additional straightforward analysis of the variational results demonstrates that the stable branch corresponds to narrow SDSs (i.e., strongly pinned ones, which prevents their mobility in the discrete direction, see below), while the short unstable branch, predicted by the VA at $k < k_{\min}$ (see Figure 1), comprises broad solitons, which might be mobile, but the instability annuls this possibility. In particular, in the limit of $k \to 0$, a very broad SDS goes over into the Townes soliton [its energy, as given by Eq. (11), $E(\alpha = 0) = 4\sqrt{3} \approx 6.93$, is different from numerical value (13) due to the approximate form of ansatz (5), cf. the other approximate value given by Eq. (14); in fact, it is shown below that more relevant characteristics of the SDSs, such as $E_{\min}$, is predicted by the VA with much better accuracy, cf. Eqs. (12) and (33)].

On the other hand, in the limit of $k \to \infty$, i.e., $E \approx 2\sqrt{2(k+1)} \to \infty$, the narrow soliton sits, chiefly, in the central core [40],

$$U_0(\tau) \approx \sqrt{2(k+1)}\,\mathrm{sech}\left(\sqrt{2(k+1)}\,\tau\right) \tag{15}$$

with small-amplitude tails in two adjacent cores determined by the linearized version of Eq. (3) at $n = \pm 1$,

$$-(k+1)U_{\pm 1} + \frac{1}{2}\frac{d^2 U_{\pm 1}}{d\tau^2} = -\frac{1}{2}U_0. \tag{16}$$

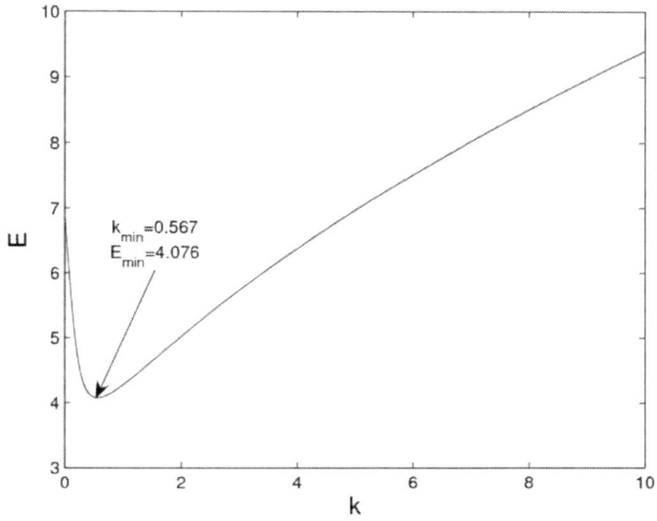

Figure 1. (Color online) The energy of the fundamental onsite-centered semi-discrete soliton versus its propagation constant, as produced by the variational approximation.

Substituting expression (15) into Eq. (16), it is straightforward to obtain an exact solution for the tail, valid under these assumptions:

$$U_{\pm 1}(\tau) = \tau e^{-\sqrt{2(k+1)}\tau} + \frac{1}{\sqrt{2(k+1)}}\cosh\left(\sqrt{2(k+1)}\tau\right)\ln\left(1+e^{-2\sqrt{2(k+1)}\tau}\right) \tag{17}$$

[in spite of its apparently asymmetric form, solution (17) is an even function of τ, and the solution is exponentially localized, although it might seem divergent at $\tau \to \infty$].

To assess the degree of the self-compression of the SDS in the transverse direction, we define R_0 as the share of the total soliton's energy concentrated in the central core. Together with Eq. (6), ansatz (5) yields

$$R_0 \equiv E_{n=0} / E = \tanh \alpha \,. \tag{18}$$

Naturally, $R_0 \to 1$ at $\alpha \to \infty$, as the soliton gets completely confined at n=0.

2.2. Twisted and Surface Solitons

It is well known that, in addition to the fundamental solitons, the discrete NLSE gives rise to *twisted solitons*, i.e., antisymmetric (odd) modes [49,19]. In the present system, twisted solitons can be naturally defined as those with $U_{-n}(\tau) = -U_n(\tau)$, which remain even functions of the continuous coordinate, τ. In fact, the shape of the stationary twisted soliton is tantamount to that of a *surface soliton* in a semi-infinite array, which is defined as a solution to the modification of Eq. (3), with the cores existing only at $n \geq 1$:

$$-kU_n + \frac{1}{2}\frac{d^2U_n}{d\tau^2} + \frac{1}{2}\left(U_{n+1} + U_{n-1} - 2U_n\right) + U_n^3 = 0, \text{ at } n \geq 2, \tag{19}$$

$$-kU_1 + \frac{1}{2}\frac{d^2U_1}{d\tau^2} + \frac{1}{2}\left(U_2 - 2U_1\right) + U_1^3 = 0, \tag{20}$$

that can be derived from the accordingly truncated version of Lagrangian (4):

$$L = \frac{1}{2}\sum_{n=-\infty}^{+\infty}\int_{-\infty}^{+\infty} d\tau\left[-(k+1)U_n^2 - \frac{1}{2}\left(\frac{dU_n}{d\tau}\right)^2 + U_nU_{n+1} + \frac{1}{2}U_n^4\right]. \tag{21}$$

Studies of surface solitons in various setting have recently attracted a great deal of attention [41,50-52]. In particular, the semi-infinite version of the present model gives rise to SDSs that may be considered as *surface light bullets* [41,52]

It is relevant to apply the VA to the description of the twisted/surface solitons too. To this end, the ansatz can be adopted in the form of Eq. (5) at $n \geq 1$, which implies that the soliton is attached to the edge of the array, $n = 1$. Solutions with a finite distance between the soliton's peak and the edge are possible too [41], but we do not consider them here, as the respective VA would be rather cumbersome. Then, a straightforward calculation yields the corresponding energy, cf. Eq. (6), and the effective Lagrangian, cf. Eq. (7):

$$E = 2A^2 / \left[\eta\left(e^{2\alpha} - 1\right)\right], \tag{22}$$

$$L_{\text{eff}} = -\frac{1}{2}(k+1)E - \frac{1}{12}\eta^2 E + \frac{1}{12}\eta E^2 \tanh\alpha + \frac{1}{2}e^{-\alpha}E. \tag{23}$$

The variational equations following from Eq. (23), $\partial L_{\text{eff}}/\partial\alpha = \partial L_{\text{eff}}/\partial\eta = \partial L_{\text{eff}}/\partial E = 0$, take the following form:

$$\eta E = 6e^{-\alpha}\cosh^2\alpha, \; \eta = (E/2)\tanh\alpha, \tag{24}$$

$$k = -1 - \frac{\eta^2}{6} + \frac{1}{3}\eta E \tanh\alpha + e^{-\alpha}.$$

A corollary of Eq. (24) is the following expression for the energy:

$$E^2 = 12e^{-\alpha}\left(\coth\alpha\right)\cosh^2\alpha, \tag{25}$$

cf. Eq. (11). It gives rise to a minimum value of the energy of the surface SDS,

$$\left(E_{\text{surf}}\right)_{\min} \approx 3.7147, \tag{26}$$

which is attained at $\alpha = (1/2)\ln(4+\sqrt{13}) \approx 1.0144$. Accordingly, the minimum energy of the twisted soliton is twice as large, $\left(E_{\text{twist}}\right)_{\min} \equiv 2\left(E_{\text{surf}}\right)_{\min} \approx 7.4294$. Finally, the dependence $E(k)$, as predicted by the VA for surface solitons is presented in Figure 2(a).

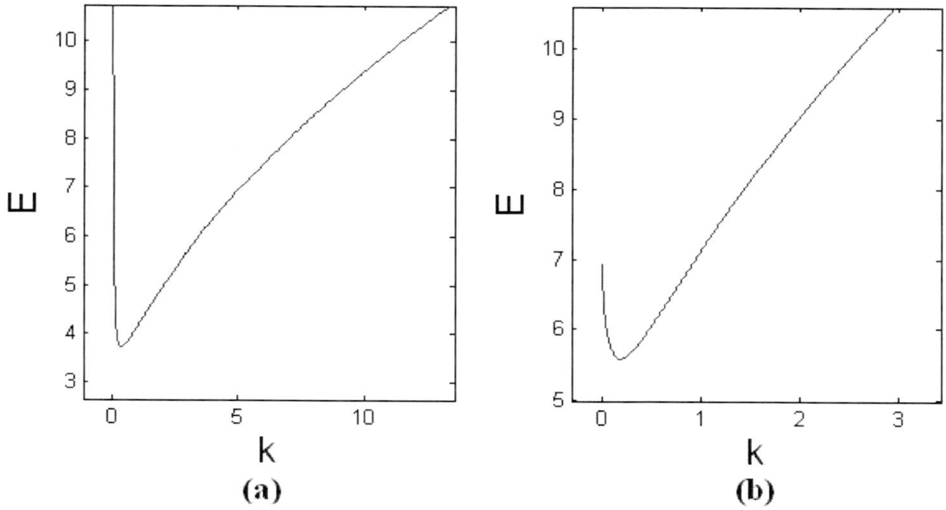

Figure 2. (Color online) The energy of the surface (a) and intersite-centered (b) semi-discrete solitons versus its propagation constant, as per the variational approximation. The energy of the twisted solitons is twice that shown in (a).

2.3. Intersite-centered Solitons

The VA can also be applied to the intersite-centered solitons (in the full lattice), which are well known as unstable stationary modes of the discrete NLSE [19]. Although the intersite SDSs are unstable too (see below), for the sake of the completeness it is relevant to briefly present the VA results for them too.

The corresponding ansatz is [cf. Eq. (5)]

$$U_n = A\exp\left(-\alpha(n+1)\right)\operatorname{sech}(\eta\tau), \text{ at } n \geq 0, \tag{27a}$$

$$U_n = A\exp\left(-\alpha|n|\right)\operatorname{sech}(\eta\tau), \text{ at } n \leq -1, \tag{27b}$$

with total energy $E = 4A^2 / \left[\eta\left(e^{2\alpha} - 1\right)\right]$, effective Lagrangian

$$L_{\text{eff}} = -\frac{1}{2}\left(k + \frac{1}{2}\right)E - \frac{1}{12}\eta^2 E + \frac{1}{24}\eta E^2 \tanh\alpha + \frac{1}{2}e^{-\alpha}E - \frac{1}{4}e^{-2\alpha}E,$$

and the variational equations

$$\eta E = 12e^{-\alpha}\left(1 - e^{-\alpha}\right)\cosh^2\alpha,\ \eta = \left(E/4\right)\tanh\alpha, \tag{28}$$

$$k = -\frac{1}{2} + \frac{1}{32}\left(E\tanh\alpha\right)^2 + e^{-\alpha} - \frac{1}{2}e^{-2\alpha}.$$

An equation for the energy, following from Eq. (27), is [cf. Eq. (25)]

$E^2 = 48e^{-\alpha}\left(1 - e^{-\alpha}\right)\left(\coth\alpha\right)\cosh^2\alpha$. This expression attains a minimum at $\alpha \approx 0.6183$,

$$\left(E_{\text{inter}}\right)_{\min} \approx 5.576. \tag{29}$$

The fact that the threshold value (29) for the intersite SDS is larger than the minimum energy (12) for its onsite counterpart complies with the fact that the intersite-centered solitons are dynamically unstable against spontaneous transformation into the onsite states. The dependence $E(k)$ predicted by the VA for the intersite modes is displayed in Figure 2(b).

Numerical Results for Single Solitons

The Numerical Method

Simulations of Eq. (1) were performed by means of an advanced version of the symmetrized split-step Fourier-transform method [28] for a 2D system, with each step separated into five computational stages. The ones handling the transverse coupling were added before and after the usual stages which implemented the longitudinal-dispersion and nonlinearity terms. This method is accurate up to the fifth order in the stepsize, $O\left(\Delta z^5\right)$. In the course of the simulations, Δz was adjusted to particular configurations, with the objective to support the necessary accuracy and computation speed. Absorbers of radiation were installed at edges of the simulation domain, in both the τ- and n- directions.

Stationary solution of Eq. (1) for the SDSs were produced by means of the imaginary-distance-propagation algorithm, built similarly to the well-known imaginary-time algorithm [53]. To this end, the simulations were started with input

$$u_n(z=0,\tau)=A\,\mathrm{sech}\left(\frac{n}{\Delta n_0}\right)\mathrm{sech}\left(\frac{\tau}{\Delta\tau_0}\right), \tag{30}$$

with some amplitude A and widths Δn_0, $\Delta\tau_0$, and were run forward in the imaginary distance, keeping the constant value of total energy E [see Eq. (2)], until the procedure would converge to a stationary SDS mode parameterized by the given value of E. Iterating this process for a range of energies, we have generated a family of the stationary SDS solutions for those energies at which they exist, $E > E_{\min}$, cf. Eq. (12). If the fixed energy was too small for the existence of the SDS ($E < E_{\min}$), the simulations lead to decay (spreading out) of the pattern. The stability of the soliton family produced by this algorithm at $E > E_{\min}$ was then tested by simulations of the perturbed evolution in the real propagation distance.

Further, the mobility of the solitons and collisions between moving ones were simulated by kicking a previously found stable SDS, $u_n(\tau)$, in the discrete or continuous direction:

$$u_n^{(n-\text{kick})}(z=0,\tau)\equiv u_n(z=0,\tau)e^{ian}, \tag{31}$$

$$u_n^{(\tau-\text{kick})}(z=0,\tau)\equiv u_n(z=0,\tau)e^{-i\omega\tau}. \tag{32}$$

Note that, due to the periodicity of $\exp(ian)$, a may be limited to interval $0<a\le\pi$. The application of the largest possible n-kick, $a_{\max}=\pi$, actually does not set the soliton into motion, but rather converts it into a *staggered* object [19], with $u_n=(-1)^n|u_n|$.

Due to the Galilean invariance of Eq. (1) in the τ-direction, the strength of the respective kick, $-\omega$, is precisely the velocity which the kick imparts to the soliton. Collisions of SDSs moving in the continuous direction were initiated by applying opposite τ-kicks to soliton pairs.

3.2. Families of Onsite-centered Semi-discrete Solitons, and Comparison to the Analytical Results

The family of stationary onsite-centered SDSs was constructed with energies exceeding a minimum value, whose numerically found value is

$$E_{\min}=4.087, \tag{33}$$

which is almost identical to the VA prediction given by Eq. (12). Real-z simulations confirm the dynamical stability of the entire soliton family found at $E > E_{\min}$.

On the other hand, the short VK-unstable branch of the broad SDSs, predicted by the VA to the left of $E = E_{\min}$ in Figure 1, is not produced by the numerical procedure. More specifically, the imaginary-z simulations, initiated by inputs corresponding to the VA-predicted shapes of the solitons which belong to the unstable branch, quickly transform them into the stationary SDSs with the same energy, but pertaining to the stable branch of $E(k)$ (at $k > k_{\min}$ in Figure 1). This outcome of the simulations may indeed imply the instability of the corresponding branch, rather than its nonexistence.

As an example, in Figure 3 we present a stable soliton obtained at E = 5.85, which exceeds $E_{\min}$, see Eq. (33). It was generated by input configuration (30) with amplitude $A = 2.8$ and widths $\Delta n_0 = 0.32$, $\Delta\tau_0 = 0.35$. The figure shows longitudinal profiles of the SDS in five central cores of the array, i.e., at $n = 0, n = \pm 1, n = \pm 2$. The comparison with the analytically predicted profiles (15) and (17) is shown as well, for $k = 3.1$, which corresponds to the sum of the energies of the U_0 and $U_{\pm 1}$ components equal to the energy of the numerical solution, i.e., $E = 5.85$. In fact, the numerically found and analytically predicted profiles are indistinguishable. To characterize the degree of the self-compression of this SDS, we note that the amplitude in the central core is larger than at $n = \pm 1$ by a factor of 11.6, and the latter amplitude is 10.2 times larger than its counterpart at $n = \pm 2$.

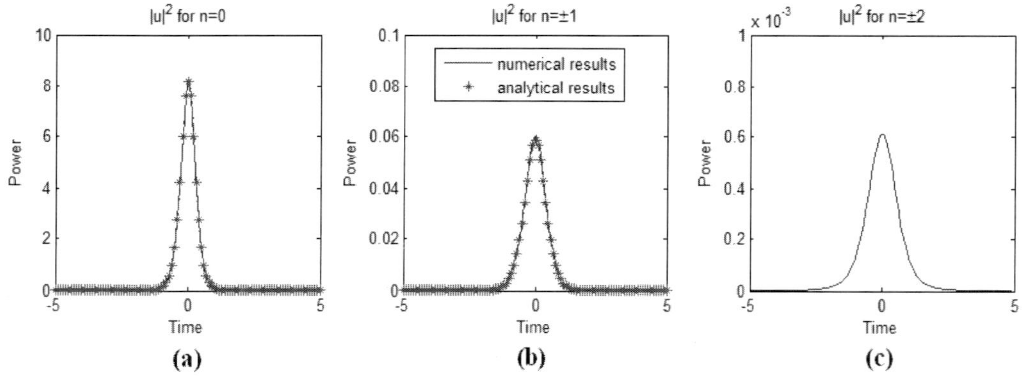

Figure 3. (Color online) Longitudinal profiles of the stable onsite-centered semi-discrete soliton with energy E=5.85 in cores n = 0 (a), n = ±1 (b), and n = ±2 (c). The blue continuous line and the chain of red stars show, severally, the numerical results and analytical predictions (15) and (17).

The entire SDS family is characterized, in Figure 4, by dependences of the amplitude, temporal width, spatial discrete width, and ratio R_0 [see Eq. (18)] on the total energy, E. The numerical widths, Δn and $\Delta\tau$, were defined in accordance with initial ansatz (30), so that $U_1(\tau = 0)/U_0(\tau = 0) \equiv \mathrm{sech}\left(1/\Delta n\right)$, and $U_0(\tau = \Delta\tau)/U_0(\tau = 0) \equiv \mathrm{sech}(1) \approx 0.648$.

The plots also display the comparison with the results produced by the VA [see Eqs. (6), (9), (11) and (18)]. It is seen that the amplitude and temporal width, as well as R_0, are predicted by the VA very accurately, while the prediction of the spatial width features a discrepancy, which is explained by the fact that variational ansatz (5) was defined with the exponential form along n, whereas the actual shape of the soliton in the discrete direction is found to be closer to sech [for this reason, the input used in the simulations was taken in the form of Eq. (30)]. Generally, we conclude that the VA, in the combination with the additional analytical results given by Eqs. (15) and (17), produces a reasonably accurate description of the SDS family.

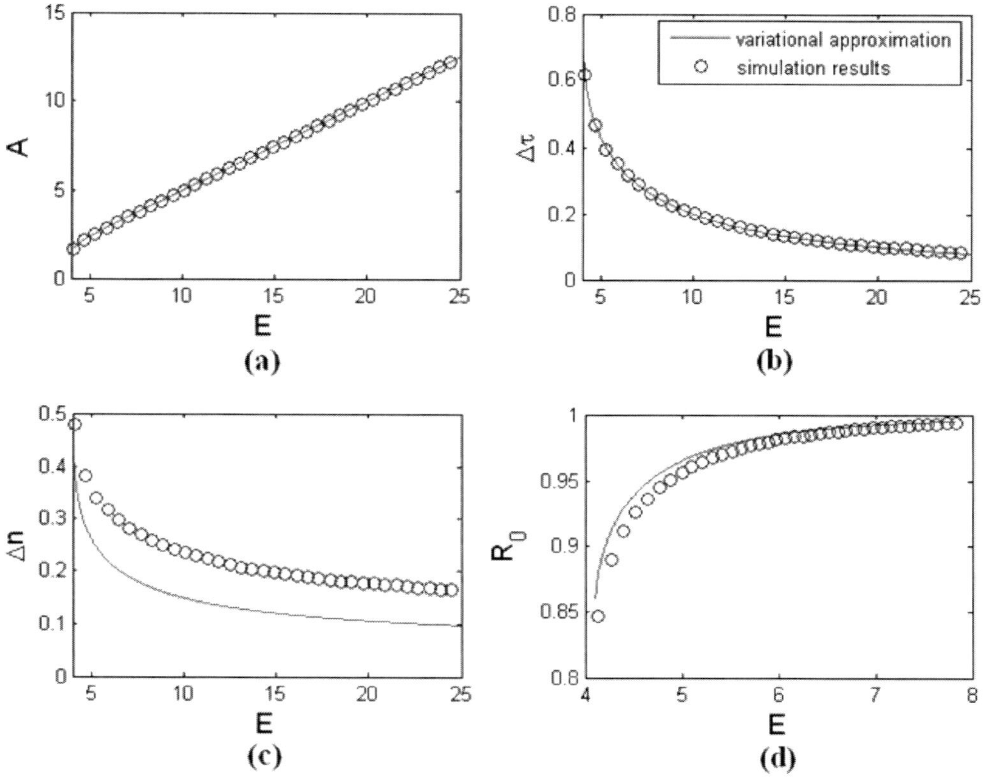

Figure 4. (Color online) The amplitude (a), temporal width (b), spatial width (c), and the share of the total energy trapped in the central core (d) of the onsite semi-discrete solitons versus the energy. The blue circles and red lines show, severally, the numerical results and predictions of the variational approximation.

3.3. Surface Solitons

The family of stationary surface SDSs attached to the edge of the semi-infinite array was found, by means if the imaginary-z simulations, for energies exceeding a minimum value,

$$E_{\min}^{(surface)} = 3.78\,,$$

which is very close to the VA prediction given by Eq. (26). Real-z simulations of the perturbed evolution confirm the stability of the entire surface-SDS family.

This family is characterized by dependences displayed in Figure 5, cf. Figure 4 for the fundamental onsite SDSs. It is seen here too that the VA predictions are reasonably accurate.

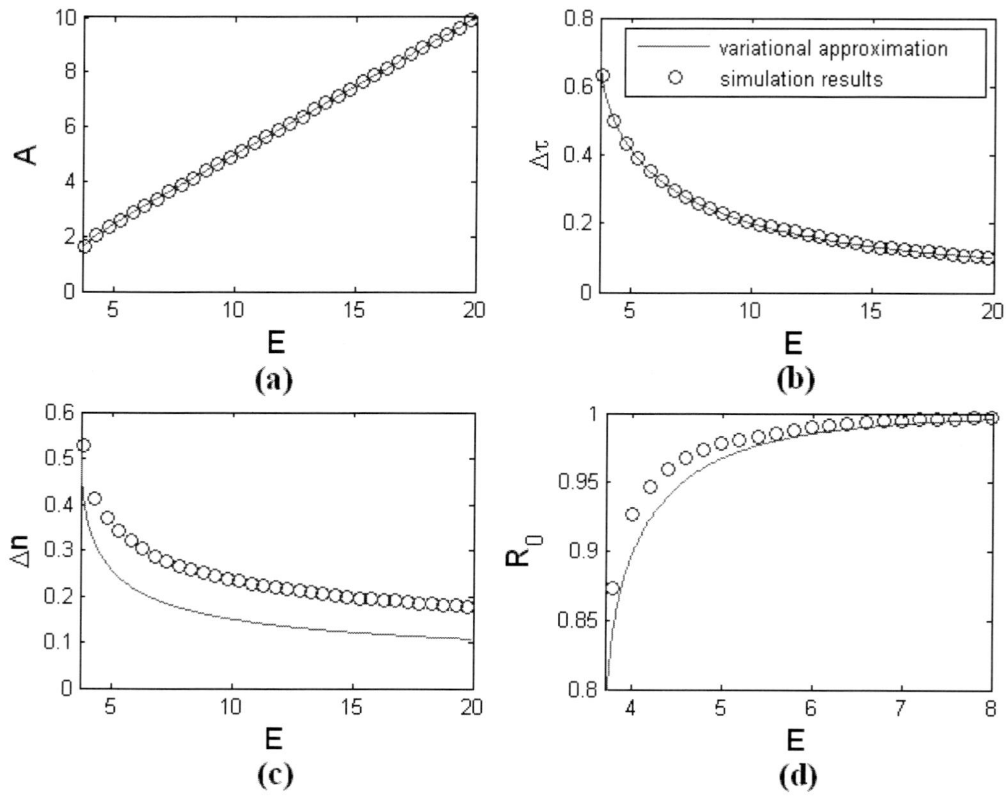

Figure 5. (Color online) The amplitude (a), temporal width (b), discrete spatial width (c), and the share of the total energy trapped in the central core (d) of the semi-discrete surface solitons versus the energy. The blue circles and red lines show, severally, the numerical results and predictions of the variational approximation.

Apart from the stable onsite-centered and surface SDSs, the possibilities of the existence of families of the intersite modes [see Eq. (27)] and twisted SDSs were also investigated. In the course of the imaginary-z propagation, the intersite-centered input spontaneously shifts in either direction, converting into a stable onsite soliton (as might be naturally expected, in view of the well-known instability of intersite-centered solitons in the discrete NLS equation [19]). On the other hand, the twisted input, in the form of $U_n = An\exp(-\alpha|n|)\cdot\mathrm{sech}(\eta\tau)$, with constants A, α, η, and the zero set at $n = 0$, is spontaneously transformed by the imaginary-z simulations into a stable fundamental SDS centered at either $n = +1$ or $n = -1$.

3.4. Transverse (Im)mobility of the Solitons

Systematic simulations of Eq. (1) in real z demonstrate that *true* stationary SDSs, produced by the imaginary-propagation-distance method, cannot be set in motion across the array by the application of the n-kick as per Eq. (31): The kick either initiates oscillations of the pinned soliton, or destroys it. A boundary between the different outcomes of the application of the kick has been identified in the plane of (E, a), as shown in Figure 6. The cause of the effective transverse immobility of the stable SDSs is that, in most cases, they all are quite narrow objects, being effectively localized, in the discrete direction, over ~3 cores, hence they are strongly pinned to the underlying discrete structure. On the other hand, essentially broader solitons, although being stable against small perturbations, turn out to be fragile against the application of the hard kick.

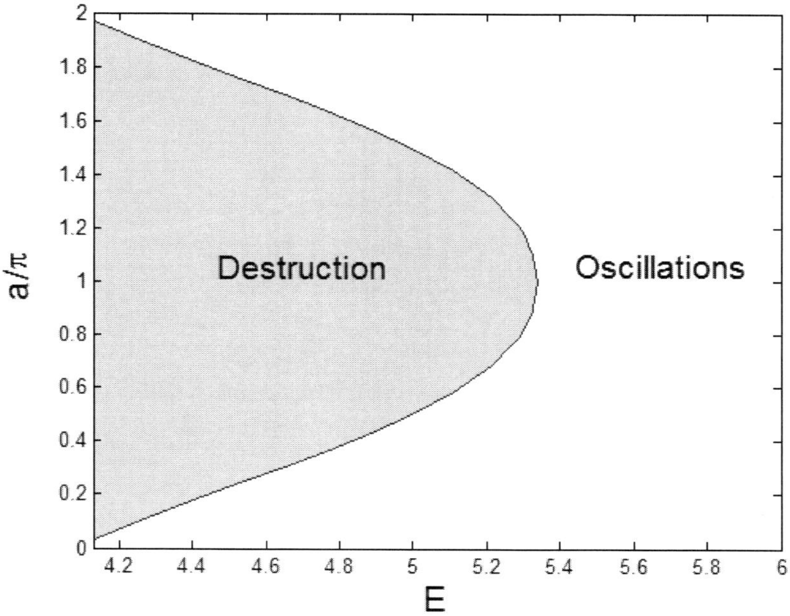

Figure 6. (Color online) Regions in the plane of the soliton's energy, E, and strength of the n-kick, a, where the stable onsite-centered soliton reacts to the kick by oscillations or destruction. Note that the picture is periodic in the vertical direction – actually, with period $\Delta a = \pi$ (taking into regard that positive and negative values of a are mutually equivalent). Values of the energy start from threshold (33). The destruction does not occur at $E > 5.3$.

On the other hand, it is possible to find non-soliton but quasi-stable broad semi-discrete localized patterns which, in the kicked form, feature effective transverse mobility, as shown in Figure 7, for a semi-discrete pulse with energy E = 4. This pattern is spread over approximately 13 cores, and, in the case shown in Figure 7, being kicked with strength a = 1.5, it travels distance $\Delta n = 10$ across the array, within propagation distance $\Delta z = 10$. It should be stressed that the energy of this pulse is below the threshold value $E_{\min}$ of the true stationary SDSs, see Eq. (33), therefore, it eventually (but quite slowly) decays after passing a sufficiently long distance: The peak power of the field falls to a half of its initial value at $z_{1/2}(a = 1.5) = 87$. For comparison, in the case when no kick is applied to the same pattern, the half-decay distance is essentially shorter, $z_{1/2}(a = 0) = 28$, i.e., the kick effectively *stabilizes* the non-soliton mode. Because the action of the nonlinearity on such low-energy modes is weak, they can collide quasi-elastically, readily passing through each other (which was verified in simulations, but is not shown here in detail).

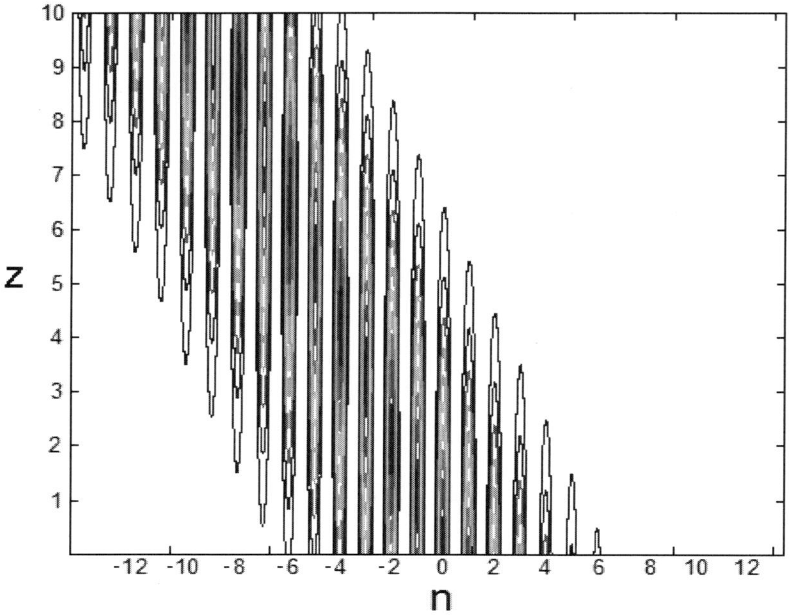

Figure 7. (Color online) Quasi-stable transverse motion of an n-kicked non-soliton semi-discrete broad mode with energy E = 4, kicked with a = 1.5. The figure shows intensity contour plots.

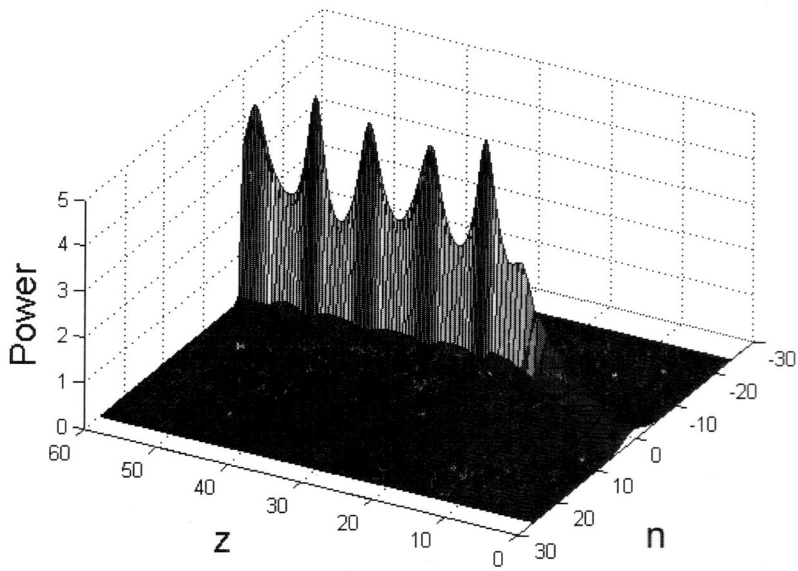

Figure 8. (Color online) The evolution of an initially broad semi-discrete pulse, spread over approximately 8 cores, with energy $E = 6.67 > E_{\min}$, which was kicked in the transverse direction by a = 1.5, see Eq. (31). The figure displays the evolution of the pulse in its central cross-section, $\tau = 0$. At first, the pulse is mobile and its center travels from n = 0 to n = -10, where it experiences the quasi-collapse and abrupt transition into the pinned mode.

A similar numerical experiment can be carried out for an initially broad semi-discrete pulse, whose total energy *exceeds* $E_{\min}$. As shown in Figure 8, the application of the transverse kick to such a "proto-soliton" sets it in motion across the array. However, the onset of the quasi-collapse causes the self-compression of the pattern, and it comes to a halt after passing a finite distance, then featuring strong oscillations in the pinned state.

4. Collisions between Solitons in the Longitudinal Direction

The mobility of SDSs in the continual (τ) direction is an obvious consequence of the Galilean invariance of Eq. (1) in this direction, which suggests to consider collisions between the solitons moving in opposite directions along axis τ , i.e., collisions of SDSs kicked as per Eq. (32), with velocities $\pm\omega$. In addition to the fundamental interest, such collisions may have potential applications to all-optical steering of nonlinear signals in photonic devices. We here focus on the onsite-centered solitons in the full array. The issue of collisions between surface SDSs is a relevant problem too, which will be considered elsewhere.

4.1. Head-on Collisions: The Merger

The most natural case corresponds to the head-on collision between identical solitons, with equal energies E, moving in opposite directions along the same central core. Systematic simulations demonstrate that there is a critical value of the kick's strength, $\omega_c(E)$, such that the collision is quasi-elastic at $\omega > \omega_c(E)$, while the two solitons *merge* into a semi-discrete bound state at $\omega < \omega_c(E)$. At energies $E < 5.2$, the inelastically colliding solitons do not merge, but rather suffer mutual destruction, see below.

In fact, the very existence of the threshold, below which the collisions lead to the merger, is a noteworthy finding. Indeed, the stable SDSs are, typically, strongly localized objects, with $\geq 85\%$ of their total energy actually concentrated in the central core, see Figure 4(d). Therefore, one might expect that the collision between them may be almost tantamount to a collision of two solitons of the integrable NLSE which describes the central core in isolation, the collisions being completely elastic in that limit. Nevertheless, the merger, which is a typical manifestation of the nonintegrability (in particular, in the 1D discrete NLSE [32]), is readily revealed by the simulations. Naturally, $\omega_c(E)$ decreases with the increase of the energy of the colliding solitons (i.e., with the strengthening of the self-compression of the semi-discrete soliton towards the single core), as shown in Figure 9.

We stress that irregularities in the $\omega_c(E)$ curve, clearly observed in Figure 9, are a genuine feature of the dependence, rather than a manifestation of numerical inaccuracies (this feature is zoomed in the inset). Qualitatively similar fine-scale irregularities (which may manifest the existence of an underlying dynamical fractal structure) in the dependence of the critical velocity, separating quasi-elastic and inelastic collisions, are known in other nonintegrable systems [54].

Apart from the irregularities, the smoothed $\omega_c(E)$ dependence at $E > 5.8$ may be fitted to a power-law curve, which is found to be

$$\omega_c(E) \approx 4.7 \cdot E^{-1.1}, \tag{34}$$

i.e., the critical velocity is almost exactly inversely proportional to the solitons' energy (at $E <$ 5.8, the dependence is strongly affected by the transition to the regime of the collision-induced destruction of broad solitons, see below). The latter dependence can be easily explained. Indeed, the Hamiltonian of system (1) is

$$H = \frac{1}{2}\sum_{n=-\infty}^{n=+\infty}\int_{-\infty}^{+\infty}\left[\left|\frac{\partial u_n}{\partial \tau}\right|^2 - u_n^*\left(u_{n+1}+u_{n-1}\right)-\left|u_n\right|^4\right]d\tau \,. \tag{35}$$

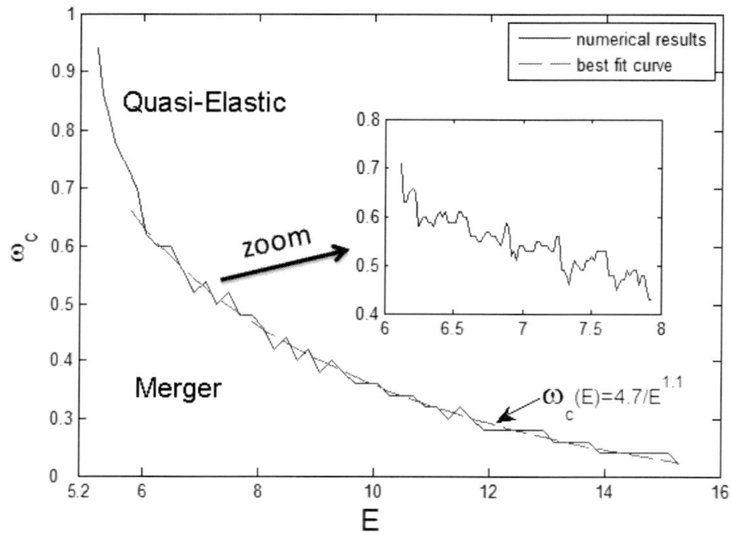

Figure 9. (Color online) The broken blue line shows the critical velocity ω_c, separating quasi-elastic head-on collisions between identical solitons with energy E, and their collision-induced merger, versus the soliton's energy, E, at, $E > 5.2$ (see Figure 10 for $E_{\min} < E < 5.2$). The inset depicts this dependence with a higher resolution for $6 < E < 8$, providing a better look at irregularities in the dependence. The red dashed line is the power-law best fit to the dependence at $E > 5.8$.

In the case of the collision between two solitons strongly confined to the central core, where their shapes are close to that of the usual NLSE soliton, given by Eq. (15), and the weaker components in adjacent cores are approximated by Eq. (17), the main contribution to the nonintegrable part of the interaction between such solitons is produced by the first term in the Hamiltonian density on the right-hand side of Eq. (35). Taking into regard that $E \approx 2\sqrt{2k}$ for such high-amplitude narrow solitons, it is easy to estimate the largest value of the interaction term, at the point of the full overlap between the colliding solitons, which plays the role of the potential barrier impeding the passage of the solitons through each other, $H_{\text{int}} \sim 1/E$ (the estimate follows from those for the width of the soliton in the τ -direction, $T \sim 1/E$, and the amplitude of the weaker component, $a \sim 1/E$). On the other hand, the net kinetic energy of the colliding solitons is $H_{\text{kin}} \approx 2E\omega^2$. Thus, the critical velocity, which is determined by equating the kinetic energy to the potential barrier, $H_{\text{int}} = H_{\text{kin}}$, is estimated as $\omega_c(E) \sim 1/E$, in agreement with the numerically generated fit (34).`

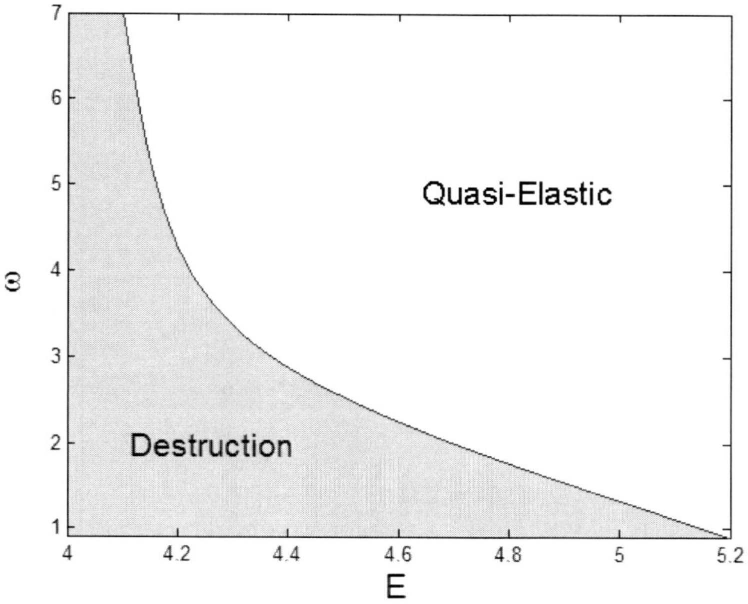

Figure 10. (Color online) Regions of the quasi-elastic and destructive head-on collisions between identical solitons, in the plane of the energy, E, and velocities, $\pm\omega$, at $E_{\min} < E < 5.2$, where $E_{\min}$ is the existence threshold, see Eq. (33).

As the energy of the SDS decreases, it becomes less confined in the central core, and ω_c exhibits in Figure 9 a transition from the approximate power-law dependence (34) to a steep rise close to $E = 5.2$. Moreover, at $E < 5.2$ inelastic collisions lead to destruction of the solitons immediately after the collision, rather than their merger. As shown in Figure 10, the critical value of the kick's strength, below which the collision results in the destruction and above which it is quasi-elastic, grows very steeply as the energy approaches $E_{\min}$, the minimum energy necessary for the existence of the onsite-centered solitons [see Eq. (33)].

To analyze the merger in more detail, we will now focus on the case of $E > 5.2$, where the collisions do not cause the destruction. At velocities sufficiently close to (but smaller than) ω_c, the merger proceeds through multiple collisions, which eventually give rise to a breather (an oscillating localized mode), as shown in Figure 11. Such multiple interactions, which demonstrate repeating attempts of the solitons to separate, are a feature known in other nonintegrable systems [54]. Note also that the collision shown in Figure 11 gives rise to a very small amount of radiative energy loss (see values of the energy indicated in the caption to the figure).

The effect of the multiple collisions disappears at small values of the velocity. Figure 12 displays the merger between the same solitons as in Figure 11, but this time kicked by $\omega = \pm 0.01$. In this case, the merger occurs immediately after the first collision, while the radiative losses remain negligible.

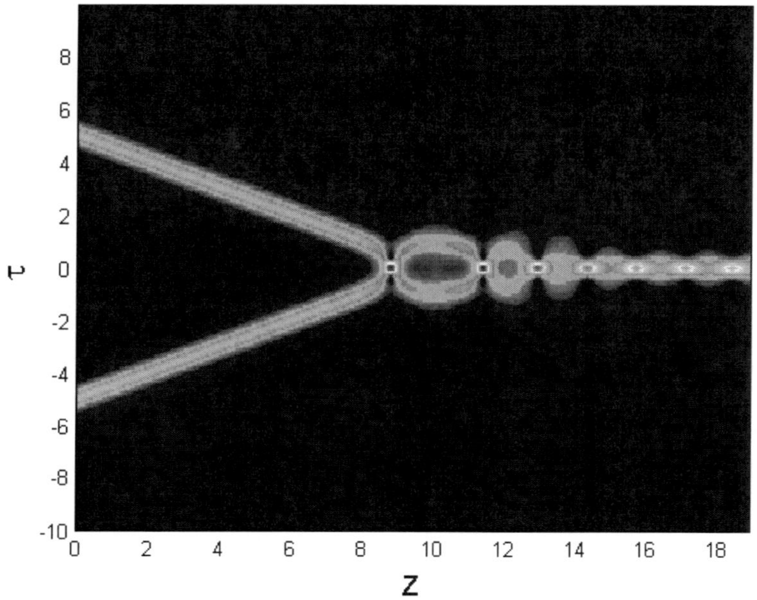

Figure 11. (Color online) The merger following multiple head-on collisions of identical solitons. The contour plot of the field amplitude is shown in the central core, $n = 0$ (brighter colors correspond to higher amplitudes). Two solitons, each with energy $E = 6$, are kicked at $z = 0$ by $\omega = \pm 0.5$, and propagate until they collide at $z = 8.8$, bouncing several times and eventually merging into the breather with energy $E = 11.9$ at $z = 15$. The critical velocity in this case is $\omega_c = 0.62$.

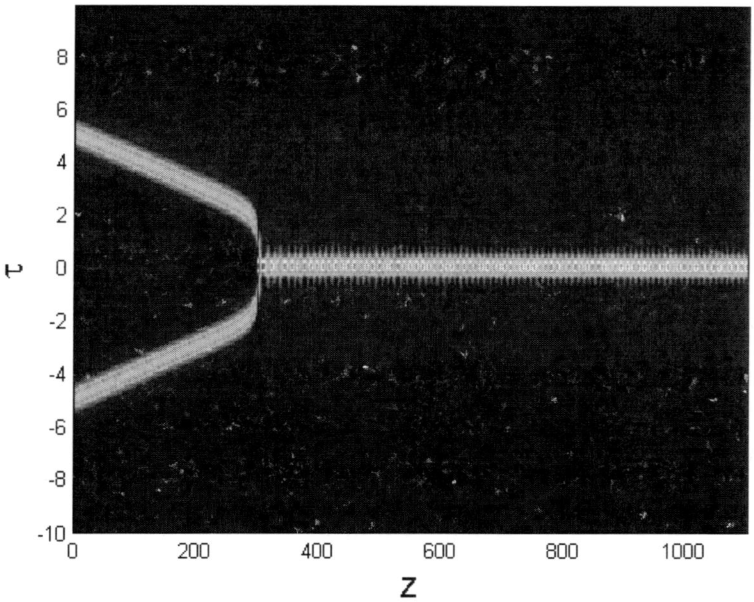

Figure 12. (Color online) An example of the merger between slowly moving solitons, produced by the single collision. The solitons are the same as in Figure 11, but kicked by $\omega = \pm 0.01$. The collision occurs at $z = 300$.

4.2. Quasi-elastic Head-on Collisions

A typical example of the quasi-elastic collision between solitons moving at velocities $\omega > \omega_c$ is shown in Figure 13. It is seen that the collision brings about a change in velocities of the colliding solitons. In Figure 13, the velocities are reduced by 55% after the collision. The collision-induced changes of the velocities are summarized in Figure 14. Here too, irregularities appearing in the plots are genuine features, rather than numerical artifacts, cf. Figure 9.

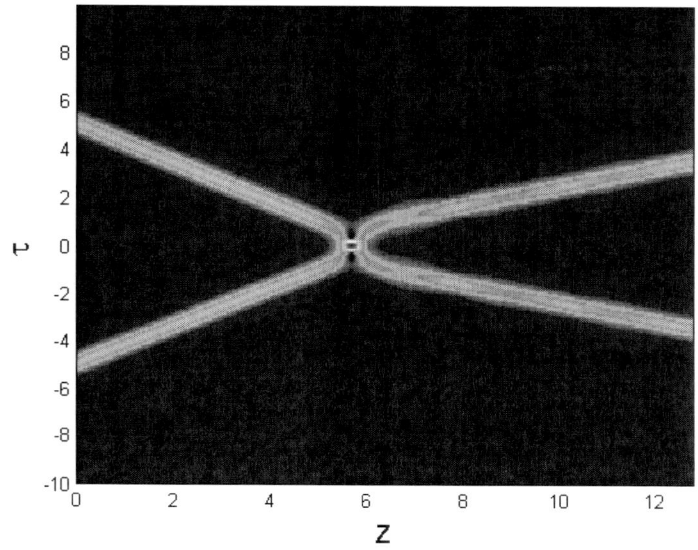

Figure 13. (Color online) An example of quasi-elastic collisions. The solitons are the same as in Figs. 10 and 11, but kicked by $\omega = \pm 0.8$. They collide at $z = 5.8$ and quickly separate afterwards.

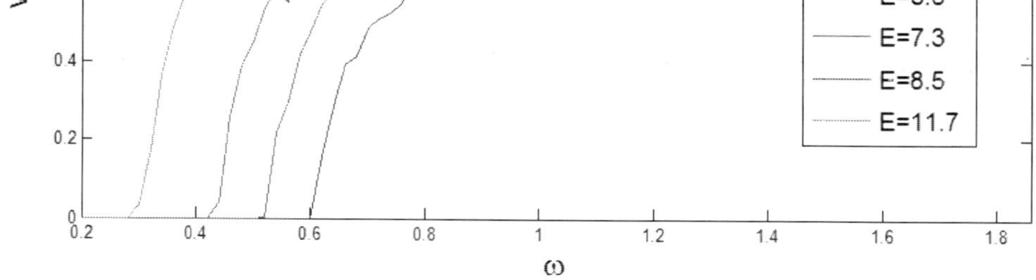

Figure 14. (Color online) The collision-induced velocity change for the colliding solitons. The plots show the ratio of the final velocity to its initial value versus the kick's strength (at $\omega > \omega_c$), for different soliton energies E.

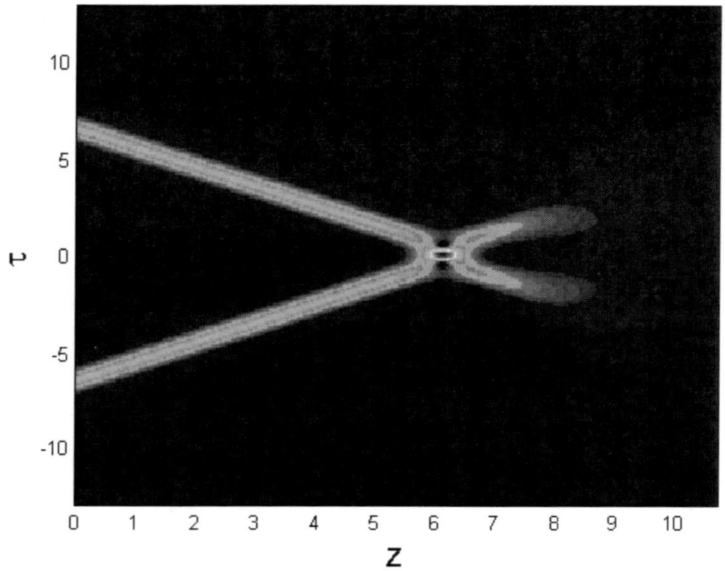

Figure 15. (Color online) An example of collision-induced soliton destruction. Two solitons, each with energy $E = 5$, are kicked at z = 0 by $\omega = \pm 1$, propagate until they collide at $z = 6$, and suffer destruction shortly afterwards.

4.3. Collision-induced Destruction of Solitons

When the kick is not strong enough, collisions between SDSs with energies $E_{\min} < E < 5.2$ result in destruction of both solitons, shortly after the collision (see Figure 10). An example n is displayed in Figure 15. This effect demonstrates that low-energy SDSs are actually fragile objects.

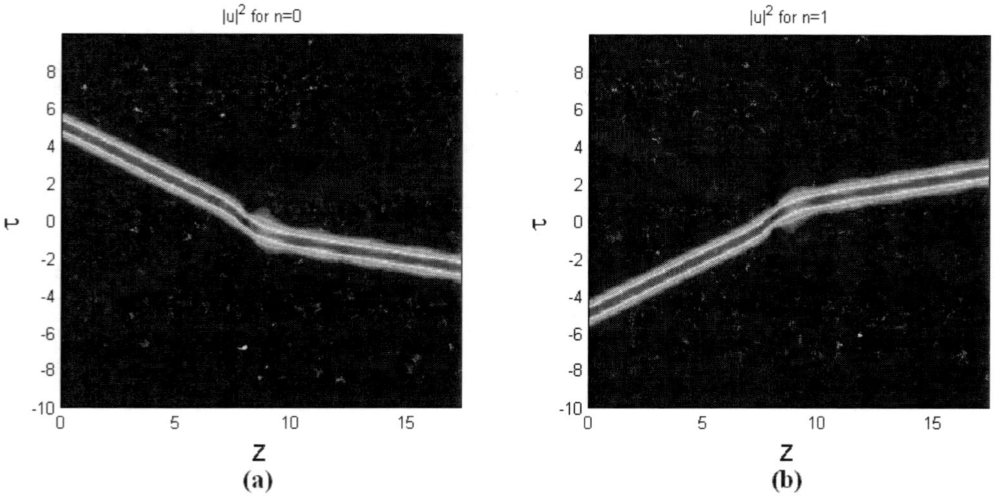

Figure 16: (Color online) The collision with the transverse shift. Shown are contour plots of the field amplitude in the central cores of the two solitons, $n = 0$ (a) $n = 1$ (b). The solitons, each with energy $E = 6$, are kicked at $z = 0$ by $\omega = \pm 0.6$. The interaction causes a significant velocity change, by 62%.

4.4. Soliton Collisions with the Transverse Shift

Collisions between counterpropagating identical solitons centered in adjacent cores, rather than in a common one, were studied too. An example is displayed in Figure 16, for the solitons centered at $n = 0$ and $n = 1$.

As in the case of the head-on collisions, the outcome of the "shifted" collision is also characterized by a critical velocity, ω_c , quasi-elastic collisions taking place at $\omega > \omega_c(E)$. However, the difference is that the interaction does not lead to merger at $\omega < \omega_c(E)$. In fact, the outcome of the interaction is not well defined in the latter case, small changes in the energy or velocity producing significantly different results. The cause of this effective instability is that, at the exact "moment" of the collision, with both solitons centered at $\tau = 0$ (in the longitudinal direction), the resulting field in the arrayed waveguide resembles that of an intersite soliton in the corresponding discrete NLSE, which is *unstable*, as discussed above (on the contrary, head-on collisions produce wave fields resembling stable onsite solitons). Below, we consider the well-defined region of quasi-elastic collisions.

The dependence of the critical velocity on the solitons' energy for the present case, produced by systematic simulations, is shown in Figure 17. As in Figs. 9 and 14, the irregular shape of the line is a true dynamical feature, rather than a result of an insufficient numerical accuracy. From Figs. 17 and 9 we conclude that ω_c is smaller for the "shifted" collisions, in comparison with the head-on configuration. This conclusion is natural, as the interactions between two solitons centered in the same core is stronger than between solitons propagating in adjacent cores. Finally, similar to the case of the head-on interactions, the "shifted" elastic collisions give rise to a change of velocities of the involved solitons (not shown here in detail, as this effect is quite similar to that displayed in Figure 14).

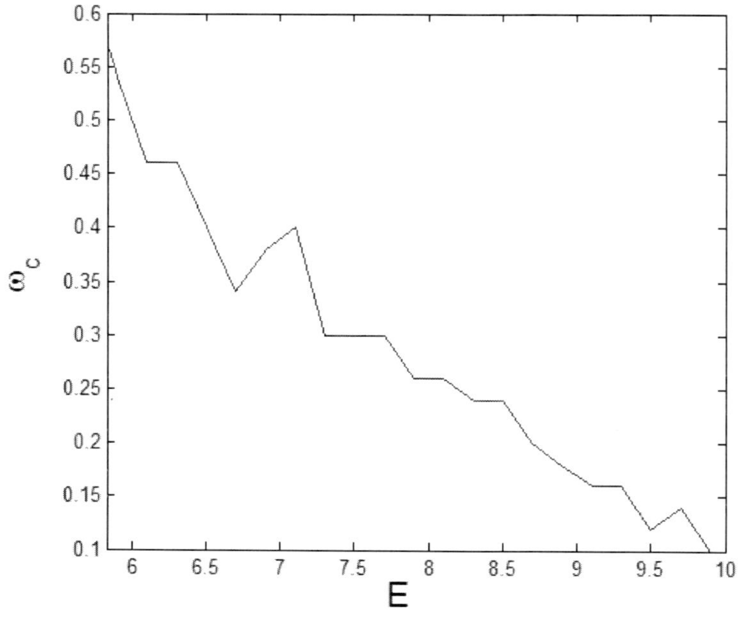

Figure 17. (Color online) The critical velocity for the collisions between solitons with the transverse shift, versus the solitons' energy, at $E > E_{\min}$.

Conclusion

In this work, we have revisited the previously known model of the tunnel-coupled array of parallel waveguides, which has realizations in optics (in both the temporal and spatial domains), as well as in BEC. It was known [21,22,24,40] that the arrest of the 2D collapse by the transverse discreteness gives rise to SDSs (semi-discrete solitons) in this system. In the present work, we aimed to develop the VA (variational approximation) and other analytical methods for the description of the SDSs, and to study, by means of systematic simulations, their mobility and collisions. It has been found that the VA yields quite accurate results for the families of stable onsite-centered and surface-mode SDSs. The exact solution of the linearized equations for the cores adjacent to the central one [see Eqs. (15) and (17)] provides a good approximation for very narrow solitons too. The simulations have shown that the true solitons have no mobility across the array (non-soliton semi-discrete patterns may be efficiently mobile in the transverse direction, and may actually be temporarily stabilized by the corresponding kick). The systematic analysis of the collisions between the onsite-centered SDSs moving in the continual direction clearly shows a sharp boundary between quasi-elastic interactions and merger. The exact form of the boundary contains irregular features, but its average shape can be explained by means of an analytical estimate. The very fact of the merger caused by the collision between the SDSs, which are strongly localized in the transverse direction, is a noteworthy finding, as, in the limit when almost all the energy is trapped in the central core, one might expect completely elastic collisions governed by the usual NLSE. Collisions between solitons centered at two adjacent cores, rather than in the same one, have been considered too.

As the next step of the analysis, it is possible to look for higher-order modes, built as N-solitons (actually, breathers), produced by multiplying the stationary SDS by integer factor N. Such exact solutions of the single-core NLSE are well-known objects [28]. We have checked that SDSs for $N = 2$ are robust breathing modes, which are similar to their counterparts in the single-core NLSE. It remains to study interactions between them in the framework of the present system.

A more challenging generalization may be a semi-discrete array in the form of a fiber bundle with the structure of a discrete 2D lattice in the transverse plane, similar to the one which was recently used to create the semi-discrete spatiotemporal solitons [25]. In particular, it would be interesting to study the corresponding SDSs with the structure of discrete vortices [55,19] in the transverse lattice.

References

[1] D. N. Christodoulides, F. Lederer, and Y. Silberberg, *Nature* 424, 817 (2003).

[2] J. W. Fleischer, O. Cohen, T. Schwartz, O. Manela, B. Freedman, M. Segev, H. Buljan, and N. K. Efremidis, *Opt. Exp.* 13, 1780 (2005).

[3] F. Lederer, G. I. Stegeman, D. N. Christodoulides, G. Assanto, M. Segev, and Y. Silberberg, *Phys. Rep.* 463, 1 (2008).

[4] Y. V. Kartashov, V. A. Vysloukh, and L. Torner, *Progr. Opt.* 52, 63 (2009).

[5] Y. V. Kartashov, V. A. Vysloukh, and L. Torner, *Eur. Phys. J. Special Topics* 173, 87 (2009).

[6] U. Röpke, H. Bartelt, S. Unger, K. Schuster, and J. Kobelke, *Appl. Phys.* B 104, 481 (2011).
[7] M. Heinrich, R. Keil, F. Dreisow, A. Tünnermann, A. Szameit, and S. Nolte, *Progr. Opt.* 104, 469 (2011).
[8] F. Eilenberger, S. Minardi, A. Szameit, U. Röpke, J. Kobelke, K. Schuster, H. Bartelt, S. Nolte, A. Tünnermann, and T. Pertsch, *Opt. Exp.* 19, 23171 (2011).
[9] Z. Chen, M. Segev, and D. N. Christodoulides, *Rep. Progr. Phys.* 75, 086401 (2012).
[10] M. A. Foster. A. C. Turner, M. Lipson, and A. L. Gaeta, *Opt. Exp.* 16, 1300 (2008).
[11] R. M. Osgood, Jr., N. C. Panoiu, J. I. Dadap, X. Liu, X. Chen, I-W. Hsieh, E. Dulkeith, W. M. J. Green, and Y. A. Vlasov, *Adv. Opt. Phot.* 1, 162 (2009).
[12] R. Yan, D. Gargas, and P. Yang, *Nature Phot.* 3, 569 (2009).
[13] P. Berini, *Adv. Opt. Phot.* 1, 484 (2009).
[14] T. Uikal, M. Hentschel, and H. Giessen, *Appl. Phys.* B 105, 51 (2011).
[15] M. I. Stockman, *Opt. Exp.* 19, 22029 (2011).
[16] J. Zhang and L. Zhang, *Adv. Opt. Phot.* 4, 157 (2012).
[17] D. N. Christodoulides and R. I. Joseph, *Opt. Lett.* 13, 794 (1988).
[18] H. S. Eisenberg, Y. Silberberg, R. Morandotti, A. R. Boyd, and J. S. Aitchison, *Phys. Rev. Lett.* 81, 3383 (1998).
[19] P. G. Kevrekidis, *The Discrete Nonlinear Schrödinger Equation: Mathematical Analysis, Numerical Computations, and Physical Perspectives* (Springer: Berlin and Heidelberg, 2009).
[20] Y. Y. M. Liu, G. Bartal, D. A. Genov, and X. Zhang, *Phys. Rev. Lett.* 99, 153901 (2007); F. Ye, D. Mihalache, B. Hu, and N.-C. Panoiu, *ibid.* 104, 106802 (2010); A. Marini, A. V. Gorbach, and D. V. Skryabin, *Opt. Lett.* 35, 3532 (2010); M. Conforti, C. De Angelis, and T. R. Akylas, *Phys. Rev.* A 83, 043822 (2011); Y. Kou, F. Ye, and X. Chen, *ibid.* 84, 033855; R. E. Noskov, P. A. Belov, and Y. S. Kivshar, *Opt. Exp.* 20, 2733 (2012).
[21] B. Aceves, C. De Angelis, A. M. Rubenchik, and S. K. Turitsyn, *Opt. Lett.* 19, 329 (1994); A. B. Aceves, C. De Angelis, G. G. Luther, and A. M. Rubenchik, *ibid.* 19, 1186 (1994).
[22] B. Aceves, G. G. Luther, C. De Angelis, A. M. Rubenchik, and S. K. Turitsyn, *Phys. Rev. Lett.* 75, 73 (1995).
[23] L. Bergé, Phys. Rep. 303, 259 (1998); E. A. Kuznetsov and F. Dias, *ibid.* 507, 43 (2011).
[24] B. Aceves, C. De Angelis, G. G. Luther, A. M. Rubenchik, and S. K. Turitsyn, *Physica* D 87, 262 (1995); S. K. Turitsyn, A. M. Rubenchik, M. P. Fedoruk, and E. Tkachenko, *Phys. Rev.* A 86, 031804 (R) (2012).
[25] S. Minardi, F. Eilenberger, Y. V. Kartashov, A. Szameit, U. Röpke, J. Kobelke, K. Schuster, H. Bartelt, S. Nolte, L. Torner, F. Lederer, A. Tünnermann, and T. Pertsch, *Phys. Rev. Lett.* 105, 263901 (2010).
[26] N.C. Panoiu, R. M. Osgood, Jr., and B. A. Malomed, *Opt. Lett.* 31, 1097 (2006).
[27] N.-C. Panoiu, B. A. Malomed, and R.M. Osgood, Jr., *Phys. Rev.* A 78, 013801 (2008).
[28] G. P. Agrawal, *Nonlinear Fiber Optics*, 4th edition (Academic Press: San Diego, 2007).
[29] D. Jaksch, C. Bruder, J. I. Cirac, C. W. Gardiner, and P. Zoller, *Phys. Rev. Lett.* 81, 3108 (1998); M. Lewenstein, A. Sanpera, V. Ahufinger, B. Damski, A. Sen(De), and U. Sen, *Adv. Phys.* 56, 243 (2007).

[30] Smerzi, A. Trombettoni, P. G. Kevrekidis, and A. R. Bishop, *Phys. Rev. Lett.* 89, 170402 (2002); V. A. Brazhnyi and V. V. Konotop, *Mod. Phys. Lett.* B 18, 627 (2004).

[31] A. Malomed and M. I. Weinstein, *Phys. Lett.* A 220, 91 (1996).

[32] E. Papacharalampous, P. G. Kevrekidis, B. A. Malomed, and D. J. Frantzeskakis, *Phys. Rev.* E 68, 046604 (2003).

[33] J. Kaup, *Math. Comput. Simul.* 69, 322 (2005); D. J. Kaup and T. K. Vogel, *Phys. Lett.* A 362, 289 (2007); B. A. Malomed, D. J. Kaup, and R. A. Van Gorder, *Phys. Rev. E* 85, 026604 (2012).

[34] R. Carretero-González, J. D. Talley, C. Chong, and B. A. Malomed, *Physica* D 216, 77 (2006); G. Herring, P. G. Kevrekidis, B. A. Malomed, R. Carretero-González, and D. J. Frantzeskakis, *Phys. Rev.* E 76, 066606 (2007); C. Chong, R. Carretero-González, B. A. Malomed, and P. G. Kevrekidis, *Physica* D 238, 126 (2009); 240, 1205 (2011).

[35] Lj. Hadžievski, G. Gligorić, A. Maluckov, and B. A. Malomed. *Phys. Rev.* A 82, 033806 (2010); M. Stojanović, A. Maluckov, Lj. Hadžievski, and B. A. Malomed, *Physica* D 240, 1489 (2011).

[36] M. Syafwan, H. Susanto, S. M. Cox, and B. A. Malomed, *J. Phys. A: Math. Gen.* 45, 075207 (2012).

[37] Chong, D. E. Pelinovsky, and G. Schneider, *Physica* D 241, 115 (2012).

[38] C. Paré and M. Florjańczyk, Phys. Rev. A 41, 6287 (1990); E. Cagliotti, S. Trillo, S. Wabnitz, B. Crosignani, and P. Di Porto, *J. Opt. Soc. Am.* B 7, 374 (1990); A. I. Maimistov, Kvant. Elektron. 18, 758 [Sov. J. Quantum Electron. 21, 687]; P. L. Chu, B. A. Malomed, and G. D. Peng, *J. Opt. Soc. Am.* B 10, 1379 (1993); I. M. Uzunov, R. Muschall, M. Gölles, Y. S. Kivshar, B. A. Malomed, and F. Lederer, *Phys. Rev.* E 51, 2527 (1995); B. A. Malomed, I. M. Skinner, P. L. Chu, and G. D. Peng, Phys. Rev. E 53, 4084 (1996); W. C. K. Mak, B. A. Malomed, and P. L. Chu, *J. Opt. Soc. Am.* B 15, 1685 (1998).

[39] A. Malomed, in: *Progr. Optics* 43, p. 71 (E. Wolf, editor: North Holland, Amsterdam, 2002).

[40] A. B. Aceves, M. Santagiustina, and C. De Angelis, *J. Opt. Soc. Am.* B 14, 1807 (1997); A. B. Aceves and M. Santagiustina, *Phys. Rev.* E 56, 1113 (1997); A. Gubeskys and B. A. Malomed, *Eur. Phys. J.* D 28, 283 (2004).

[41] Mihalache, D. Mazilu, F. Lederer, and Y.S. Kivshar, *Opt. Exp.* 15, 589 (2007).

[42] T. Peschel, U. Peschel, and F. Lederer, Phys. Rev. E 57, 1127 (1998); D. A. Semagin, S. V. Dmitriev, T. Shigenari, Y. S. Kivshar, and A. A. Sukhorukov, *Physica* B 316, 136 (2002); S.V. Dmitriev, D. A. Semagin, A.A. Sukhorukov, and T. Shigenari, *Phys. Rev.* E 66, 046609 (2002).

[43] B. Duncan, J.C. Eilbeck, H. Feddersen, and J.A.D. Wattis, *Physica* D 68, 1 (1993); S. Flach and K. Kladko, *ibid.* 127, 61 (1999); S. Flach, Y. Zolotaryuk, and K. Kladko, *Phys. Rev.* E 59, 6105 (1999); M. J. Ablowitz, Z. H. Musslimani, and G. Biondini, *ibid.* 65, 026602 (2002).

[44] O. Bang, J. J. Rasmussen and P. L. Christiansen, Nonlinearity 7, 205 (1994); E. W. Laedke, K. H. Spatschek, and S. K. Turitsyn, *Phys. Rev. Lett.* 73, 1055 (1994).

[45] R. A. Vicencio and M. Johansson, *Phys. Rev.* E 73, 046602 (2006).

[46] H. Susanto, P. G. Kevrekidis, R. Carretero-González, B. A. Malomed, and D. J. Frantzeskakis, *Phys. Rev. Lett.* 99, 214103 (2007).

[47] M. Desaix, D. Anderson and M. Lisak, *J. Opt. Soc. Am.* B 8, 2082 (1991).

[48] M. Vakhitov, and A. Kolokolov, Radiophys. *Quantum Electron.* 16, 783 (1973).

[49] S. Darmanyan, A. Kobyakov, and F. Lederer, *J. Exp. Theor. Phys.* 86, 682 (1998).

[50] K. G. Makris, S. Suntsov, D. N. Christodoulides, G. I. Stegeman, and A. Hache, Opt. Lett. 30, 2466 (2005); S. Suntsov, K. G. Makris, D. N. Christodoulides, G. I. Stegeman, A. Hache, R. Morandotti, H. Yang, G. Salamo, and M. Sorel, *Phys. Rev. Lett.* 96, 063901 (2006); Y. V. Kartashov, L.Torner, and V. A. Vysloukh, *ibid.* 96, 073901 (2006); C. R. Rosberg, D. N. Neshev, W. Królikowski, A. Mitchell, R. A. Vicencio, M. I. Molina, and Y. S. Kivshar, *ibid.* 97, 083901 (2006); M. I. Molina, R. A. Vicencio, and Y. S. Kivshar, *Opt. Lett.* 31, 1693 (2006); E. Smirnov, M. Stepić, C. E. Rueter, D. Kip, and V. Shandarov, *ibid.* 31, 2338 (2006); K. G. Makris, J. Hudock, D. N. Christodoulides, G. I. Stegeman, O. Manela, and M. Segev, *ibid.* 31, 2774 (2006); Y. V. Kartashov, A. A. Egorov, V. A. Vysloukh, and L. Torner, *Opt. Exp.* 14, 4049 (2006); G. A. Siviloglou, K. G. Makris, R. Iwanow, R. Schiek, D. N. Christodoulides, G. I. Stegeman, Y. Min, and W. Sohler, *ibid.* 14, 5508 (2006); A. Szameit, Y. V. Kartashov, F. Dreisow, T. Pertsch, S. Nolte, A. Tünnermann, and L. Torner, *Phys. Rev. Lett.* 98, 173903 (2007); X. Wang, A. Bezryadina, Z. Chen, K. G. Makris, D. N. Christodoulides, and G. I. Stegeman, *ibid.* 98, 123903 (2007); B. Alfassi, C. Rotschild, O. Manela, M. Segev, and D. N. Christodoulides *ibid.* 98, 213901 (2007).

[51] Y. S. Kivshar, Laser Phys. Lett. 5, 703 (2008); Y.-J. He and B. A. Malomed, in: *The Discrete Nonlinear Schrödinger Equation: Mathematical Analysis, Numerical Computations, and Physical Perspectives*, pp. 259-276 (P. G. Kevrekidis, editor; Springer: Berlin and Heidelberg, 2009).

[52] Mihalache, D, Mazilu, F. Lederer, and Y. S. Kivshar, *Opt. Lett.* 32, 3173 (2007).

[53] M. L. Chiofalo, S. Succi and M. P. Tosi, *Phys. Rev.* E 62, 7438 (2000).

[54] D. K. D. K. Campbell, J. S. Schonfeld, and C. A. Wingate, *Physica* D 9, 1 (1983); D. K. Campbell and M. Peyrard, *Physica* D 18, 47 (1986); 19, 165 (1986); P. Anninos, S. Oliveira, and R. A. Matzner, *Phys. Rev.* D 44, 1147 (1991); Y. Tan and J. Yang, *Phys. Rev.* E 64, 056616 (2001); B. Piette and W. J. Zakrzewski, *J. Phys.* A 40, 329 (2007); R. H. Goodman and R. Haberman, *Phys. Rev. Lett.* 98, 104103 (2007).

[55] A. A. Malomed and P. G. Kevrekidis, *Phys. Rev.* E 64, 026601 (2001); P. G. Kevrekidis, B. A. Malomed, Z. Chen, and D. J. Frantzeskakis, *ibid.* 70, 056612 (2004).

In: Solitons
Editor: Matthew C. Williams

ISBN: 978-1-62618-234-9

Chapter 4

SOLITON RELATED EFFECTS IN SUPERCONTINNUM GENERATION

S. Rodrigues, M. Facão, S. Latas and M. Ferreira
I3N – Institute of Nanostructures, Nanomodelling and Nanofabrication,
Department of Physics, University of Aveiro, Aveiro, Portugal

Abstract

The role of the solitonic effects in the context of supercontinuum generation in optical fibers is reviewed. Such effects become important whenever light propagates in the anomalous dispersion regime. When ultrashort pulses with large enough power are launched into highly nonlinear fibers, soliton fission gives origin to multiple fundamental solitons of different widths and peak powers. Among the host of soliton related effects, there are two which become particularly important: the soliton self-frequency shift and the emission of dispersive radiation in the normal dispersion region. If long duration pulses or a continuous wave are used to pump the supercontinuum, the modulation instability becomes the main initial mechanism, breaking the pump wave into multiple fundamental solitons. In this case, both the soliton self-frequency shift and inelastic collisions among such solitons play an important role.

1. Introduction

The supercontinuum generation (SCG) is essentially a nonlinear phenomenon and corresponds to an extremely wide spectrum achieved by an optical pulse while propagating in a nonlinear medium, namely an optical fiber [1]. It results generally from the synergy between several fundamental nonlinear processes, such as self-phase modulation (SPM), cross-phase modulation (XPM), stimulated Raman scattering (SRS), and four-wave mixing (FWM) [2]. The spectral locations and powers of the pumps, as well as the nonlinear and dispersive characteristics of the medium determine the relative importance and the interaction between these nonlinear processes. Soliton related effects also play a significant role whenever light with sufficient power propagates in the anomalous dispersion regime [1].

One aspect of the effort to enhance the supercontinuum generation in optical fibers is to increase the fiber nonlinear parameter by decreasing the effective mode area, together with the suitable adjustment of its dispersion characteristics. These developments have led to extensive research on microstructured optical fibers (MOFs) with optimized designs, as well as on suspended core fibers using compound glasses [3-7]. The so-called jacketed air-suspended rod fiber [8] design has the smallest possible effective mode area, but it is mainly a theoretical concept, which can only be approximately realized in practice. On of such approximations is provided by the triangular core fiber [4].

The supercontinuum generation can be broadly divided in two categories, depending on the duration of the pump pulses. The first category corresponds to long (picosecond, nanosecond and continuous wave) pump pulses, whereas the second category is obtained with short (femtosecond) pump pulses.

The key to long pulse and continuous wave (CW) SCG is the occurrence of modulation instability (MI), which is inherent to any anomalously dispersive, nonlinear medium [2]. Through this process the continuous pump wave breaks up into fundamental solitons. The rate of breakup and the duration of these solitons depend on the pump conditions and fibre parameters such as dispersion and nonlinearity.

In some cases, the soliton duration can be short enough to have the spectral width necessary for Raman self-scattering, leading to the soliton self-frequency shift (SSFS) [9]. As MI is a noise seeded process, the solitons will have variable bandwidths and durations and thus a range of SSFS occurs, creating a smooth red-shifted continuous. Another, main contributor to the CW continuum is inelastic inter-soliton collisions, which also enhances the red-shift of the continuum [10, 11].

When ultrashort pulses are launched into anomalously dispersive highly nonlinear fibers with large enough power, soliton fission gives origin to multiple fundamental solitons of different widths and peak powers [12]. Among the host of soliton related effects, there are two which become then particularly important: the SSFS induced by the Raman scattering and the emission of dispersive (or "Cherenkov") radiation, which comes from the overlap of a soliton spectrum with the normal GVD range [13, 14]. Solitons dominate the long-wavelength edge of the supercontinuum, while the dispersive radiation determines the short wavelength expansion [15].

In this paper we discuss the role that solitons can have in the supercontinuum generation process in optical fibers. Section 2 provides a review of the main characteristics of highly nonlinear fibers suitable for supercontinuum generation. In Section 3 we discuss the modeling of supercontinuum generation, while in Section 4 we present several simulation results for different situations. Finally, Section 5 summarizes the main conclusions.

2. Highly Nonlinear Fibers for Supercontinuum Generation

Both the nonlinear and the dispersive properties of the optical fiber play an important role on the supercontinuum generation process. In the following we discuss both aspects, concerning particularly two types of highly nonlinear fibers: the jacketed air-suspended rod and the microstructured optical fiber.

2.1. The Nonlinear Parameter

The fiber nonlinearity is commonly characterized by the nonlinear parameter, γ, which is given by [2]

$$\gamma = \frac{2\pi}{\lambda}\frac{n_2}{A_{eff}} \tag{1}$$

where λ is the light wavelength, n_2 is the nonlinear-index coefficient of the fiber core and A_{eff} is the effective mode area, given by:

$$A_{eff} = \frac{\left(\iint |F(x,y)|^2 dxdy\right)^2}{\iint |F(x,y)|^4 dxdy} \tag{2}$$

$F(x, y)$ representing the spatial distribution of the fiber mode.

Eq. (1) shows that, for a fixed wavelength, since n_2 is determined by the material from which the fiber is made, the most practical way of increasing the nonlinear parameter γ is to reduce the effective mode area A_{eff}. An approach to achieve this objective is to reduce the core radius, while increasing the relative index difference between the core and the cladding of the fiber, given by

$$\Delta = (n_1 - n_c)/n_1 \tag{3}$$

where n_1 and n_c represent the core and the cladding refractive index, respectively.

The maximum value of Δ is achieved when the cladding material is replaced by air ($n_c \approx 1$), in which case the fiber becomes the so-called jacketed air-suspended rod (JASR). Narrow JASRs have been produced by tapering standard optical fibers, with an original cladding diameter of 125 μm [16-18]. Tapering a fiber involves heating and stretching it to form a narrow waist connected to the untapered fiber by taper transitions. If the transitions are gradual, light propagating along the fiber suffers very little loss.

In recent years there has been a great interest in tapered fibers with a core diameter below 1 μm [19-21]. Numerical results, obtained without using the standard Gaussian approximation, have shown that the mean-field diameter of the optical mode of such a nanoscale JASR achieves a minimum value when the rod diameter is 0.74λ [20]. The nonlinear parameter γ attains then a maximum value, which scales with λ^{-3}.

Figure 1 shows the effective mode area of a jacketed air-suspended rod against its diameter (dashed curve). We observe that, as the diameter of the rod is reduced, the effective mode area decreases, since the mode becomes more confined. However, when such diameter is further reduced, the core can no more confine adequately the light and the mode broadens again. As a consequence, there is a minimum effective mode area that, for a given optical wavelength, depends on the refractive index of the fiber [8]. For a silica fiber, the minimum effective mode area is $A_{eff} = 1.48\ \mu m^2$ and it occurs for a rod diameter $\approx 1.2\ \mu m$.

The jacket-air suspended rod considered above can be used to model the behavior of the fundamental mode of air-silica microstructured optical fibers with a high air-filling fraction in their cladding [22, 23]. Microstructured optical fibers (MOFs) represent a new class of optical fibers that are characterized by the fact that the silica cladding presents an array of embedded air holes. They are also referred as photonic crystal fibers (PCFs), since they were first realized in 1996 in the form of a photonic crystal cladding with a periodic array of air holes [24].

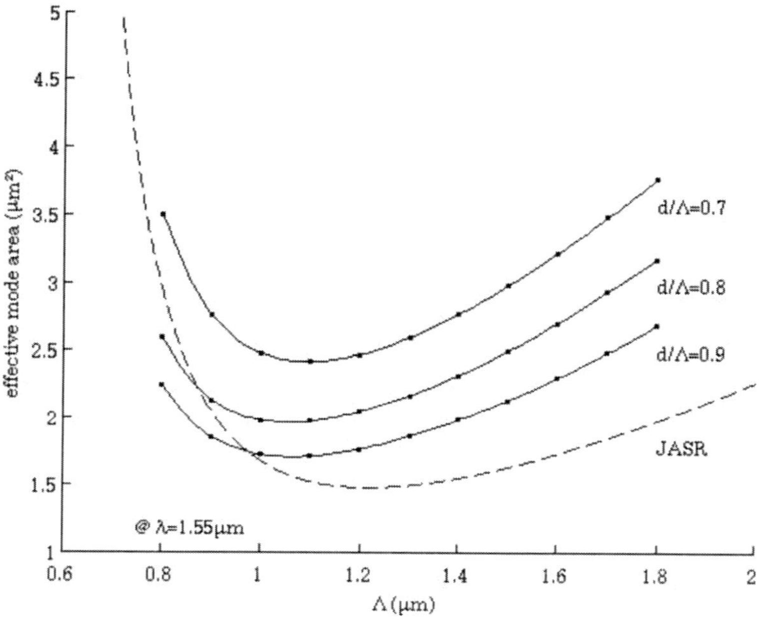

Figure 1. Effective mode area of a jacketed air-suspended rod (dashed curve) and of a microstructured optical fiber (full curves), as a function of the rod diameter or the hole-to-hole spacing, respectively (see the text for details).

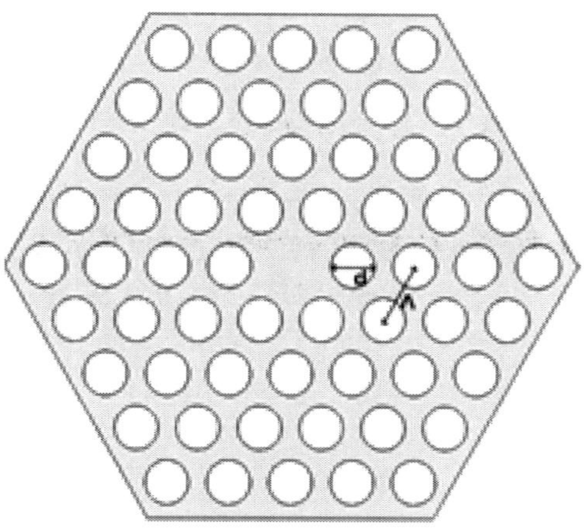

Figure 2. Schematic of a microstructured optical fiber with a hexagonal pattern of holes.

A MOF with an hexagonal pattern of holes is represented in Figure 2; it has two main structural parameters: the hair-hole diameter d and the hole-to-hole spacing Λ. In such fibers the periodic nature of air holes in the cladding is not important as long as they provide an effective reduction of its refractive index below that of the silica core [25, 26]. This helps to concentrate the mode field in a very small area, which is particularly the case in MOFs with small-scale cladding features and large air-filling fractions.

The effective mode area of a MOF is shown in Figure1 against the hole-to-hole spacing Λ, for different hole-pitch ratios d/Λ (full curves). As observed for the JASR, MOFs also exhibit a minimum effective mode area. As expected, the smallest effective mode area is achieved by use of the largest air-filling fraction, which corresponds to the case $d/\Lambda = 0.9$. A minimum effective mode area $A_{eff} = 1.7\ \mu\text{m}^2$ is obtained in this case, which occurs for a hole-to-hole spacing $\Lambda \approx 1\ \mu\text{m}$.

2.2. Fiber Dispersion

The mode-propagation constant $\beta(\omega)$ is related with the refractive index $n(\omega)$ in the form

$$\beta(\omega) = \frac{\omega n(\omega)}{c} \tag{4}$$

Mathematically, the effects of fiber dispersion are accounted for by expanding $\beta(\omega)$ in a Taylor series about the carrier frequency ω_0 at which the pulse spectrum is centered:

$$\beta(\omega) = \beta_0 + \beta_1(\omega - \omega_0) + \frac{1}{2}\beta_2(\omega - \omega_0)^2 + \ldots \tag{5}$$

where

$$\beta_j = \left(\frac{d^j\beta}{d\omega^j}\right)_{\omega=\omega_0} \quad (j = 0, 1, 2, \ldots) \tag{6}$$

The group velocity dispersion (GVD) is characterized by the parameter β_2

$$\beta_2 = \frac{d^2\beta}{d\omega^2} = -\frac{dv_g/d\omega}{v_g^2} \tag{7}$$

where $v_g = c/n_g$ and $n_g = n + \omega(dn/d\omega)$ represent the group velocity and the group refractive index, respectively. In practice, the GVD is often characterized by another parameter, D, given by

$$D = -\frac{2\pi c \beta_2}{\lambda^2} = -\frac{\lambda}{c}\frac{d^2 n}{d\lambda^2}. \tag{8}$$

The parameter D gives the delay of arrival time in ps unit for two wavelength components separated by 1 nm over a distance of 1 km. The GVD of pure silica is zero around 1.3 μm, being negative (normal dispersion) at shorter wavelengths and positive (anomalous dispersion) at longer wavelengths.

The overall GVD of a guided mode in an optical fiber depends not only on the material dispersion, as given above, but also on the waveguide dispersion [2]. The last contribution becomes especially important in the case of fibers using a large difference in refractive index between the core and cladding materials. Besides affecting the dispersion characteristics, increasing such refractive index difference provides also an enhancement of the nonlinear parameter of the fiber, as discussed previously.

Figure 3 shows the dispersion curves for the fundamental mode of a silica jacketed air-suspended rod considering different rod diameters. The dispersive properties are shown to be very sensitive to the rod diameter. By decreasing such diameter, the zero dispersion wavelength (ZDW) is shifted towards shorter wavelengths. Moreover, the strongly normal waveguide GVD causes a decrease of the overall GVD at long wavelengths and a second ZDW appears, defining a wavelength window with anomalous GVD. The GVD values are quite large in the anomalous dispersion regime, ranging in the order of 200 ps/km/nm. Meanwhile, the first zero dispersion points do not exceed a wavelength of 800 nm for a fiber diameter below 3 μm.

SC generation generally requires that the pump wavelength be close to the ZDW. If the dispersion slope is positive at this point it means that the dispersion is anomalous at longer wavelengths, where soliton propagation is possible.

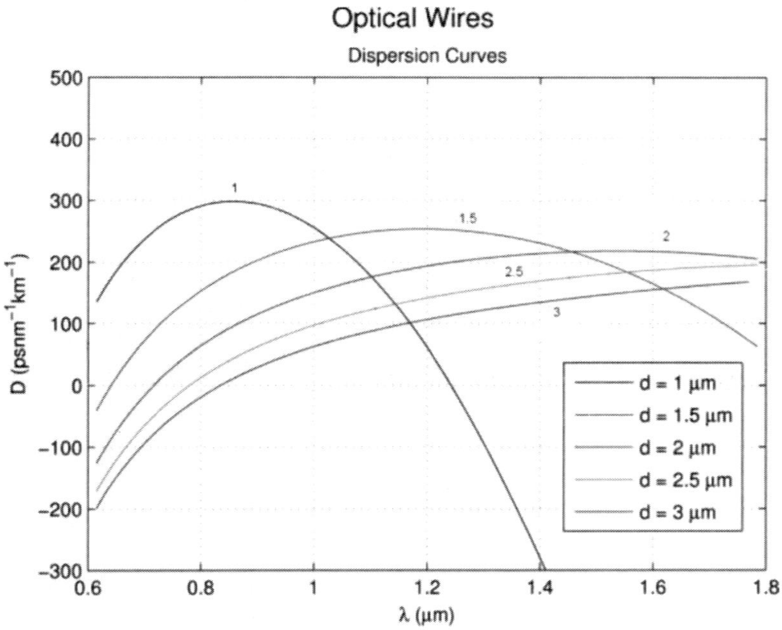

Figure 3. Dispersion properties of a jacketed air-suspended rod, for several values of the diameter.

Mode-locked Ti:sapphire lasers, mode-locked krypton ion lasers at 647 nm and frequency doubled Nd^{3+}-based lasers at 532 nm can be considered for this purpose, using tapered silica fibers with diameters sufficiently reduced.

Microstructured optical fibers offer greatly enhanced design flexibility. In this case, the dispersion characteristics can be manipulated by controlling structural parameters such as the hair-hole diameter d and the hole-to-hole spacing Λ. In fact, the dispersive properties of MOFs are quite sensitive to these parameters and can be tailored by changing appropriately each of them.

In a MOF, as the holes get larger, the core becomes more and more isolated, until it resembles a JASR. The MFs with larger cores exhibit semi-infinite anomalous dispersion above the ZDW. By decreasing the core size, the ZDW tends to be shifted to a shorter wavelength, leading to the anomalous dispersion at near infrared and visible wavelengths. When the core size is decreased further, a second ZDW arises in the longer wavelength side, such that the GVD is anomalous in the spectral window between the two ZDWs and normal outside it.

Figure 4 shows the dispersion characteristics of a MOF with a hexagonal pattern of holes spaced by i) $\Lambda = 1\,\mu m$ and ii) $\Lambda = 2.5\,\mu m$, considering a wavelength $\lambda = 1.55\,\mu m$, for different hole-pitch ratios. In the case i), two ZDWs are observed, defining a wavelength window with anomalous GVD. The separation between the two ZDWs increases and the anomalous GVD becomes larger when the hole-pitch ratio increases. For $d/\Lambda = 0.9$ this is essentially the same as a JASR with a diameter $\sim 2\Lambda - d$. On the other hand, for $\Lambda = 2.5\,\mu m$ and $d/\Lambda = 0.2$ the waveguide dispersion is relatively weak and the dispersion approaches that of the bulk silica. In the case $d/\Lambda = 0.3$ a nearly zero dispersion and flattened behaviour is achieved. This is particularly important for supercontinuum generation and other applications making use of the four-wave mixing effect.

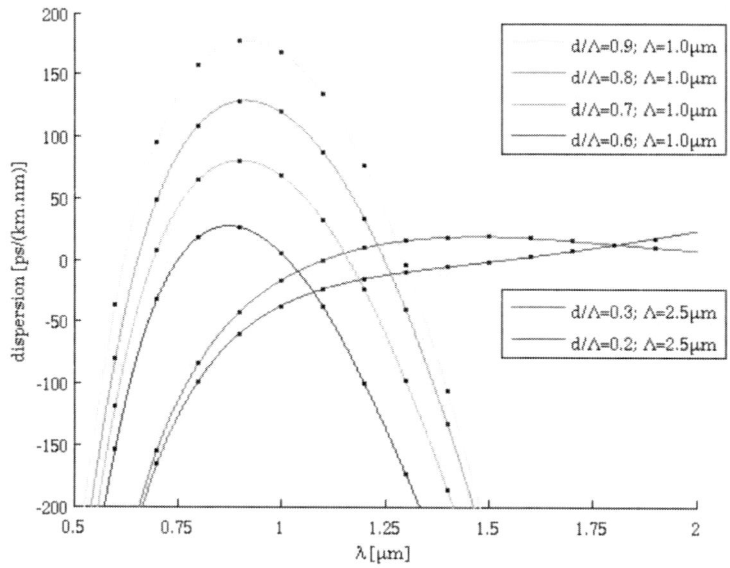

Figure 4. Dispersion properties of a microstructured fiber with a hexagonal pattern of holes spaced by i) $\Lambda = 1\,\mu m$ and ii) $\Lambda = 2.5\,\mu m$, considering a wavelength $\lambda = 1.55\,\mu m$, for different hole-pitch ratios.

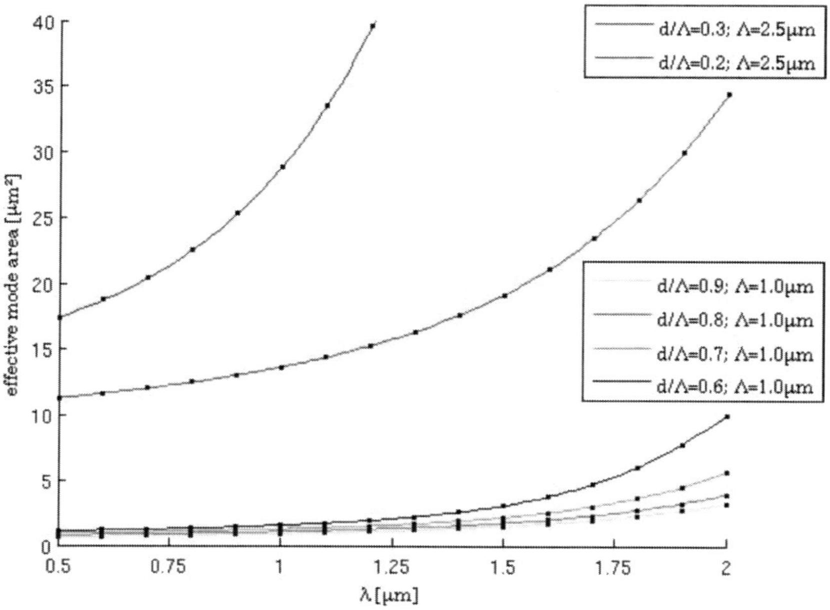

Figure 5. Effective mode area for the same cases of Figure 3.

Figure 5 shows the effective mode area for the same cases of Figure 4. We observe that the effective area is significantly smaller in the case $\Lambda = 1\,\mu\text{m}$, compared with the case $\Lambda = 2.5\,\mu\text{m}$. In both cases, increasing the hole-pitch ratio provides a reduced effective area. Consequently, the nonlinear parameter is increased, according with Eq. (1).

3. Modelling of the Supercontinuum

Modelling the supercontinuum generation can be realized considering a generalized NLSE that includes higher-order dispersion effects, as well as intrapulse Raman scattering. Such equation can be written as [1, 2, 27].

$$\frac{\partial U}{\partial z} - i\sum_{k\geq 2}\frac{i^k\beta_k}{k!}\frac{\partial^k U}{\partial \tau^k} + \frac{\alpha(\omega)}{2}U = i\gamma\left(1+\frac{i}{\omega_0}\frac{\partial}{\partial t}\right)\left(U(z,t)\int_{-\infty}^{+t} R(t')\left|U(z,t-t')\right|^2 dt'\right) \tag{9}$$

where $U(z,t)$ is the electric field envelope, ω_0 is the center frequency, β_k are the dispersion coefficients at the center frequency, $\alpha(\omega)$ is the frequency-dependent fiber loss, and γ is the nonlinear parameter.

The nonlinear response function $R(t)$ in Eq. (9) can be written as

$$R(t) = (1-f_R)\delta(t) + f_R h(t), \tag{10}$$

where the δ -function represents the instantaneous electron response (responsible for the Kerr effect), $h(t)$ represents the delayed ionic response (responsible for the Raman scattering) and

f_R is the fractional contribution of the delayed Raman response to the nonlinear polarization, in which a value $f_R = 0.18$ is often assumed [28]. It is common to approximate $h(t)$ in the form [27, 29]:

$$h(t) = \frac{\tau_1^2 + \tau_2^2}{\tau_1 \tau_2^2} \exp(-t/\tau_2) \sin(t/\tau_1) \tag{11}$$

where $\tau_1 = 12.2$ fs and $\tau_2 = 32$ fs. More accurate forms of the response function $h(t)$ have also recently been investigated [30].

Eq. (9) can be used to describe the propagation of femtosecond pulses in optical fibers, in both the normal and anomalous dispersion regimes. When such pulses have enough power, their spectra undergo extreme broadening. In the anomalous dispersion regime, this process is mainly influenced by the phenomenon of soliton fission, which occurs whenever a higher-order soliton is affected by third- or higher-order dispersion. The soliton order N is given by [2, 27]:

$$N = \sqrt{\frac{L_D}{L_{NL}}} \tag{12}$$

where $L_D = T_0^2 / |\beta_2|$ is the dispersion distance and $L_{NL} = 1/\gamma P_0$ is the nonlinear length.

In the presence of higher-order dispersion, an Nth-order soliton gives origin to N fundamental solitons whose widths and peak powers are given by [31]:

$$T_k = \frac{T_0}{2N + 1 - 2k} \tag{13}$$

$$P_k = \frac{(2N + 1 - 2k)^2}{N^2} P_0 \tag{14}$$

where $k = 1$ to $\tilde{N}$, where $\tilde{N}$ is the integer closest to N when N is not an integer. Soliton fission occurs generally after a propagation distance $L_{fiss} \sim L_D / N$, at which the injected higher-order soliton attains its maximum bandwidth. The fission distance L_{fiss} is a particularly significant parameter in the context higher-order soliton effect compression [32, 33]

Besides higher-order dispersion, another main effect affecting the dynamics of a higher-order soliton fission is the intrapulse Raman scattering (IRS). This phenomenon leads to a continuous downshift of the soliton carrier frequency, an effect known as the soliton self-frequency shift (SSFS) [34]. Such effect was observed for the first time by Mitschke and Mollenauer in 1986 [9]. The origin of SSFS can be understood by noting that for ultrashort solitons the pulse spectrum becomes so broad that the high-frequency components of the pulse can transfer energy through Raman amplification to the low-frequency components of the same pulse. Such an energy transfer appears as a red shift of the soliton spectrum, with shift increasing with distance.

Since its discovery in conventional single-mode fibers, the SSFS effect has been also observed in other types of fibers, including tapered fiber [35], solid-core MOFs [36], and air-core MOFs [37]. The SSFS effect can be significantly enhanced in some highly nonlinear fibers, where it has been used during the recent years for producing femtosecond pulses [38-41], as well as to realize several signal processing functions [42, 43].

The rate of frequency shift per propagation length is given by [2]

$$\frac{df}{dz} = -\frac{4t_R|\beta_2|}{15\pi T_0^4} = -\frac{4t_R(\gamma P_0)^2}{15\pi|\beta_2|} \tag{15}$$

where $t_R \equiv f_R \int_{-\infty}^{+\infty} th(t)dt \approx 5\,\text{fs}$ is the Raman parameter and P_0 is the soliton peak power. Since the SSFS effect is proportional to $(\gamma P_0)^2$, it will be enhanced if short pulses with high peak pulses are propagated in highly nonlinear fibers.

The higher-order soliton evolution and soliton fission as described above become progressively less important when pumping in the anomalous dispersion regime with long duration and high peak power pulses. This is because the fission distance L_{fiss} also increases with the pump pulse duration.

Under these conditions, the modulation instability causes the temporal breakup of the pump pulse into a large number of fundamental solitons. The subsequent evolution of these solitons then leads to additional spectral broadening and supercontinuum formation through a variety of mechanisms, namely soliton dynamical effects such as dispersive wave generation and the Raman-induce frequency shift.

The general features observed in the case of long duration pulses can also be extended to the continuous wave pumping in the anomalous dispersion regime. In this case, the supercontinuum generation also starts with the modulation instability of the continuous pump wave. This instability evolves from background noise and leads to the splitting of the continuous wave into fundamental solitons.

In the frequency domain, the modulation instability causes the spread of energy from the central wave frequency to side-bands with a frequency separation given by [2]:

$$\Omega_s = \sqrt{\frac{2\gamma P_p}{|\beta_2|}} \tag{16}$$

where P_p is the pump power. The MI period can be obtained from the above frequency separation as

$$T_s = \frac{2\pi}{\Omega_s} = \sqrt{\frac{2\pi^2|\beta_2|}{\gamma P_p}} \tag{17}$$

This period determines the number of solitons generated per unit time, as well as the energy of each soliton, given by:

$$E_{sol} = 2P_0T_0 = P_pT_s \tag{18}$$

where P_0 and T_0 are the soliton peak power and duration, respectively.

In some circumstances, the fundamental solitons arising from MI can be short enough in order to enable the occurrence of intrapulse Raman scattering, leading to the soliton self-frequency shift. As the modulation instability evolves from noise, there will be some jitter on the period and hence also on the duration of the generated solitons. In this case, a range of soliton self-frequency shifts occurs, creating a smooth red-shifted continuum evolution.

Soliton collisions constitute also an important mechanism for the spectral expansion in CW pumped supercontinua [10, 11]. In this case, the number of solitons can be particularly large and as redshifted solitons travel more slowly than those at higher frequencies, they can eventually collide while propagating through a fibre. During such collisions, some energy will be transferred to the most red-shifted soliton, if it is within the spectral range of Raman scattering. The energy increase leads to a corresponding compression, which results in a further red-shift for that soliton, according with Eq. (15). The increased red-shift also slows down the soliton, leading it to pass by more solitons and hence experience more collision events. Soliton collisions can considerably increase the extent of a supercontinuum compared to Raman self-scattering alone.

4. Simulation Results

In this Section we present and discuss several simulation results for the supercontinuum generation in highly nonlinear microstructured fibers. Pumping in the anomalous dispersion regime using femtosecond pulses is assumed, such that the soliton dynamical effects mentioned in the previous section play a prominent role. Figure 6 shows the dispersion curves for two different microstructured fibers with a hexagonal pattern of holes: fiber A, with holes of diameter $d = 1.4$ μm spaced by $\Lambda = 1.6$ μm, and fiber B with holes of diameter $d = 0.8$ μm spaced by $\Lambda = 1$ μm. Fiber A has is a single zero dispersion wavelength at 735 nm, whereas fiber B has two ZDWs: one at 654 nm and the other at 1248 nm.

Figure 7 shows the temporal and spectral evolution of an optical pulse along the microstructured fiber A. This result was obtained by numerically solving the Eq. (9), taking into account third- and higher-order dispersion but neglecting the Raman contribution. We consider a pump located in the anomalous dispersion regime at 790 nm, where the nonlinear parameter is $\gamma = 117\,\text{W}^{-1}\text{km}^{-1}$. An input pulse $U(0,\tau) = \sqrt{P_0}\,\text{sech}(\tau / T_0)$, is assumed, where $P_0 = 5$ kW, and the width $T_0 = 14.2$ fs, which corresponds to an intensity full width at half maximum FWHM of 25 fs.

For the assumed pulse and fiber parameters, we have $N = 3.01$ and a fission distance $L_{fiss} \sim 0.51$ cm. The individual solitons resulting from the fission process are not apparent in Figure 7 because they do not separate much from each other when the Raman contribution is neglected. However, a clear signature of soliton fission is the appearance of a new spectral peak in Figure 7a at $z = L_{fiss}$. Regarding the time domain, Figure 7b shows that a dispersive wave spreads with propagation and lags behind the main soliton.

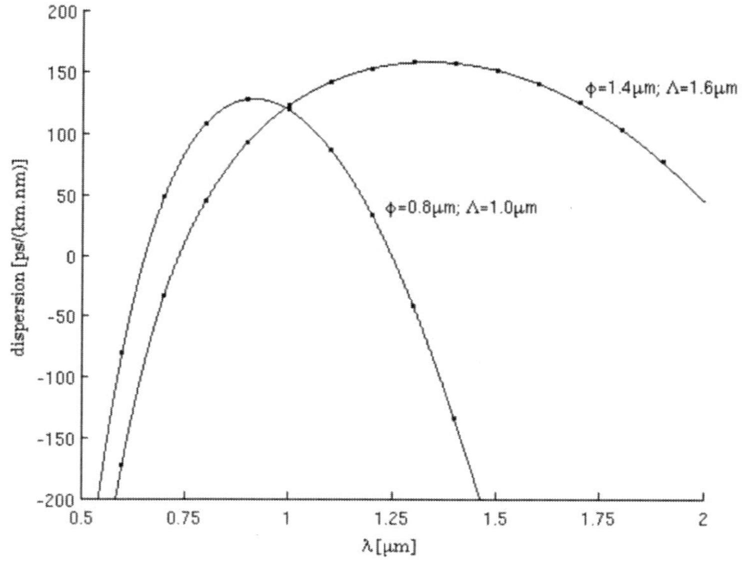

Figure 6. Dispersion curves for two different microstructured fibers with a hexagonal pattern of holes: fiber A, with holes of diameter $d = 1.4$ μm spaced by $\Lambda = 1.6$ μm, and fiber B with holes of diameter $d = 0.8$ μm, spaced by $\Lambda = 1$ μm.

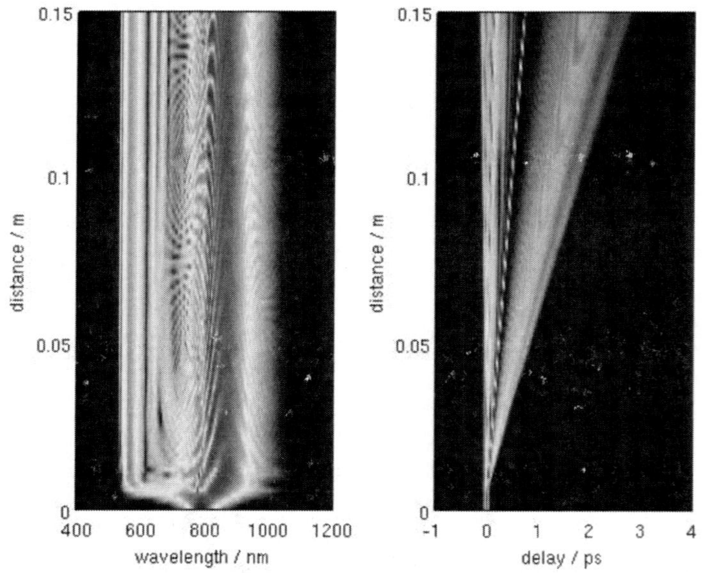

Figure 7. (a) Spectral and (b) temporal evolution of an optical pulse along a microstructured fiber obtained from Eq. (9) by neglecting the Raman effect. The input sech pulse at 790 nm has a peak power of 5 kW and 25 fs FWHM.

The new spectral peak observed in Figure 7a corresponds to the so-called *nonsoliton radiation* (NSR) [44] or *Cherenkov radiation* [14] which is emitted by the solitons resulting from the fission process in the presence of higher-order dispersion. The NSR is emitted at a frequency such that its phase velocity matches that of the soliton. Neglecting the fourth- and higher-order dispersions, the frequency shift $\Omega_d = \omega_d - \omega_s$ between the frequency of dispersive waves (ω_d) and that of the soliton (ω_s) is given approximately by [14]:

$$\Omega_d \approx -\frac{3\beta_2}{\beta_3} + \frac{\gamma P_0 \beta_3}{3\beta_2^2} \tag{16}$$

where β_j (j=2, 3) are the dispersion parameters, calculated at the soliton central frequency ω_s. For solitons propagating in the anomalous-GVD regime we have $\beta_2 < 0$. In these circumstances, Eq. (15) shows that $\Omega_d > 0$ when $\beta_3 > 0$, in which case the NSR is emitted at wavelengths shorter than that of the soliton, as observed in Figure 7a. On the contrary, if $\beta_3 < 0$ the NSR is emitted at a longer wavelength than that of the soliton.

The impact of the intrapulse Raman scattering on the soliton fission process can be appreciated by solving numerically Eq. (9), taking into account both the full dispersion curve and the Raman effect. The result of such simulation is shown in Figure 8, assuming the same fiber and pumping conditions as for Figure 7. Figure 9 shows the initial and final pulse profiles, both in frequency and time domains.

The most noticeable difference of Figure 8 in comparison with Figure 7 is the red-shift of the solitons created by the fission process. As a consequence of the SSFS induced by intrapulse Raman scattering, such solitons separate from each other. Since the SSFS is the largest for the shortest soliton, its spectrum shifts the most toward the red side in Figure 8a. The change of the soliton´s frequency determines a reduction in the soliton´s speed because of dispersion. This deceleration appears as a bending of the soliton trajectory in the time domain, as observed in Figure 8b.

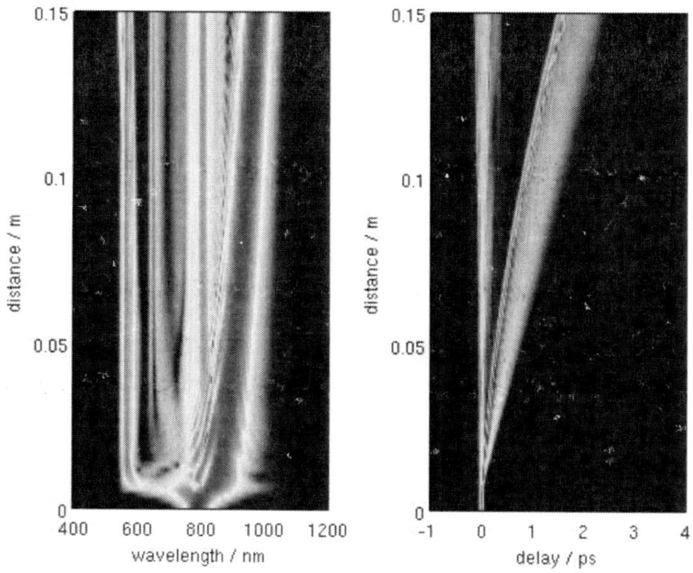

Figure 8. (a) Spectral and (b) temporal evolution of an optical pulse along a microstructured fiber obtained from Eq. (9) taking into account both the higher dispersion and the Raman effect. The fiber and pump conditions are the same as for Figure 7.

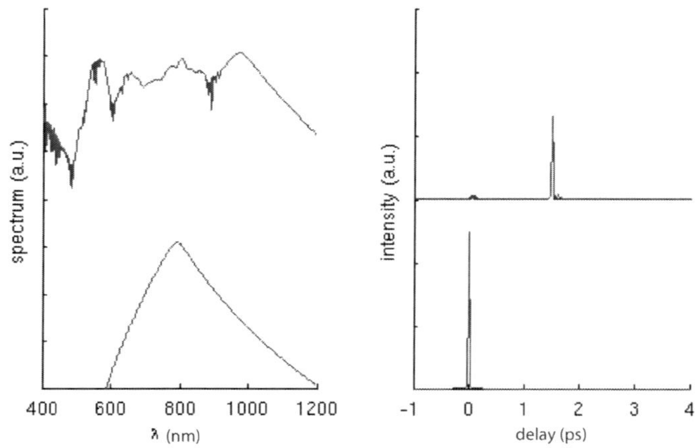

Figure 9. Initial and final pulse profiles in (a) frequency and (b) time domains for the case of Figure 8.

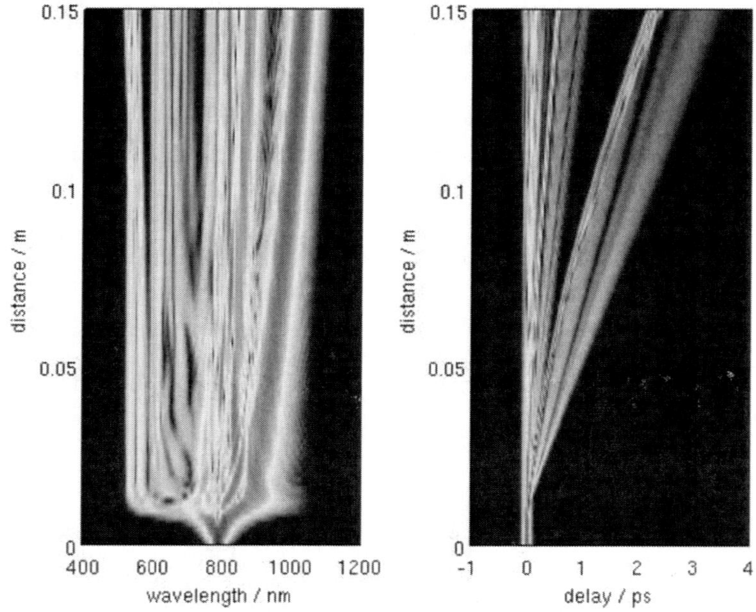

Figure 10. (a) Spectral and (b) temporal evolution of an optical pulse along a microstructured fiber obtained from Eq. (9). The input sech pulse at 790 nm has a peak power of 5 kW and 50 fs FWHM.

Keeping the input pulse peak power at 5 kW but increasing its width to 50 fs provides a soliton order N = 6.02. Figure 10 shows that, in this case, we achieve a wider and more uniform supercontinumm in comparison with that of Figure 8. Figure 11 shows the initial and final pulse profiles, both in frequency and time domains. Since the power of the first ejected soliton is greater than in the case of Figure 8, it experiences a larger SSFS, according with Eq. (15). At the same time, the corresponding nonsolitonic radiation is generated at a lower wavelength in the normal region.

As seen in Figure 6, fiber B has two ZDWs, such that the GVD is anomalous in the spectral window between them. The third-order dispersion parameter β_3 is positive near the first ZDW and negative near the second one. Near the first ZDW the NSR is always blue-shifted, as observed in Figures 8 and 10, so that the recoil and SSFS act in the same spectral direction. However, the opposite behaviour must occur near the second ZDW. This situation is illustrated in Figure 12, which shows the supercontinuum generated in fiber B, when it pumped at 860 nm using 100 fs pulses with a peak power of 1 kW. In this case, the GVD at the pump wavelength is $\beta_2 = -0.0487 \ \mathrm{ps^2/m}$ and the nonlinear parameter is $\gamma = 196.2 \ \mathrm{W^{-1}km^{-1}}$, which provide a soliton order N = 3.6. As the soliton loses energy to the red-shifted NSR, which is emitted in the normal-GVD regime, it becomes slightly wider in the time domain, as a result of energy conservation. On the other hand, since the detuning between the radiation and soliton gets smaller with propagation (Figure 12), the radiation is exponentially amplified.

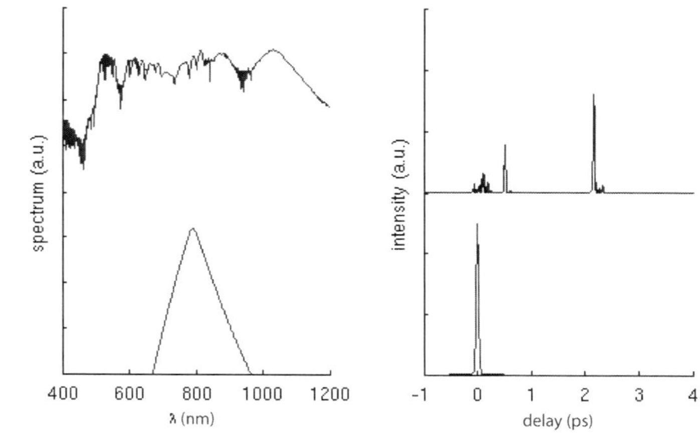

Figure 11. Initial and final pulse profiles in (a) frequency and (b) time domains for the case of Figure 10.

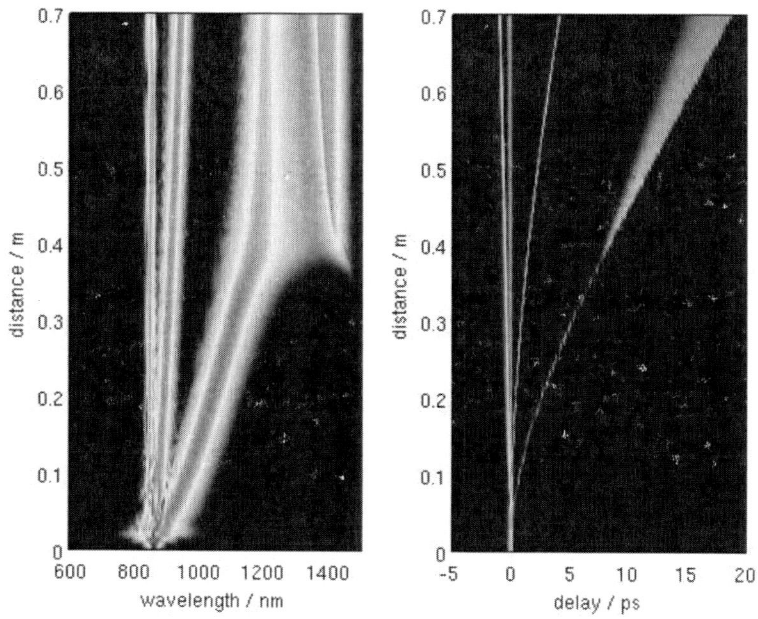

Figure 12. (a) Spectral and (b) temporal evolution of the supercontinuum generated in fiber B of Figure 6, pumped at 860 nm using 100 fs pulses with a peak power of 1 kW.

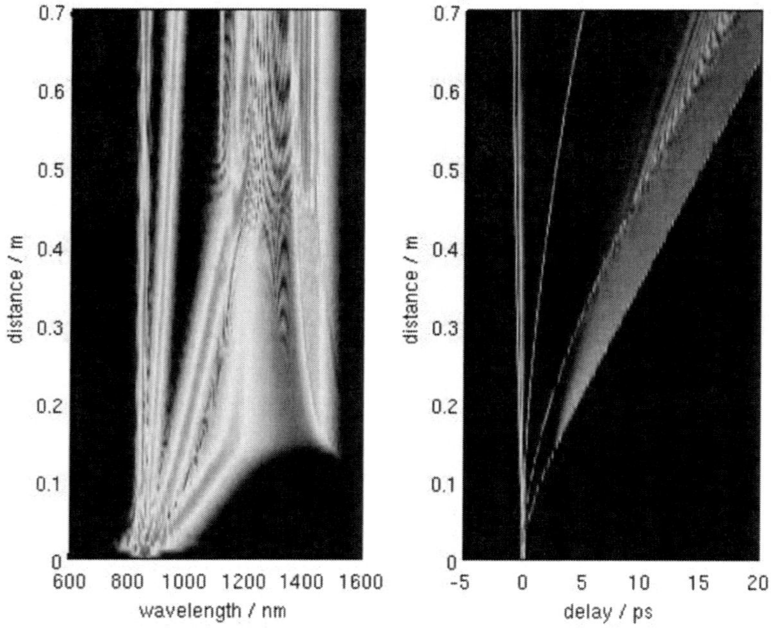

Figure 13. (a) Spectral and (b) temporal evolution of the supercontinuum generated in fiber B of Figure 6, pumped at 860 nm using 100 fs pulses with a peak power of 2 kW.

This amplification leads to the strong spectral recoil on the soliton into the anomalous-GVD regime, providing the suppression of SSFS [45-47] These effects are clearly visible in Figure 12, regarding the evolution of the first ejected soliton. We observe that the frequency shift of this soliton is suppressed after Z ~ 0.4 m.

One can observe in Figure 12 that the NSR is emitted ahead of the soliton and has no chance of interacting with it. However, the radiation emitted by one soliton can interact with other solitons present in the fibre [48, 49]. This is illustrated in Figure 13, which was obtained assuming the same conditions of Figure 12, but considering a peak power of 2 kW for the input pulse. In this case, the soliton order is N = 5.1. The suppression of the frequency shift is observed in Figure 13a for two ejected solitons: the first after Z = 0.15 m and the second after Z = 0.5 m. Beyond these positions, the trajectories of the ejected solitons in the time domain become linear, as shown in Figure 13b.

Conclusion

In this paper we discussed the main solitonic effects contributing to the supercontinuum generation process in highly nonlinear fibers. Pumping with femtosecond pulses in the anomalous dispersion region of the fiber, the soliton fission process gives origin to multiple fundamental solitons of different widths and peak powers. Among the host of soliton related effects, there are two which become then particularly important: the SSFS induced by the Raman scattering and the emission of dispersive (or "Cherenkov") radiation. Solitons dominate the long-wavelength edge of the supercontinuum, while the dispersive radiation determines the short wavelength expansion.

Concerning the case of long pulse and continuous wave pumping, the initial key mechanism for supercontinuum generation is the occurrence of modulation instability. Through this process the pump wave breaks up into fundamental solitons. The duration of such solitons can be short enough to have the spectral width necessary for Raman self-scattering, leading to the soliton self-frequency shift. Another main contributor to the supercontinuum in this case is inelastic inter-soliton collisions.

Acknowledgments

We acknowledge FCT (Fundação para a Ciência e Tecnologia) for supporting through the Projects PTDC/EEA-TEL/105254/2008, PTDC/FIS/ 112624/2009, and PEst-C/CTM/LA0025/2011.

References

[1] Dudley, J.; Taylor, J. (Eds) (2010) *Supercontinuum Generation in Optical Fibers.* Cambridge University Press, Cambridge, UK.

[2] Ferreira, M. F. (2011) *Nonlinear Effects in Optical Fibers*, John Wiley and Sons.

[3] Russel, P. (2006). Photonic-Crystal fibers. *J. Lightwave Technol.* 24, 4729-4749.

[4] Leong, J.; Petropoulos, P.; Price, J.; Ebendorff-Heidepriem, H.; Asimakis, S.; Moore, R.; Frampton, K.; Finazzi, V.; Feng, X.; Monro, T.; Richardson, D. (2006) High-nonlinearity dispersion-shifted lead-silicate holey fibers for efficient 1-μm pumped supercontinuum generation. *J. Lightwave Technol.* 24, 183-190.

[5] Ebendorff-Heidepriem, H.; Petropoulos, P.; Asimadis, S.; Finazzi, V.; Moore, R.; Frampton, K.; Koizumi, F.; Richardson, D.; Monro, T. (2002). Bismuth glass holey fibers with high nonlinearity. *Opt. Express* 12, 5082-5087.

[6] Ravi, V.; Kumar, K.; George, A.; Reeves, W.; Knight, J.; Russell, P.; Omenetto, F.; Taylor, A. (2002). Extruded soft glass photonic crystal fiber for ultrabroad supercontinuum generation. *Opt. Express* 10, 1520-1525.

[7] Ravi, V.; Kumar, K.; George, A.; Knight, J.; and Russel, P. (2003). Tellurite photonic crystal fiber", *Opt. Express* 11, 2641-2645.

[8] Finazzi, V.; Monro, T.; and Richardsom, D. (2003). Small-core silica holey fiber: Nonlinearity and confinement loss trade-offs. *J. Opt. Soc. Am. B.* 20, 1427-1436.

[9] Mitschke, F; Mollenauer, L. (1986). Discovery of the soliton self-frequency shift. *Opt. Lett.* 11, 659-661.

[10] Korneev N., Kuzin E. A., Ibarra-Escamilla B., Flores-Rosas M. B. J. A. (2008). Initial development of supercontinuum in fibers with anomalous dispersion pumped by nanosecond–long pulses. *Opt. Express* 16, 2636–2645.

[11] Frosz M. H., Bang O., Bjarklev A. (2006). Soliton collision and Raman gain regimes in continuous-wave pumped supercontinuum generation. *Opt. Express* 14, 9391–9407.

[12] Kodama,Y. and Hasegawa, A. (1987) Nonlinear pulse propagation in a monomode dielectric waveguide. *IEEE J. Quant. Elect.* QE23, 510–524.

[13] Wai P. K. A., Menyuk C. R., Lee Y. C., Chen H. H. (1986). Nonlinear pulse propagation in the neighborhood of the zero-dispersion wavelength of monomode optical fibers. *Opt. Lett.* 11, 464–4.

[14] Akhmediev N., Karlsson M. (1995). Cherenkov radiation emitted by solitons in optical fibers. *Phys. Rev. A* 51, 2602–2607.

[15] Dudley, J.M., Genty G. and Coen, S. (2006) Supercontinuum generation in photonic crystal fiber. *Rev. Mod. Phys.* 78, 1135–1184.

[16] Harbold, J.; Ilday, F.; Wise, F.; Birks, T.; Wadsworth, W.; and Chen, Z. (2002) Long-wavelength continuum generation about the second dispersion zero of a tapered fiber. *Opt. Lett.* 27, 1558-1560.

[17] Lu F. and Knox, W. (2004) Generation of a broadbend continuum with high specral coherence in tapered single-mode optical fibers. *Opt. Express* 12, 347-353.

[18] Foster M. and Gaeta, A. (2004) Ultra-low threshold supercontinuum generation in sub-wavelength waveguides. *Opt. Express* 12, 3137-3143.

[19] Tong, L.; Lou, J.; and Mazur, E. (2004). Single-mode guiding properties of subwavelength-diameter silica and silicon wire waveguides. *Opt. Express* 12, 1025-1035.

[20] Foster, M.; Moll, K.; and Gaeta, A. (2004) Optimal waveguide dimensions for nonlinear interactions. *Opt. Express* 12, 2880-2887.

[21] Cordeiro, C.; Wadsworth, W.; Birks, T.; and Russel, P. (2005). Engineering the dispersion of tapered fibers for supercontinuum generation with a 1064 nm pump laser. *Opt. Lett.* 30, 1980-1982.

[22] Birks, T.A., Mogilevtsev, D., Knight, J.C. and Russell, P.St.J. (1999). Dispersion compensation using single-material fibers. *IEEE Photon. Technol. Lett.* 11 674–676.

[23] Knight, J.C., Arriaga, J., Birks, T.A., Ortigosa-Blanch, A., Wadsworth, W.J. and Russell, P.St.J. (2000). Anomalous dispersion in photonic crystal fiber. *IEEE Photon. Technol. Lett.* 12 807–809.

[24] Knight, J.; Birks, T.; Russel, P.; Atkin, D. (1996). All-silica single-mode optical fiber with photonic crystal cladding. *Opt. Lett.* 21, 1547-1549.

[25] Broeng, J.; Mogilevstev, D.; Barkou, S.; Bjarklev, A. (1999). Photonic crystal fibers: a new class of optical waveguides. *Opt. Fiber Technol.* 5, 305-330.

[26] Eggleton, B.; Kerbage, C.; Westbrook, P.; Windeler, R.; Hale, A. (2001). Microstructured optical fiber devices. *Opt. Express* 9, 698-713.

[27] Agrawal G. P. (2007). *Nonlinear Fiber Optics*, 4th edn. San Diego: Academic Press.

[28] Stolen, R. H., Gordon, J. P.,Tomlinson,W. J.andHaus, H.A. (1989) Raman response function of silica-core fibers. *J. Opt. Soc. Am. B*, 6, 1159–1166.

[29] Blow K. J., Wood D. (1989). Theoretical description of transient stimulated Raman scattering in optical fibers. *IEEE J. Quantum Electron. 25*, 12, 2665–2673.

[30] Lin Q., Agrawal G. P. (2006). Raman response function for silica fibers. *Opt. Lett. 31*, 21, 3086–3088.

[31] Kodama Y and Hasegawa A. (1987). Nonlinear pulse propagation in a monomode dielectric guide. *IEEE J. Quantum Electron.* 23, 510-524.

[32] Chen C.-M., Kelley P. L. (2002). Nonlinear pulse compression in optical fibers: scaling laws and numerical analysis. *J. Opt. Soc. Am. B* 19, 1961–1967.

[33] Dianov E. M., Nikonova Z. S., Prokhorov A. M., SerkinV. N. (1986). Optimal compression of multi-soliton pulses in optical fibers. *Pis'ma Zh. Tekh. Fiz.* 12, 756–760 [Sov. Tech. Phys. Lett. 12, 311–313 (1986)].

[34] Gordon J. P. (1987). Theory of the soliton self-frequency shift. *Opt. Lett.* 11, 659-661.

[35] Liu, X.; Xu, C.; Knox, W.; Chandalia, J.; Eggleton, B.; Kosinski, S.; Windler, R. (2001). Soliton self-frequency shift in a short tapered air-silica microstructure fiber. *Opt. Lett.* 26, 358-360.

[36] Cormack, I.; Reid, D.; Wadsworth, W.; Knight, J.; Russel, P. (2002). Observation of soliton self-frequency shift in photonic crystal fibre. *Electron. Lett.* 38, 167-168.

[37] Ouzounov, D.; Ahmad, F.; Muller, D.; Venkataraman, N.; Gallagher, M.; Thomas, M.; Silcox, J.;Koch, K.; Gaeta, A. (2003). Generation of Megawatt Optical Solitons in Hollow-Core Photonic Band-Gap Fibers. *Science* 301, 1702-1704.

[38] Nishizawa, N.; Ito, Y.; Goto, T. (2002). 0.78-0.90-μm wavelength-tunable femtosecond soliton pulse generator using photonic crystal fiber. *IEEE Photon. Technol. Lett.* 14, 986-988.

[39] Efimov, A.; Taylor, A.; Omenetto, F.; Vanin, E. (2004). Adaptive control of femtosecond soliton self-frequency shift in fibers. *Opt. Lett.* 29, 271-273.

[40] Abedin, K.; Kubota, F. (2004). Wavelength tunable high-repetition-rate picosecond and femtosecond pulse sources based on highly nonlinear photonic crystal fiber. *IEEE J. Sel. Topics Quantum Electron.* 10, 1203-1210.

[41] Lee, J.; Howe, J.; Xu, C.; and Liu, X. (2008). Soliton self-frequency shift: experimental demonstrations and applications. *J. Sel. Quantum Electron.* 14, 713-723.

[42] Nishizawa N.; Goto, T. (2003). Ultrafast all optical switching by use of pulse trapping across zero-dispersion wavelength. *Opt. Express* 11, 359-365.

[43] Oda, S.; Maruta, A. (2006). All-optical tunable delay line based on soliton self-frequency shift and filtering broadened spectrum due to self-phase modulation. *Opt. Express* 14, 7895-7902.

[44] Husakou, A.; Herrmann, J. (2001). Supercontinuum generation of higher-order solitons by fission in photonic crystal fibers. *Phys. Rev. Lett.* 87, 203901.

[45] Skryabin D. V., Luan F., Knight J. C., Russell P. S. (2003). Soliton self-frequency shift cancellation in photonic crystal fibers. *Science* 301, 1705–1708.

[46] Biancalana F., Skryabin D. V.,Yulin A. V. (2004). Theory of the soliton self-frequency shift compensation by the resonant radiation in photonic crystal fibers. *Phys. Rev. E* 70, 016615.

[47] Tsoy E., deSterke C. M. (2006). Dynamics of ultrashort pulses near zero dispersion wavelength. *J. Opt. Soc. Am. B* 23, 2425–2433.

[48] Efimov A., Yulin A. V., Skryabin D. V. et al. (2005). Interaction of an optical soliton with a dispersive wave. *Phys. Rev. Lett.* 95, 213902.

[49] Gorbach A. V., Skryabin D. V. (2007). Bouncing of a dispersive pulse on an accelerating soliton and stepwise frequency conversion in optical fibers. *Opt. Express* 15, 14560–14565.

In: Solitons
Editor: Matthew C. Williams

ISBN: 978-1-62618-234-9

Chapter 5

ION-ACOUSTIC SUPER- SOLITONS IN PLASMA

A.E. Dubinov* and D.Yu. Kolotkov†
Sarov Institute for Physics and Technology of National Research Nuclear Institute "MEPhI" (SarFTI-MEPhI), Sarov, Russia

Abstract

New forms of nonlinear waves in plasmas – super-solitons are proposed. They are different from ordinary solitons by nontrivial topology of their phase portraits. Phase trajectories of the super-nonlinear super-solitons envelope one or several separatrixes on the phase portrait.

A number of models of non-magnetized multicomponent plasmas are considered in the paper. These are: dusty electron-ion plasma, electron-positron-ion plasma, multi-ion plasma. Possibility of existence of the super-solitons in such plasmas is shown. Profiles of physical values in the specified waves are calculated. Examples of experimental observation of the super-solitons in the multi-ion plasmas are presented.

Introduction

As is known to those skilled in the art, any phase trajectory of some certain η value, changing in the periodic wave, represents itself on the phase plane ($\acute{\eta}$, η) as the closed curve without self-crossings, whereas the phase portrait of the wave, treated as the dynamic system, is the continuum ensemble of all the phase trajectories embedded into each other. It is notable, that phase trajectories, as a rule, embrace the equilibrium point.

Closed low-swing phase trajectories, concentrated near equilibrium point, correspond to harmonic linear waves of low amplitude. Their characteristics, frequency and wavelength can be derived from the dispersion relation, which results from linearization and subsequent solution of the original wave equations.

The above closed phase trajectories are the testimony to the fact, that some invariant of energy type is present in the system, the total potential energy of which has the minimum, coinciding with equilibrium point.

* E-mail address: dubinov-ae@yandex.ru.
† E-mail address: molotok300@gmail.com.

But such a simple topology of the phase wave portrait is not the only one. In paragraph 8.2 [1] the example of the phase portrait is presented, in which the dedicated phase trajectory or separatrix can be seen. The closed phase trajectories, which embrace the separatrix, correspond to the specific non-linear periodic waves, which cannot have the span, which is minor to that of the separatrix. The waves featuring the big span will be referred to as super-linear waves, as is proposed in [1].

To further explain the idea of super-linear waves, a more complicated phase portrait can be imagined, in which one of separatrices embraces wholly only one or several interior separatrices and a number of equilibrium points. Then the new form of an solitary wave, never described anywhere before – super-soliton will correspond to the exterior separatrix.

For the solitary waves to exist in the form of super-solitons the accumulated potential energy of the system has to have not less than three minima, separated by two maxima. Then the minima will correspond to equilibrium points and the maxima – to the saddle points (i.e. the points of self-crossing of separatrix). Several minima and maxima of the potential energy present in the system indicate that the system may also have super-solitons.

Can super-soliton exist in real media such as plasma? As it turned out, under certain conditions, for instance in some specially-composed plasmas, such solitary waves may really exist.

In spite of the above, this class of waves in plasma had never been studied anywhere in literature; as it turned out, it has even never been mentioned before.

The extensive literature search and analysis revealed that a series of papers exists, in which non-linear waves of different type in plasma were considered, and it was shown there, that the cumulative potential energy (further referred to as the pseudo-potentials of Sagdeev or Bernoulli) features two or even three maxima. According to papers [2, 3], for instance, pseudo-potential for ion-acoustic waves in electron-ion plasma under certain conditions has two minima. It seemed to be of the highest importance here that for the two minima to occur the electron component has to have a substantially non-Maxwell distribution, allowing for separation of electrons into two or more non-mixing energy sub-groups. The similar situation springs up in case of electrostatic waves in electron-positron plasma [4], in which both types of particles were split into two non-mixing energy sub-groups: the hot ones and the cold ones, and in case of electron-ion plasma containing hot and cold electrons and heavy and light ions [5].

In papers [6-10] electrostatic waves in dusty electron plasmas of different models are considered; pseudo-potential in these papers also has two minima. The same pseudo-potentials with two minima were observed in out recent studies [11, 12], where ion-acoustic waves in dusty plasma were considered. Two minima are obtained as well when analyzing electrostatic waves in magnetized electron-positron-ion plasma [13] and electron-ion plasma, containing two types of ions with the negative and positive charge [14]. In [15], where non-linear dusty-acoustic waves are considered in dusty electro-ion plasma with two types of dust particles having different charges, pseudo-potential has three minima. Two separatrices, one of which totally embraces to other one, correspond to this pseudo-potential. Consequently, super-solitons may possibly exist in model [15].

However, several minima in pseudo-potential were observed not only in case of electrostatic waves. For instance, in [16] Alfven waves are studied in electron-ion plasma containing two types of ions, and in [17, 18] the same waves are studied on dusty plasma. Two minima were also revealed for this sort of waves.

Unfortunately, not even a short reference to the fact, that there is a possibility to exist for the waves with an excursion wider than that of separatrix is present in the above mentioned papers.

It was not occasionally that we've mentioned the compositions and the models of plasmas from [2-18] in such a detail. As a matter of fact, no waves of this kind are detectable in the ordinary Maxwell two-component plasma. Thus, based on the analysis of the papers found, a very important conclusion can be made: for super-solitons to exist the multi-component nature of plasma is highly essential, and the more components it has, the more minima can the effective potential have, and the more complicated and diverse the structure of the waves can be. Recognition of this fact allowed us to reasonably select the plasma models, in which solitary super-nonlinear waves can be found.

Thus, super-solitons form a new class of waves in plasma. Presented below are several analytical solutions in the form of super-solitons for different plasma models. The compliance of theoretical solutions to experimental facts will be shown, confirming probability of existence of super-solitons in plasma.

Super-Solitons in Epii-Plasma

As the first example we would like to show, that the solutions in the form of super-solitons are practicable even in the simplest model – a non-magnetized collision-free epii-plasma, containing four types of charged particles: the light ones – electrons (e) and positrons (p) and the heavy ones – positively charged (1) and negatively charged (2) ions.

Let us designate electron charge as $e < 0$, that of positron $(-e) > 0$; the charge and the mass of positively charged ions as $(-Z_1 e) > 0$ and m_1 and the charge and the mass of negatively charged ions as $Z_2 e < 0$ and m_2. It will be further assumed that the masses of the light particles are negligibly low. The condition of quasi-neutrality for the non-disturbed epii-plasma of the above composition will look like: $Z_1 e n_{0i1} - Z_2 e n_{0i2} - e n_{0e} + e n_{0p} = 0$, where index "0" labels concentrations (and, also, lower: another variables of the value) for non-disturbed plasma components.

Let us think, that the temperatures of the light components are equal to T_e and T_p, and disregard the temperatures of heavy components. Now we introduce dimensionless parameters $\alpha = n_{0i2}/n_{0i1}$ and $\delta = n_{0p}/n_{0i1}$. With consideration to the proposed normalizations quasi-neutrality conditions will take the following form: $n_{0e} = n_{0i1}(Z_1 - Z_2\alpha + \delta)$

Let us write the basic equations, describing the dynamics of the ion components 1 and 2:

$$\frac{\partial n_{i1,2}}{\partial t} + \frac{\partial (n_{i1,2} v_{i1,2})}{\partial x} = 0, \tag{1}$$

$$\frac{\partial v_{i1,2}}{\partial t} + v_{i1,2}\frac{\partial v_{i1,2}}{\partial x} = \pm \frac{Z_{1,2} e}{m_{1,2}} \frac{\partial \varphi}{\partial x}, \tag{2}$$

$$\frac{\partial^2\varphi}{\partial x^2} = 4\pi e\left(Z_1 n_{i1} - Z_2 n_{i2} - n_e + n_p\right), \tag{3}$$

where in the equation of motion the "plus" sign on the right stands for the positive ions and the "minus" sign – for the negative ones.

Since we have disregarded the mass of the light particles, the following Boltzmann distributions will be true for them:

$$n_e = n_{0e} \exp\left(-\frac{e\varphi}{kT_e}\right), \tag{4}$$

$$n_p = n_{0p} \exp\left(\frac{e\varphi}{kT_p}\right). \tag{5}$$

Let us introduce a new variable $\xi = x - Vt$ into the equation of ion dynamics and transfer from the laboratory reference frame to the system, related to the wave according to Galileo rule:

$$\frac{\partial}{\partial t} = -V\frac{d}{d\xi},\ \frac{\partial}{\partial x} = \frac{d}{d\xi},\ u_{i1,2} = v_{i1,2} - V,$$

where V takes the sense of phase wave speed.

Then, the transformed system of equations will look like:

$$\frac{d\left(n_{i1,2}u_{i1,2}\right)}{d\xi} = 0, \tag{6}$$

$$u_{i1,2}\frac{du_{i1.2}}{d\xi} = \pm\frac{Z_{1,2}e}{m_{i1,2}}\frac{d\varphi}{d\xi}, \tag{7}$$

$$\frac{d^2\varphi}{d\xi^2} = 4\pi e\left(Z_1 n_{i1} - Z_2 n_{i2} - n_e + n_p\right). \tag{8}$$

With motion equation (7) and continuity equation (6) integrated at $\lim\limits_{u_i \to -V} n_{i1,2} = n_{0i1,2}$ and $\lim\limits_{u_i \to -V} \varphi = 0$, the explicit dependence of ion component from electro-static potential will be revealed:

$$n_{i1,2}(\varphi) = n_{0i1,2}\frac{Z_{1,2}}{\sqrt{1 \pm \dfrac{2Z_{1,2}e\varphi}{m_{1,2}V^2}}}. \tag{9}$$

Let us introduce (4), (5) and (9) into equation of Poisson (8), thus having differential equation of the second order relative to $\varphi(\xi)$:

$$\frac{d^2\varphi}{d\xi^2} = 4\pi\rho(\varphi) = 4\pi\left[n_{0i1}\frac{Z_1 e}{\sqrt{1+\frac{2Z_1 e\varphi}{m_{i1}V^2}}} - n_{0i2}\frac{Z_2 e}{\sqrt{1-\frac{2Z_2 e\varphi}{m_{i2}V^2}}} - n_{0e}e\exp\left(-\frac{e\varphi}{kT_e}\right) + n_{0p}e\exp\left(\frac{e\varphi}{kT_p}\right)\right]. \quad (10)$$

Since super-solitons, as opposed to ordinary solitons, cannot have low amplitude, i.e. lower than the excursion of the interior separatrix, then the well-known method of non-linear weak disturbances, resulting in the equations similar to Korteweg-de-Vries, is not applicable to describe the same. That is why we will analyze equation (10) using the method of mechanical analogy – the Sagdeev pseudo-potential.

It is worth noting that equation (10) has the form of the motion equation for some pseudo-particle, where potential φ plays the role of pseudo-coordinate of pseudo-particle, ξ - the role of pseudo-time and expression $4\pi\rho(\varphi)$ – the role of pseudo-power.

Poisson equation, such as equation (10), has the following integral of motion:

$$\frac{1}{2}\left(\frac{d\varphi}{d\xi}\right)^2 + U(\varphi) = \text{const}, \quad (11)$$

in which $U(\varphi)$ has the name of the Sagdeev pseudo-potential. For equation (10) specifically it has the following form:

$$U(\varphi) = -4\pi\int_0^{\varphi}\rho(\varphi)d\varphi = 4\pi m_1 V^2\left\{ n_{0i1}\left(1-\sqrt{1+\frac{2Z_1 e\varphi}{m_1 V^2}}\right) + n_{0i2}\frac{m_2}{m_1}\left(1-\sqrt{1-\frac{2Z_2 e\varphi}{m_{i2}V^2}}\right) + n_{0e}\frac{kT_e}{m_1 V^2}\left[1-\exp\left(-\frac{e\varphi}{kT_e}\right)\right] + n_{0p}\frac{kT_p}{m_1 V^2}\left[1-\exp\left(\frac{e\varphi}{kT_p}\right)\right]\right\}. \quad (12)$$

In Figure 1 the plot for the Sagdeev pseudo-potential $U(\varphi)$ is presented for the following set of parameters: $\alpha = 4.8$; $\delta = 0.045$; $m_2/m_1 = 0.495$; $T_p/T_e = 1.02$; $\sqrt{m_1 V^2/kT_e} = 11.35$; $Z_1 = 48$; $Z_2 = 4$ and the integration constant, chosen such, that $U(0) = 0$. The figure also presents the corresponding phase portrait of pseudo-particle (or the phase portrait of ion-acoustic wave).

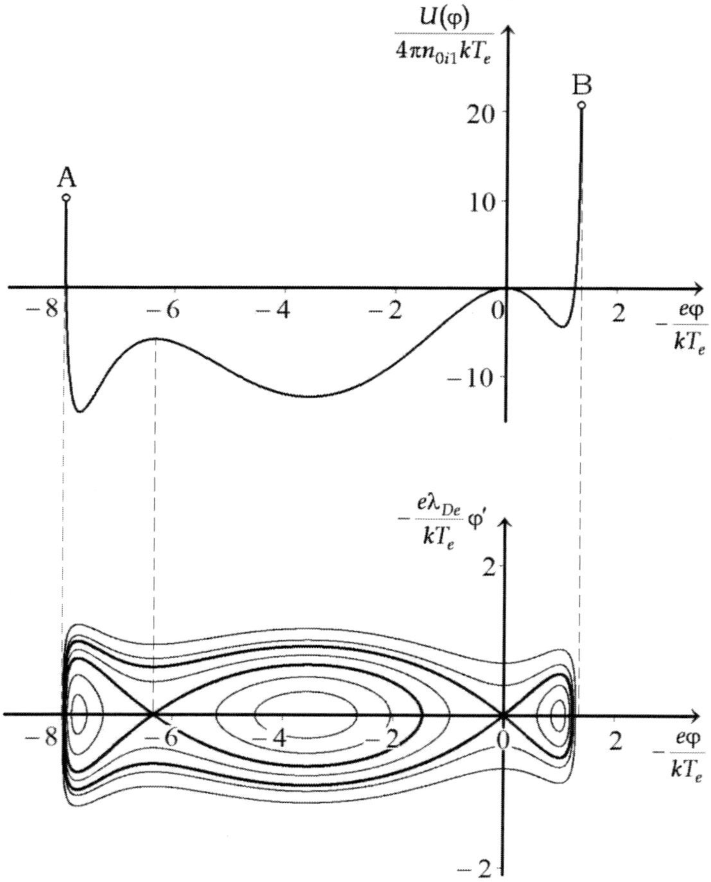

Figure 1. Pseudo-potential (top) and phase portrait of ion-acoustic wave (bottom, separatrices are outlined with solid line, λ_{De} is Debye electron length).

The pseudo-potential, obviously, has three minima and two maxima, where the right maximum in the origin of coordinates coincides with the point of unstable equilibrium of pseudo-particle, and corresponds to quasi-neutrality of non-disturbed plasma. This point at the phase portrait gives rise to two separatrices, one going to the positive area of potential φ values, another – to the negative area. Both separatrices comply with soliton solutions. It has to be noted, that paper [15] also presents the example of the Sagdeev pseudo-potential, having three local minima and two local maxima for the model of four-component epiid-plasma, comprised by electrons, ions and two types of dust particles with different signs of charge. In this example the parameters of plasma were specially chosen to make the height of maxima equal, – that is the extreme case, where solitons are impossible, but the running double layers are possible. That is why super-solitons in [15] were not identified.

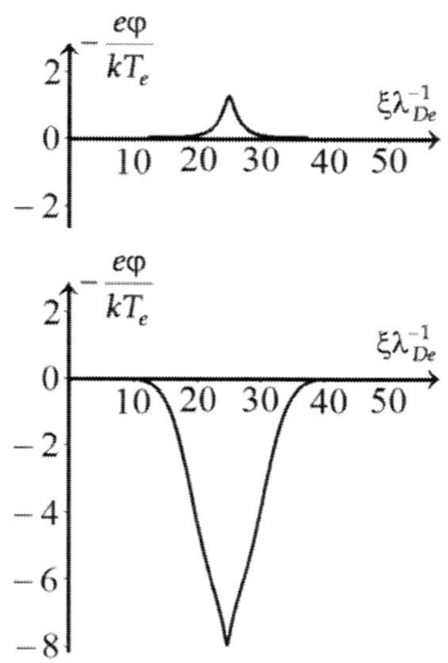

Figure 2. Profiles of the ordinary soliton (top) and super-soliton (bottom).

As it can be seen in Figure 1, the separatrix, going leftwards, embraces another interior separatrix. This geometry of a phase portrait complies with the solution sought for, which is super-soliton. The rightward separatrix corresponds to an ordinary soliton. In Figure 2 for comparison purposes the profiles of potential $\varphi(\xi)$ are presented both for an ordinary and super-soliton.

Since super-solitons cannot be described by Korteweg-de-Vries equation, their profile differs strongly from the well-known bell-shaped solution with one maximum, described in the range of low amplitudes by the function $\sec h^2(x - Vt)$. Super-solitons may also be much wider, and they always have more points of inflection at their profile then the ordinary solitons – not less than four.

It is worth mentioning, that super-solitons may exist in plasma, comprised with not less than four charged components. In simpler two or three-component plasmas the Sagdeev pseudo-potential has not more than one or two minima, correspondingly. In this case the situation, when a separatrix is embraced by an exterior separatrix is impossible, thus super-solitons cannot be produced.

Several other obligatory conditions for existence of super-solitons follow, when the profile of pseudo-potential $U(\varphi)$ is analyzed. The second maximum $U(\varphi)$ has to be strictly lying lower than the first maximum (compare with [15]), located in the origin of coordinates, that is lower, than the point of quasi-neutrality of plasma, whereas the last point, located at the side, where the second maximum is located (in our case point A in Figure 1, corresponding to the point of ion reflection from the potential barrier in the wave), on the contrary, has to lie beyond.

These conditions can be easily written in the form of inequalities, which are skipped here due to their bulky nature.

And, once more, let us show, how super-solitons can be identified among the ordinary experimental solitons. It can be seen, that the form of the signals, presented in Figure 2 is hardly differentiated visually.

However, if a differentiating circuit is included into the instrumental channel, the difference between an ordinary soliton and super-soliton becomes evident (see Figure 3): the derivative of an ordinary soliton always has one maximum sharp (or one minimum sharp with another signal polarity) before the middle of the pulse, whereas the derivative of super-soliton has not less than two of them before the middle of the pulse.

The number of extremes of the derivative is the measure of the number of equilibrium points at the phase portrait, which is embraced by separatrix, corresponding to super-soliton.

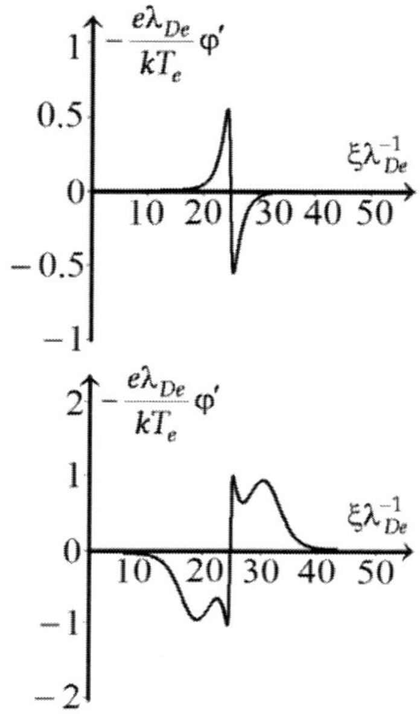

Figure 3. Profiles of derivatives for the ordinary soliton (top) and super-soliton (bottom).

Super-Solitons in Epiid-Plasma

Let us take a look at the dusty multi-component epiid-plasma, containing electrons, positrons, two types of ions with the opposite signs of charge and immobile negatively charged dust particles. The plasma will be assumed to be homogeneous, shock-free and non-magnetized one.

Quasi-neutrality condition for non-disturbed state of such plasma can be described as: $Z_1 e n_{0i1} - Z_2 e n_{0i2} - e n_{0e} + e n_{0p} - q_d n_d = 0$, where $e < 0$ is electron charge; $(-e) > 0$ is positron charge; $(-Z_1 e) > 0$ is the charge of positive ions; $Z_2 e < 0$ is the charge of negative ions; $q_d < 0$ is the charge of dust particles.

To describe the value of spatial charge of the dust component the dimensionless parameter $\alpha = q_d n_d / e n_{0i1}$ is introduced.

Then, the condition of quasi-neutrality can be used to obtain the concentration of electron component in the non-disturbed state: $n_{0e} = n_{0i1}(Z_1 - Z_2\gamma + \delta - \alpha)$, where the additional dimensionless parameters $\gamma = n_{0i2}/n_{0i1}$ and $\delta = n_{0p}/n_{0i1}$ are introduced.

Let us write the system of equations, describing the dynamics of cold ions in the wave:

$$\frac{\partial n_{i1,2}}{\partial t}+\frac{\partial\left(n_{i1,2}V_{i1,2}\right)}{\partial x}=0, \tag{13}$$

$$\frac{\partial V_{i1,2}}{\partial t}+V_{i1,2}\frac{\partial V_{i1,2}}{\partial x}=\pm\frac{Z_{1,2}e}{m_{i1,2}}\frac{\partial\varphi}{\partial x}, \tag{14}$$

$$\frac{\partial^2\varphi}{\partial x^2}=4\pi e\left(Z_1 n_{i1}-Z_2 n_{i2}-n_e+n_p-\alpha n_{0i1}\right), \tag{15}$$

where in the equations of motion (14) for the positive ions the "plus" sign stands on the right and the "minus" sign stands on the left. Assuming inertia-free character of the light particles, let us use Boltzmann distribution for the same:

$$n_e=n_{0e}\exp\left(-\frac{e\varphi}{k_B T_e}\right), \tag{16}$$

$$n_p=n_{0p}\exp\left(\frac{e\varphi}{k_B T_p}\right). \tag{17}$$

If we admit, that all the variables from equations (13) – (15) depend only on $\xi = x - Vt$ and by providing computations similar to that made for the previous model, we will have the following expression for the Sagdeev pseudo-potential:

$$U(\varphi)=4\pi m_{i1}V^2\left[n_{0i1}\left(1-\sqrt{1+\frac{2Z_1 e\varphi}{m_{i1}V^2}}\right)+n_{0i2}\frac{m_{i2}}{m_{i1}}\left(1-\sqrt{1-\frac{2Z_2 e\varphi}{m_{i2}V^2}}\right)+\right.$$
$$\left.+n_{0e}\frac{k_B T_e}{m_{i1}V^2}\left(1-e^{-\frac{e\varphi}{k_B T_e}}\right)+n_{0p}\frac{k_B T_p}{m_{i1}V^2}\left(1-e^{\frac{e\varphi}{k_B T_p}}\right)-\alpha n_{0i1}\frac{e\varphi}{m_{i1}V^2}\right]. \tag{18}$$

In Figure 4 the plot for the Sagdeev pseudo-potential $U(\varphi)$ in form (18) is presented for the following set of parameters: $\alpha = 0.6$; $\gamma = 4.35$; $\delta = 0.04$; $T_p/T_e = 1$; $\sqrt{m_1V^2/kT_e} = 11.36$; $m_{i2}/m_{i1} = 0.49$; $Z_1 = 41$; $Z_2 = 4$ and the integration constant, chosen such, that $U(0) = 0$. The figure also presents the corresponding phase portrait of pseudo-particle (or the phase portrait of ion-acoustic wave).

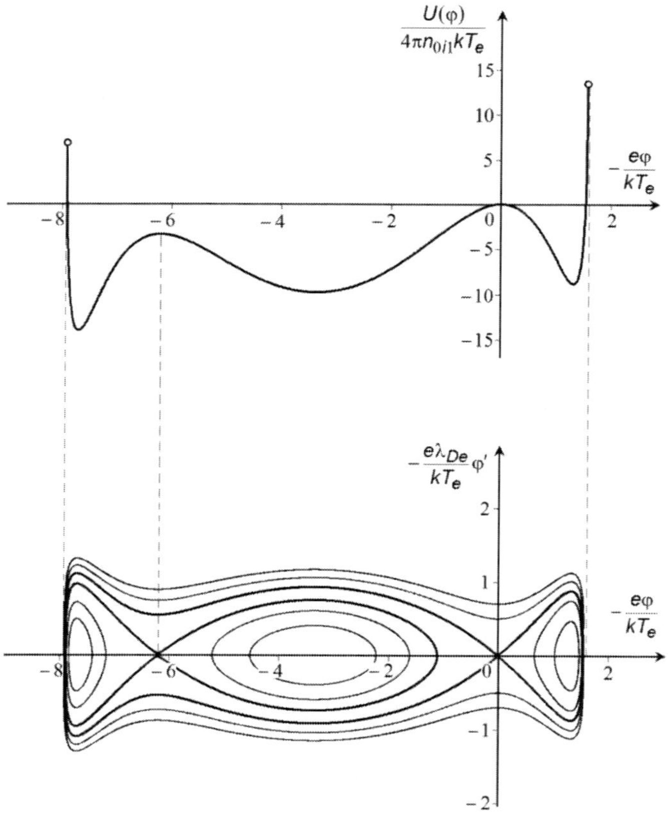

Figure 4. Pseudo-potential (top) and phase portrait of ion-acoustic wave (bottom, separatrices are outlined with solid line, λ_{De} is Debye electron length).

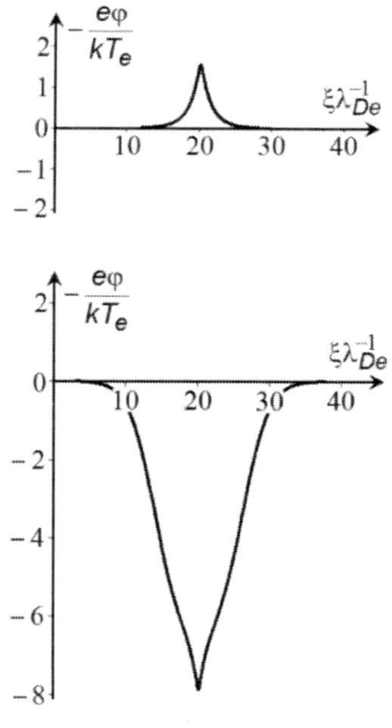

Figure 5. Profiles of the ordinary soliton (top) and super-soliton (bottom).

Pseudo-potential, obviously, has the geometry, corresponding to the solution in the form of super-soliton: the separatrix going leftward embraces another interior separatrix.

In Figure 5 for comparison purposes the profiles of potential $\varphi(\xi)$ are plotted for the ordinary soliton and super-soliton; in Figure 6 time dependencies of the derivative of the ordinary soliton versus super-soliton are presented.

Super-Solitons in Warm Epiid-Plasma

Considered further is the plasma with the same composition as in the previous paragraph, however more complicated to a certain extent due to the temperature of heavy particles.

We will use the denominations already introduced and the dimensionless parameters.

The system of equations, describing the dynamics of hot ions in the wave looks like:

$$\frac{\partial n_{i1,2}}{\partial t}+\frac{\partial(n_{i1,2}v_{i1,2})}{\partial x}=0, \tag{19}$$

$$\frac{\partial v_{i1,2}}{\partial t}+v_{i1,2}\frac{\partial v_{i1,2}}{\partial x}=\pm\frac{Z_{1,2}e}{m_{i1,2}}\frac{\partial\varphi}{\partial x}-\frac{1}{m_{i1,2}n_{i1,2}}\frac{\partial P_{i1,2}}{\partial x}, \tag{20}$$

$$\frac{\partial^2\varphi}{\partial x^2}=4\pi e\left(Z_1n_{i1}-Z_2n_{i2}-n_e+n_p-\alpha n_{0i1}\right), \tag{21}$$

Note that in the equation of motion the "plus" sign on the right stands for the positive ions and the "minus" sign – for the negative ones.

Let us assume, that heavy particles (ions of both types) and the light particles (electrons and positrons) obey Maxwell distribution.

Then, the equations of state for ions can be added to system (19) – (21):

$$P_{i1,2}=n_{i1,2}kT_{i1,2}. \tag{22}$$

With inertia-free character of the light particles assumed, their equations of state and motion can be easily used to derive Boltzmann distributions:

$$n_e=n_{0e}\exp\left(-\frac{e\varphi}{kT_e}\right), \tag{23}$$

$$n_p=n_{0p}\exp\left(\frac{e\varphi}{kT_p}\right). \tag{24}$$

Let us introduce a new variable $\xi = x - Vt$ into the equation of ion dynamics and transfer from the laboratory reference frame to the system, related to the wave according to Galileo rule.

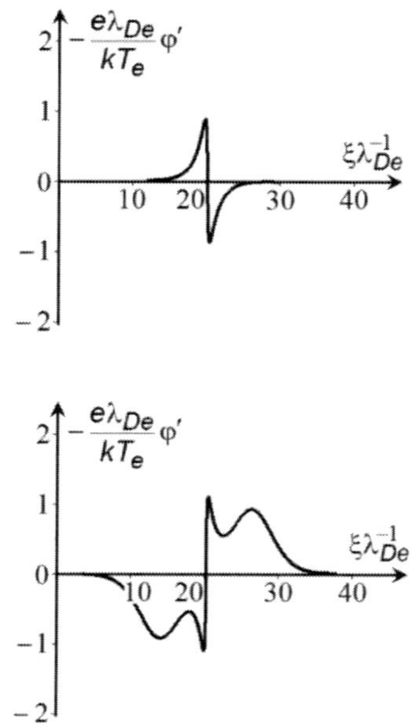

Figure 6. Profiles of the derivatives of ordinary soliton (top) and super-soliton (bottom).

Then we integer motions equations (20) and continuity equation (19) with the new variable at $\lim_{u_i \to -V} n_{i1,2} = n_{0i1,2}$ and $\lim_{u_i \to -V} \varphi = 0$, and as a result we obtain the connection between concentration of ion constants and electrostatic potential:

$$\varphi(n_{i1}) = \frac{m_{i1}V^2}{2e}\left[\left(\frac{n_{0i1}}{n_{i1}}\right)^2 - 1\right] + \frac{kT_{i1}}{e}\ln\left(\frac{n_{i1}}{n_{0i1}}\right), \tag{25}$$

$$\varphi(n_{i2}) = -\frac{m_{i2}V^2}{2e}\left[\left(\frac{n_{0i2}}{n_{i2}}\right)^2 - 1\right] - \frac{kT_{i2}}{e}\ln\left(\frac{n_{i2}}{n_{0i2}}\right). \tag{26}$$

Equations (25) and (26) can be solved analytically with relation to $n_{i1}(\varphi)$ and $n_{i2}(\varphi)$; these dependencies look like as follows:

$$n_{i1}(\varphi) = n_{0i1}\sqrt{-\frac{m_{i1}V^2}{kT_{i1}} \frac{1}{W_{0,-1}\left[-\frac{m_{i1}V^2}{kT_{i1}}\exp\left(-\frac{m_{i1}V^2}{kT_{i1}} - \frac{2Z_1 e\varphi}{kT_{i1}}\right)\right]}}, \tag{27}$$

$$n_{i2}(\varphi)=n_{0i2}\sqrt{-\frac{m_{i2}V^2}{kT_{i2}}\frac{1}{W_{0,-1}\left[-\frac{m_{i2}V^2}{kT_{i2}}\exp\left(-\frac{m_{i2}V^2}{kT_{i2}}+\frac{2Z_2e\varphi}{kT_{i2}}\right)\right]}} \tag{28}$$

The new transcendental function the Lambert W function [19, 20] is used in the above, having two actual branches, designated by indices "0" and "-1".

Dependencies (27) and (28) are very important for analytical purposes – they contain the information about the law of preservation of impulse and the number of particles in ion motion. As it can be seen, they are double-valued functions and what we have to do first, is to exclude non-physical branches from them. We will do the same as in [21, 22], i.e. use the condition of quasi-neutrality for non-disturbed plasma.

From the two branches of dependency (27) we have to choose the one, which in the plot $n_{i1}(\varphi)$ goes through the point of quasi-neutrality of plasma with coordinate n_{0i1} at $\varphi=0$. As the analysis revealed, at $m_{i1}V^2>kT_{i1}$ in (27) the negative real branch of the Lambert W function is to be taken $W_{-1}(\ldots)$, and at $m_{i1}V^2<kT_{i1}$ – the principal real branch of the Lambert W function $W_0(\ldots)$.The similar situation is for (28): at $m_{i2}V^2>kT_{i2}$ the negative real branch of the Lambert W function is to be taken $W_{-1}(\ldots)$, and at $m_{i2}V^2<kT_{i2}$ – the principal real branch of the Lambert W function $W_0(\ldots)$. In Figure 7 the plots for dependencies (27) and (28) are presented for the case, when the bottom branches, calculated with $W_{-1}(\ldots)$ go through their own points of quasi-neutrality, whereas the upper branches, calculated with $W_0(\ldots)$ and shown by a dotted line, are disregarded.

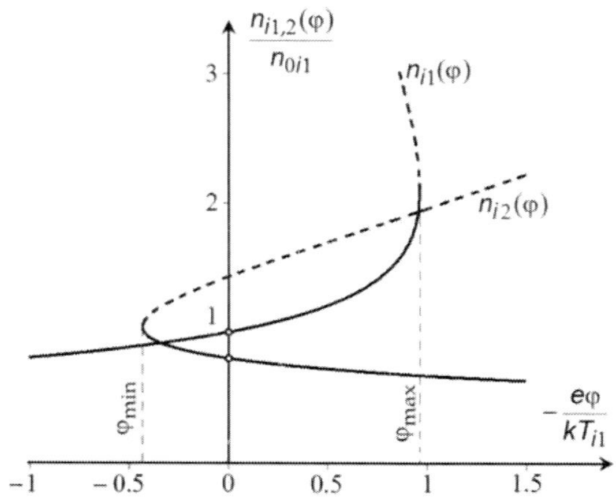

Figure 7. Dependencies (27) and (28) plotted with non-physical branches designated by a dotted line and the points of quasi-neutrality of non-disturbed plasma shown by circles.

It can be seen from Figure 7, that knowing coordinates of conjugation points of branches (in conjugation point the argument of the Lambert W function has to be equal to $-\exp(-1)$), one can refer from (27) and (28) the maximum and the minimum values of electrostatic potential in the wave:

$$\varphi_{\min} = -\frac{kT_{i2}}{2Z_2 e}\left[\ln\left(\frac{m_{i2}V^2}{kT_{i2}}\right) - \frac{m_{i2}V^2}{kT_{i2}} + 1\right], \tag{29}$$

$$\varphi_{\max} = \frac{kT_{i1}}{2Z_1 e}\left[\ln\left(\frac{m_{i1}V^2}{kT_{i1}}\right) - \frac{m_{i1}V^2}{kT_{i1}} + 1\right]. \tag{30}$$

Having inserted relations (23), (24), (27) and (28) into the Poisson equation (21), we will have differential equation of the second order with relation to $\varphi(\xi)$:

$$\frac{d^2\varphi}{d\xi^2} = 4\pi\rho(\varphi), \tag{31}$$

where

$$\rho(\varphi) = n_{0p}e\exp\left(\frac{e\varphi}{k_B T_p}\right) + Z_1 e n_{0i1}\sqrt{-\frac{m_{i1}V^2}{kT_{i1}}\frac{1}{W_{-1}\left[-\frac{m_{i1}V^2}{kT_{i1}}\exp\left(-\frac{m_{i1}V^2}{kT_{i1}} - \frac{2Z_1 e\varphi}{kT_{i1}}\right)\right]}}$$

$$-Z_2 e n_{0i2}\sqrt{-\frac{m_{i2}V^2}{kT_{i2}}\frac{1}{W_{-1}\left[-\frac{m_{i2}V^2}{kT_{i2}}\exp\left(-\frac{m_{i2}V^2}{kT_{i2}} + \frac{2Z_2 e\varphi}{kT_{i2}}\right)\right]}} - n_{0e}e\exp\left(-\frac{e\varphi}{k_B T_e}\right) - \alpha e n_{0i1}.$$

For equation (31) the Sagdeev pseudo-potential has the following representation:

$$U(\varphi) = -4\pi\int_0^{\varphi}\rho(\varphi)d\varphi\,. \tag{32}$$

Let us present the explicit expression for $U(\varphi)$, obtained as a result of integration:

$$\begin{aligned}\frac{U(\varphi)}{4\pi kT_{i1}} &= n_{0i1}\frac{W_{-1}\left[-\frac{m_{i1}V^2}{kT_{i1}}\exp\left(-\frac{m_{i1}V^2}{kT_{i1}}\right)\right]+1}{\sqrt{-\frac{m_{i1}V^2}{kT_{i1}}W_{-1}\left[\frac{m_{i1}V^2}{kT_{i1}}\exp\left(-\frac{m_{i1}V^2}{kT_{i1}}\right)\right]}} - n_{0i1}\frac{W_{-1}\left[-\frac{m_{i1}V^2}{kT_{i1}}\exp\left(-\frac{m_{i1}V^2}{kT_{i1}} - \frac{2Z_1 e\varphi}{kT_{i1}}\right)\right]+1}{\sqrt{-\frac{m_{i1}V^2}{kT_{i1}}W_{-1}\left[-\frac{m_{i1}V^2}{kT_{i1}}\exp\left(-\frac{m_{i1}V^2}{kT_{i1}} - \frac{2Z_1 e\varphi}{kT_{i1}}\right)\right]}} + \\ &+ n_{0i2}\frac{\frac{T_{i2}}{T_{i1}}W_{-1}\left[-\frac{m_{i2}V^2}{kT_{i2}}\exp\left(-\frac{m_{i2}V^2}{kT_{i2}}\right)\right]+1}{\sqrt{-\frac{m_{i2}V^2}{kT_{i2}}W_{-1}\left[-\frac{m_{i2}V^2}{kT_{i2}}\exp\left(-\frac{m_{i2}V^2}{kT_{i2}}\right)\right]}} - n_{0i2}\frac{\frac{T_{i2}}{T_{i1}}W_{-1}\left[-\frac{m_{i2}V^2}{kT_{i2}}\exp\left(-\frac{m_{i2}V^2}{kT_{i2}} + \frac{2Z_2 e\varphi}{kT_{i2}}\right)\right]+1}{\sqrt{-\frac{m_{i2}V^2}{kT_{i2}}W_{-1}\left[-\frac{m_{i2}V^2}{kT_{i2}}\exp\left(-\frac{m_{i2}V^2}{kT_{i2}} + \frac{2Z_2 e\varphi}{kT_{i2}}\right)\right]}} + \\ &+ n_{0e}\frac{T_e}{T_{i1}}\left[1 - \exp\left(-\frac{e\varphi}{kT_e}\right)\right] + n_{0p}\frac{T_p}{T_{i1}}\left[1 - \exp\left(\frac{e\varphi}{kT_p}\right)\right] + \alpha n_{0i1}\frac{e\varphi}{kT_{i1}}.\end{aligned} \tag{33}$$

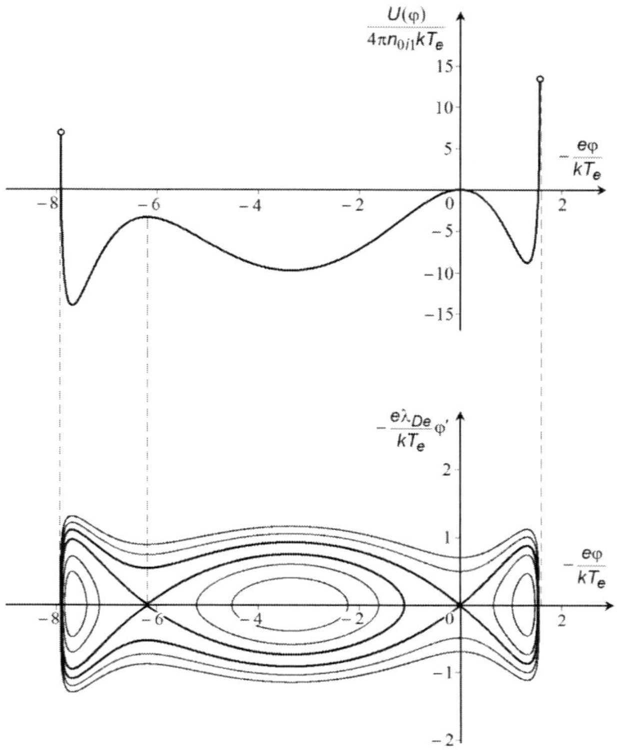

Figure 8. Pseudo-potential (top) and phase portrait of ion-acoustic wave (bottom, separatrices are outlined with solid line, λ_{De} – Debye electron length).

In Figure 8 the Sagdeev pseudo-potential $U(\varphi)$ is plotted for the following set of parameters: $\alpha = 0.1852$; $\gamma = 0.8$; $\delta = 0.13$; $T_p/T_e = 0.87$; $T_{i1}/T_e = 1.76$; $T_{i2}/T_e = 9.2$; $\sqrt{m_1 V^2/kT_e} = 2.1$; $m_{i2}/m_{i1} = 2$; $Z_1 = Z_2 = 1$ and the integration constant, chosen such, that $U(0) = 0$. The figure also presents the corresponding phase portrait of imaginary particle (or the phase portrait of ion-acoustic wave). As in the previous models, the geometry of pseudo-potential, corresponding to the solution in the form of super-soliton was found. The separatrix embraces another interior separatrix is present in the phase portrait. In Figure 9 for comparison purposes the profiles of potential $\varphi(\xi)$ are plotted for the ordinary soliton and super-soliton; in Figure 10 time dependencies of the derivative of the ordinary soliton versus super-soliton are presented.

Super-Solitons in the Experiments with SF6-Ar Plasma

Further the probability for super-solitons to exist in chemically complex plasma will be considered. In [23-28] and some other papers ion-acoustic solitons were studied experimentally in the complex, as to its ion composition, plasma, i.e. the plasma, containing the mixture of SF_6 and Ar gases. There, apart from electrons, the highest significance belonged to ions Ar^+, F^-, SF_5^-, SF_5^+ (according to the data from [24], although some other ions could have been present in small quantities).

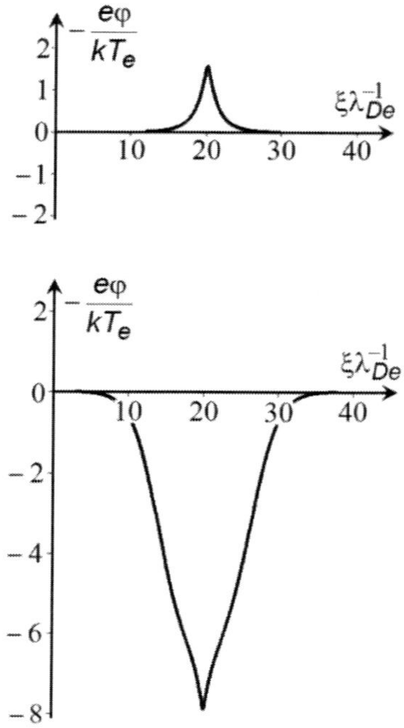

Figure 9. Profiles of the ordinary soliton (top) and super-soliton (bottom).

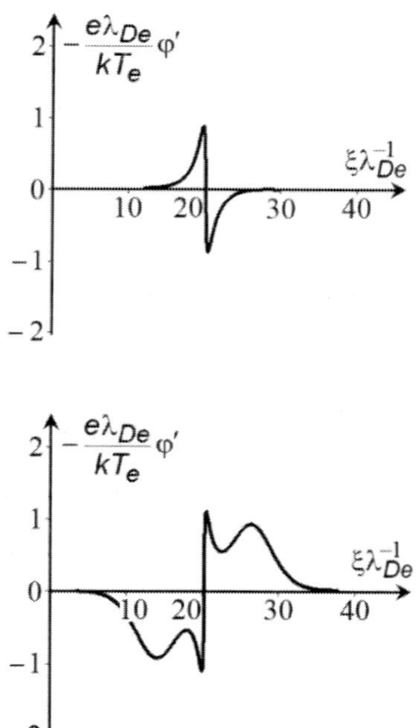

Figure 10. Profiles of the derivatives of the ordinary soliton (top) and super-soliton (bottom).

Almost in all the papers [23-28] the interesting circumstances were detected, although they were never discussed anywhere and cannot be fished out from the earlier theoretical papers. These circumstances are: the recorded during experiments profiles of solitons, as a rule, strongly differ from the bell-shaped functions, since they have either angularity with one maximum (or minimum) and several bending points, or non-monotonous form with several maxima (or minima). The "angular" solitons, recorded during experiments, are presented, for instance, in [25, 26].

It is possible, that the unordinary as to their form solitons in multi-component plasma, shown in [25, 26], are super-solitons, and the profile of non-monotonous soliton with several maxima or minima is obtained from the profile of super-soliton using differentiation by the argument.

For us to be able to prove that, let us consider multi-component plasma, consisting of the cold single-charged ions Ar^+, F^-, SF_5^-, SF_5^+ and electrons. The plasma will be assumed to be homogeneous, collision-free and non-magnetized.

The condition of quasi-neutrality for the non-disturbed state of plasma has the following expression:

$$en_{0\mathrm{Ar}^+} - en_{0\mathrm{F}^-} + en_{0\mathrm{SF}_5^+} - en_{0\mathrm{SF}_5^-} - en_{0e} = 0, \tag{34}$$

where $e < 0$ is the charge of electrons and negative ions, $(-e) > 0$ is the charge of positive ions; all the non-disturbed variables here and further will be designated by index "0".

Now we introduce dimensionless parameters: $\alpha = n_{0F^-}/n_{0\mathrm{Ar}^+}$, $\beta = n_{0\mathrm{SF}_5^-}/n_{0\mathrm{Ar}^+}$, $\gamma = n_{0\mathrm{SF}_5^+}/n_{0\mathrm{Ar}^+}$. With consideration to them the condition of quasi-neutrality can be written as follows: $n_{0e} = n_{0\mathrm{Ar}^+}(1 - \alpha - \beta + \gamma)$.

The numerical values for plasma can be determined in accordance with experimental conditions, described in [24], in particular: it will be assumed that $\frac{n_{0\mathrm{SF}_6}}{n_{0e}} << 1$. The initial concentrations of ion gases in this case F^- and SF_5^- are to be even lower, than n_{0e}, which means, that inequality $\alpha << \beta$ has to be observed for the dimensionless parameters earlier introduced. To further estimate the parameters numerically we will make use of one more fact, taken from [24]: for the measured relative ion concentrations the following conditions are observed: $0.65 < \frac{n_{0\mathrm{SF}_5^-}}{n_{0\mathrm{SF}_5^-} + n_{0\mathrm{F}^-}} < 0.95$ and $0.11 < \frac{n_{0\mathrm{SF}_5^+}}{n_{0\mathrm{SF}_5^+} + n_{0\mathrm{Ar}^+}} < 0.77$, from which, at $\beta = 0.52$, it may be deduced that $0.027 < \alpha < 0.28$ and $0.12 < \gamma < 3.35$. With the above circumstances considered, the following numbers will be chosen: $\alpha = 0.035$, $\beta = 0.52$, $\gamma = 0.2$.

Owing to the fact, that the mass of electron is several orders lower than that of ions, we will assume, that electrons are inertia-free and isothermal. Now they comply with Boltzmann distribution condition:

$$n_e = n_{0e} \exp\left(-\frac{e\varphi}{k_B T_e}\right), \tag{35}$$

For the same reason all the ion components of plasma can be deemed cold.

For the above assumptions the system of 1D electro-hydrodynamic equations, describing the dynamics of ion components in the wave can be written as follows:

$$\frac{\partial n_l}{\partial t} + \frac{\partial (n_l V_l)}{\partial x} = 0, \tag{36}$$

$$\frac{\partial V_l}{\partial t} + V_l \frac{\partial V_l}{\partial x} = \pm \frac{e}{m_l}\frac{\partial \varphi}{\partial x}, \tag{37}$$

$$\frac{\partial^2 \varphi}{\partial x^2} = 4\pi e \left(n_{Ar^+} - n_{F^-} - n_{SF_5^-} + n_{SF_5^+} - n_e\right), \tag{38}$$

where $l = Ar^+, F^-, SF_5^-, SF_5^+$ and in the motion equation (37) the "plus" sign stands on the right for the positive ions and the "minus" sign stands for the negative ones.

Similar to the previous models let us introduce a new wave variable $\xi = x - Vt$ into the equation of ion dynamics and transfer from the laboratory reference frame to the system, related to the wave according to Galileo rule. Then, the Sagdeev pseudo-potential can be written in a standard way for the converted system:

$$U(\varphi) = -4\pi \int_0^{\varphi} \rho(\varphi) d\varphi = 4\pi m_{Ar^+} V^2 \left\{ n_{0Ar^+}\left(1 - \sqrt{1 + \frac{2e\varphi}{m_{Ar^+} V^2}}\right) + \right.$$

$$+ n_{0F^-}\frac{m_{F^-}}{m_{Ar^+}}\left(1 - \sqrt{1 - \frac{2e\varphi}{m_{F^-} V^2}}\right) + n_{0SF_5^-}\frac{m_{SF_5^-}}{m_{Ar^+}}\left(1 - \sqrt{1 - \frac{2e\varphi}{m_{SF_5^-} V^2}}\right) +, \tag{39}$$

$$\left. + n_{0SF_5^+}\frac{m_{SF_5^+}}{m_{Ar^+}}\left(1 - \sqrt{1 + \frac{2e\varphi}{m_{SF_5^+} V^2}}\right) + n_{0e}\frac{k_B T_e}{m_{Ar^+} V^2}\left(1 - e^{-\frac{e\varphi}{k_B T_e}}\right) \right\}$$

where

$$\rho(\varphi) = 4\pi e\left\{\frac{n_{0\,\mathrm{Ar}^+}}{\sqrt{1+\dfrac{2e\varphi}{m_{\mathrm{Ar}^+}V^2}}} - \frac{n_{0\,\mathrm{F}^-}}{\sqrt{1-\dfrac{2e\varphi}{m_{\mathrm{F}^-}V^2}}} - \frac{n_{0\,\mathrm{SF}_5^-}}{\sqrt{1-\dfrac{2e\varphi}{m_{\mathrm{SF}_5^-}V^2}}} + \right.$$

$$\left. + \frac{n_{0\,\mathrm{SF}_5^+}}{\sqrt{1+\dfrac{2e\varphi}{m_{\mathrm{SF}_5^+}V^2}}} - n_{0e}\exp\left(-\frac{e\varphi}{k_{\mathrm{B}}T_e}\right)\right\}$$

Now we specify the velocity $v = \sqrt{m_{\mathrm{Ar}^+}V^2/kT_e} = 1.7$ in such a way, that it would be deliberately supersonic and estimate the relations between ion masses $\dfrac{m_{\mathrm{SF}_5^+}}{m_{\mathrm{Ar}^+}} = 3.18$, $\dfrac{m_{\mathrm{SF}_5^-}}{m_{\mathrm{Ar}^+}} = 3.18$, $\dfrac{m_{\mathrm{F}^-}}{m_{\mathrm{Ar}^+}} = 0.476$. Making use of these values the plot of pseudo-potential (39) can be drawn together with the corresponding phase portrait of oscillations of pseudo-particle, i.e. the phase portrait of ion-acoustic wave (Figure 11).

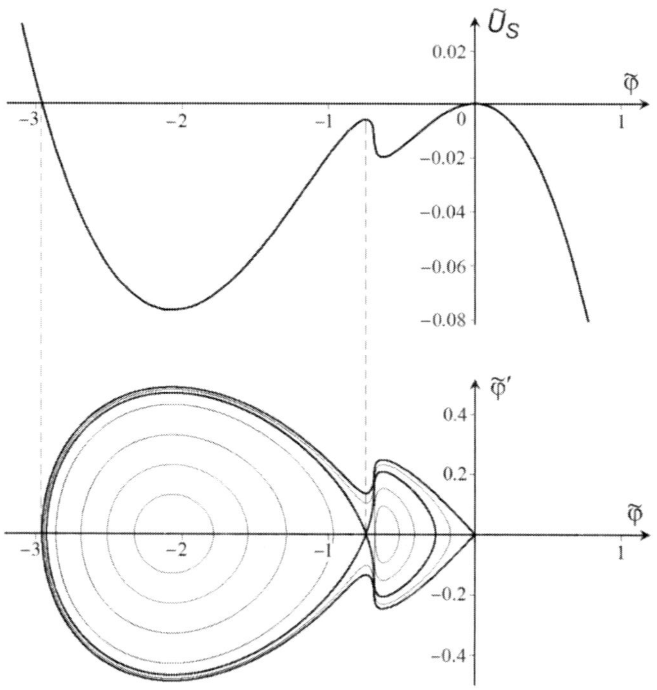

Figure 11. The normalized Sagdeev pseudo-potential (top) and the phase portrait of ion-acoustic wave in normalized coordinates, on which separatrices are designated by solid lines (top); normalizations: $\tilde{U}_S = \dfrac{U_S}{4\pi n_{0\,\mathrm{Ar}^+}k_{\mathrm{B}}T_e}$, $\tilde{\varphi} = -\dfrac{e\varphi}{k_{\mathrm{B}}T_e}$, $\tilde{\varphi}' = -\dfrac{e\lambda_{\mathrm{D}}}{k_{\mathrm{B}}T_e}\dfrac{d\varphi}{d\xi}$, λ_{D} is Debye electron length.

Analysis of the same reveals that in the plasma being considered the ion-acoustic soliton may exist. The phase trajectory of the plasma is represented by the exterior separatrix in the form of a guitar, which embraces the interior separatrices. Let us now draw the profiles of certain physical values in super-soliton: electrostatic potential, concentration of electrons and concentration of argon ions (Figure 12).

It is clearly seen, that they have angular shape. Now let us draw the profiles of the derivatives of the same values in the wave (Figure 13), which have non-monotonous shape with several minima and maxima.

Thus, the angular-shaped and non-monotonous ion-acoustic solitons in multi-component plasma, like those recorded in papers [23-28], can be essentially super-solitons and may have topologically nontrivial phase portraits, in which one separatrix embraces one or several other interior separatrices.

Occurrence of this type of phase portraits is, as we see, related to multi-component nature of the plasma involved.

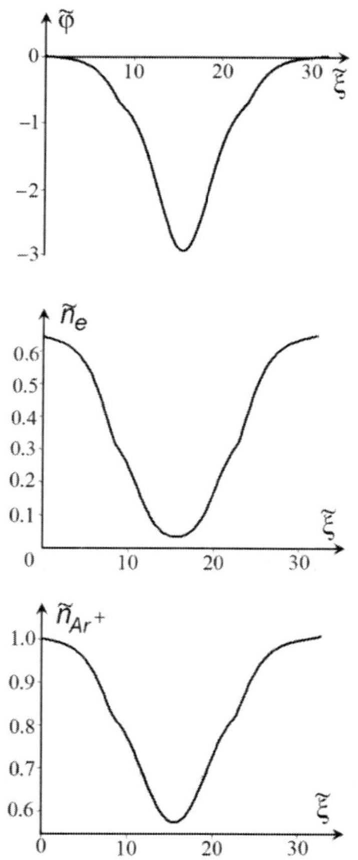

Figure 12. Normalized profiles of physical values in angular-shaped super-soliton: electrostatic potential, concentration of electrons and argon ions (from the top down); normalizations: $\tilde{\varphi} = -\dfrac{e\varphi}{k_B T_e}$

$\tilde{n}_e = \dfrac{n_e}{n_{0\,\mathrm{Ar}^+}}$, $\tilde{n}_{\mathrm{Ar}^+} = \dfrac{n_{\mathrm{Ar}^+}}{n_{0\,\mathrm{Ar}^+}}$, $\tilde{\xi} = \xi\lambda_D^{-1}$.

Conclusion

It is being proved in the article, that in multi-component plasmas of different compositions ion-acoustic solitons may live in the new form of ion-acoustic super-solitons, which feature the nontrivial shape of their phase portraits.

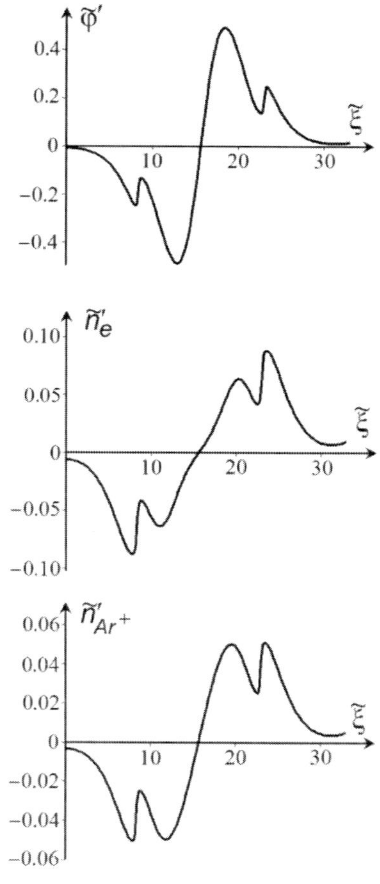

Figure 13. Normalized profiles of the derivatives of physical values in super-soliton: electrostatic potential, concentration of electrons and argon ions (from the top down); normalizations: $\tilde{\varphi}' = -\dfrac{e\lambda_D}{k_B T_e}\dfrac{d\varphi}{d\xi}$, $\tilde{n}'_e = \dfrac{\lambda_D}{n_{0\,\mathrm{Ar}^+}}\dfrac{dn_e}{d\xi}$, $\tilde{n}'_{\mathrm{Ar}^+} = \dfrac{\lambda_D}{n_{0\,\mathrm{Ar}^+}}\dfrac{dn_{\mathrm{Ar}^+}}{d\xi}$, $\tilde{\xi} = \xi\lambda_D^{-1}$.

References

[1] N. M. Ryskin and D. I. Trubetskov, *Nonlinear Waves*, Nauka, Moscow (2000) (in Russian).

[2] H. H. Kuehl, K. Imen, *IEEE Trans. Plasma Sci.* 13, 37 (1985).

[3] R. A. Cairns, A. A. Mamun, R. Bingham et al., *Geophys. Res. Lett.* 22, 2709 (1995).

[4] F. Verheest, M. A. Hellberg, G. J. Gray, R. L. Mace, *Astrophys. and Space Sci.* 239, 125 (1996).

[5] S. S. Ghosh, G. S. Lakhina, *Nonlinear Processes in Geophysics.* 11, 219 (2004).

[6] C. R. Choi, C.-M. Ryu, D.-Y. Lee et al., *Phys. Lett. A.* 364, 297 (2007).

[7] T. K. Baluku, M. A. Hellberg, I. Kourakis, N. S. Saini, *Phys. Plasmas.* 17, 053702 (2010).

[8] M. Akbari-Moghanjoughi, *Phys. Plasmas.* 17, 123709 (2010).

[9] A. Mamun, P. K. Shukla, *Phys. Lett. A.* 374, 4238 (2010).

[10] F. Verheest, *Phys. Plasmas.* 18, 083701 (2011).

[11] E. Dubinov and M. A. Sazonkin, *Tech. Phys.* 53, 1129 (2008).

[12] E. Dubinov, D. Yu. Kolotkov and M. A. Sazonkin, *Plasma Phys. Rep.* 37, 64 (2011).

[13] M. A. Mahmood, S. Mahmood, A. M. Mirza, H. Saleem, *Chin. Phys. Lett.* 22, 632 (2005).

[14] H. Kuar, T. S. Gill, *J. Phys.: Conf. Series.* 208, 012080 (2010).

[15] F. Verheest, *Phys. Plasmas.* 16, 013704 (2009).

[16] L. Yang, D.-J. Wu, *Chin. Phys. Lett.* 23, 2155 (2006).

[17] R. Choi, D.-Y. Lee, *Phys. Plasmas.* 14, 052304 (2007).

[18] W. Masood, H. A. Shah, N. L. Tsintsadze, M. N. S. Qureshi, *Eur. Phys. J. D.* 59, 413 (2010).

[19] S. R. Valluri, D. J. Jeffrey and R. M. Corless, *Canadian J. Phys.* 78, 823 (2000).

[20] E. Dubinov, I. D. Dubinova, *J. Plasma. Phys.* 71, 715 (2005).

[21] E. Dubinov, I. D. Dubinova, V. A. Gordienko, *Phys. Plasmas* 13, 082111 (2006).

[22] V. A. Gordienko, A. E. Dubinov, *Techn. Phys.* 53, 43 (2008).

[23] Y. Nakamura, *IEEE Trans. on Plasma Sci.* 10, 180 (1982).

[24] G. O. Ludwig, J. L. Ferreira, Y. Nakamura, *Phys. Rev. Lett.* 52, 275 (1984).

[25] Y. Nakamura, J. L. Ferreira, G. O. Ludwig, *J. Plasma Phys.* 33, 237 (1985).

[26] Y. Nakamura, I. Tsukabayashi, *J. Plasma Phys.* 34, 401 (1985).

[27] J. L. Cooney, M. T. Gavin, K. E. Lonngren, *IEEE Trans. on Plasma Sci.* 19, 545 (1991).

[28] J. L. Cooney, M. T. Gavin, J. Tao, K. E. Lonngren, *IEEE Trans. on Plasma Sci.* 19, 1259 (1991).

In: Solitons
Editor: Matthew C. Williams

ISBN: 978-1-62618-234-9

Chapter 6

DIAMOND-CONTROLLED SOLITONS AND TURBULENCE IN THE EXTRACELLULAR MATRIX AND LYMPHATIC DYNAMICS

T. Ivancevic, B. Jovanovic, H. Greenberg, R. Greenberg and L. Lukman*
Tesla Science Evolution Institute, Adelaide, Australia
Sports Academy, Belgrade, Serbia
Diamond International Corporation, Adelaide, Australia

Abstract

This paper presents local and global turbulence, shock-waves and solitons occurring in human lymphatic and matrix dynamics controlled by Diamond Biomesotherapy. We start with a basic description of the extracellular matrix (ECM), which is followed by an outline of Diamond ECM therapy. The next section presents modern research in lymphatic and ECM turbulent dynamics, from both a global-macroscopic perspective and local-microscopic (i.e., quantum) one. After that, we explore related Diamond therapies and fat-burning mechanics. We conclude with two biomedical applications: (i) a nonstandard Diamond Biomesotherapy for motor neuron diseases, and (ii) a Diamond fat-burning mechanism.

Keywords: Extracellular matrix, lymphatic turbulence, solitons and shock-waves, Diamond Biomesotherapy

1. The Extracellular Matrix (ECM)

Alfred Pischinger, Austrian professor of Histology and Embryology at the University of Vienna, researched the extracellular matrix (ECM) in 1975 and presented his work in the book 'System of Matrix Regulation'. The most important aspect of Pischinger's research was that it was based on nonlinear thermodynamics and Wiener's cybernetic theory

*E-mail address: tijana.ivancevic@alumni.adelaide.edu.au, CEO, Tesla Science Evolution Institute, Adelaide, Australia (Corresponding Author)

of highly connected open systems, rather than the linear and Newton mechanicistics cause and effect theory. He discovered that for chronic inflammation and complex diseases including tumour cell distribution, the classical medical approach did not give acceptable outcomes [Pis07].

An ECM is a functional network of capillaries, connective tissue cells and autonomic nerve endings closely connected to the lymphatic system. The lymphatic system is connected with the ECM via anchoring filaments [LB68]. The matrix system together with the lymphatic system is responsible for cell nutrition, as well as the elimination of the toxins. If the lymphatic system is not functional then lymphedema formation starts and immune function is compromised [SSP11]. According to Pischinger, 'the concept of a cell is strictly speaking, only a morphological abstraction. Seen from a biological viewpoint, a cell cannot be considered by itself without taking its environment into account' (1975).

One of the main components of the ECM is Hyaluronic Acid (HA). HA is present on and inside cells [TDT01]. It plays an important role in the supply and movement of water and is essential for water homeostasis [FLL97]. Other HA functions include the regulation of water balance, skin moisture and osmotic pressure; elimination of particular molecules; lubrication and shock absorption [Ste04]; and cell proliferation and migration. This leads to the connection between cancer and the ECM: cancers attack the ECM at the primary cancer location, emitting enzymes which are incorporated into the ECM at that location (e.g. breasts). At the same time, this allows metastatic cancer cells to migrate to other vital parts of the body (e.g. liver or lungs). But high levels of HA can reduce the risk of metastatic cancer. There is an interesting story about the people living in Japanese village Yuzuri Hara, commonly known as 'The Village of Long Life'. There, the villagers live to an average age of 86 and have a virtually nonexistent history of chronic or malignant diseases. This is believed to be due to the fact that the local diet consists mainly of vegetables very rich in HA, particularly potatoes.

So the newly discovered importance of the ECM offers many explanations regarding the bodys immune system during disease. It has been found that the ECM contributes to important body mechanisms such immune cell migration into and within inflamed tissues, immune cell activation and propagation, and cell separation processes. The ECM itself is voluminous and highly insoluble, with non-collagenous glycoproteins which are negatively charged and consequently interact with other charged molecules, such as growth factors and chemokines (proteins secreted by cells) [Sor10]. It also consists of 3 major groups of biomolecules: structural proteins (collagen and elastin), specialised proteins (fibrillin, fibronectin and laminin) and proteoglycans [Kin12].

2. Diamond Medisculpt ECM Therapy

Diamond Biomesotherapy treats the extracellular matrix, where cellulite is formed by lymphatic stagnation. Diamond Medisculpt increases lymphatic dynamics in the ECM by combining ancient techniques with 21st century technology.

The build up of cellulite is directly related to the purity and efficiency of processes that regulate the ECM. Cellulite is the first stage of lymphatic stagnation and blockage in the ECM, leading to breakdown of the geloid infrastructure, which includes Hyaluronic acid (HA). HA, when broken down, leads to increased acidity. As HA attracts water, this leads to the local fluid build-up.

The ECM in its 3D structure is a filter for entry and exit of nutrients gases in and out of the cells. It is a molecular sieve of capillaries and nerve endings for the release of defense cells, neurotransmitters and hormones [DI12].

Diamond Contours Biomesosculpture uses a system which creates a reduction of visible inflammation and congestion anywhere on the body. It stimulates the extra cellular matrix or the fluid substance just under the skin, to treat the first stage of stagnation, and the second stage which is inflammation. Diamond Contours reduces stagnation by improving tissue circulation and infusing pure plant botanical herbal and homeopathic extracts into the skin. This is a comfortable treatment that effectively reduces inflammation and fatty build up on the face and body. The accumulation of toxins in the extra-cellular matrix disrupts normal cellular function and is believed to be a major cause of inflammation. The Diamond Contours machine is effective in the reduction of stagnation and promotion of lymphatic drainage.

3. Lymphatic Turbulence versus Quantum ECM Turbulence

From physical perspective, *lymphatic flow* obeys the laws of classical fluid dynamics, of which the extreme case is *lymphatic turbulence*. Lymphatic turbulence is the only way to prevent *lymphatic stagnation*, which is the main cause of many global human diseases.

On the other hand, the general *ECM flow/behavior* obeys the laws of *quantum fluid dynamics*, of which the extreme case is *ECM quantum turbulence*. ECM quantum turbulence is the only way to prevent *ECM stagnation*, which is the main cause of many local human diseases.

When Diamond Medisculpt's hand piece is connected to any part of a human body (in either gliding or pulsing movements, either fast or slow), it stimulates the local ECM by causing microscopic (quantum) turbulence of the extracellular fluid. After a significant area of the body is covered in this way, it stimulates the global lymphatic flow by causing its macroscopic (classical) turbulence. Therefore, lymphatic turbulence is a cumulative effect of a number of Diamond contouring movements that cause local quantum ECM turbulence. This is the physical mechanism behind the extraordinary efficiency of Diamond Medisculpt contouring, body shaping and cleansing therapy.

In summary, turbulence (first quantum on the local ECM level, then global lymphatic one) is the way to break both ECM stagnation and lymphatic stagnation and prevent all potential diseases caused by the respective stagnations.

3.1. Classical Lymphatic Fluid Dynamics and Lymph Turbulence

The so-called *turbulent flow* is a fluid flow regime characterized by low momentum diffusion, high momentum convection, and rapid variation of pressure and velocity in space and time. Flow that is not turbulent is called *laminar flow*. Also, recall that the *Reynolds number Re* characterizes whether flow conditions lead to laminar or turbulent flow. The structure of turbulent flow was first described by A. Kolmogorov (for an overview, see e.g. [II09b]). Consider the flow of water over a simple smooth object, such as a sphere. At very low speeds the flow is laminar, i.e., the flow is locally smooth (though it may involve vortices on a large scale). As the speed increases, at some point the transition is made to turbulent (or, chaotic) flow. In turbulent flow, unsteady *vortices* appear on many scales and interact with each other. Drag due to boundary layer skin friction increases. The structure and location of boundary layer separation often changes, sometimes resulting in a reduction of overall drag.

Applied to lymph dynamics, *vorticity* $\omega = \omega(\mathbf{x}, t)$ is defined as the circulation per unit area at a point in the lymph flow field, that is as the curl of the the lymph flow velocity: $\omega = \nabla \times \mathbf{u}$. It is a vector quantity, whose direction is approximately along the axis of the swirl. The movement of a lymph flow can be said to be vortical if the fluid moves around in a helix (which from the top looks like a circle), or if it tends to spin around some axis. Such motion can also be called *solenoidal motion*.

Because laminar-turbulent transition in lymph dynamics is governed by Reynolds number, the same transition occurs if the size of the lymph is gradually increased, or the viscosity of the lymph is decreased, or if the density of the lymph is increased.

In particular, in a turbulent lymph flow, there is a range of scales of the lymph flow motions, called *eddies*. A single packet of lymph flow moving with a bulk velocity is called an 'eddy'. The size of the largest scales (eddies) are set by the overall geometry of the lymph flow.

Such turbulent lymphatic flow shows characteristic statistical behavior (compare with [Kol41a, Kol41b]). For simplicity, we will assume a steady state of fully developed turbulence of an incompressible classical lymph flow. Energy is injected into the lymph flow at a rate ε in an energy-containing range. In an inertial range, this lymph energy is transferred to smaller length scales without dissipation. In this range, the lymph is locally homogeneous and isotropic, leading to energy spectral statistics described by the *Kolmogorov law*,

$$E(k) = C\,\varepsilon^{2/3}\,k^{-5/3}. \qquad (1)$$

The lymph energy spectrum $E(k)$ is defined by $E = \int E(k)\,d\mathbf{k}$, where E is the kinetic energy of the lymph per unit mass and k is the lymph wave-number from the Fourier transform of the velocity field (compare with the last subsection on quantum lymph waves). The spectrum of equation (1) is derived by assuming that $E(k)$ is locally determined only by the lymph energy flux ε and by k. The lymph energy transferred to smaller scales is dissipated at the Kolmogorov wave-number $k_K = (\epsilon/\nu^3)^{1/4}$ in an energy-dissipative range via the viscosity of the lymph flow at a dissipation rate ε in equation (1), which is equal to the lymph energy flux Φ in the inertial range. The Kolmogorov constant C is a dimensionless parameter of order unity. The inertial range is thought to be sustained by a self-similar *Richardson cascade* in which large lymph eddies break up into smaller eddies through lymph vortex reconnections.

In order for two lymph flows to be similar they must have the same geometry and equal Reynolds numbers. When comparing lymph flow behavior at homologous points in a lymph model and a full-scale lymph flow, we have $Re^* = Re$, where quantities marked with * concern the flow around the lymph model and the other the real lymph flow.

Fluid dynamicists believe that *Navier-Stokes PDEs* accurately describe turbulence (see, e.g. [II06c]). Therefore, we can assume that viscous lymph flows evolve according to nonlinear Navier-Stokes equations

$$\dot{\mathbf{u}} + \mathbf{u}\cdot\nabla\mathbf{u} + \nabla p/\rho = \nu\Delta\mathbf{u} + \mathbf{f}, \qquad (2)$$

where $\mathbf{u} = \mathbf{u}(\mathbf{x}, t)$ is the 3D velocity of a lymph flow, $\dot{\mathbf{u}} \equiv \partial_t \mathbf{u}$ is the 3D acceleration of a lymph flow, $p = p(\mathbf{x}, t)$ is the lymph pressure field, while $\mathbf{f} = \mathbf{f}(\mathbf{x}, t)$ is the external nonlinear energy source to the lymph, while ρ, ν are the lymph flow density and viscosity coefficient, respectively. Such a lymph flow can be characterized by the ratio of the second term on the left-hand side of equation (2), $\mathbf{u} \cdot \nabla \mathbf{u}$, referred to as the lymph inertial term, and the second term on the right-hand side, $\nu \Delta \mathbf{u}$, that we call the lymph viscous term. This ratio defines the *Reynolds number* $Re = \bar{v}D/\nu$, where $\bar{v}$ and D are a characteristic velocity and length scale, respectively. When $\bar{v}$ increases and the Reynolds number Re exceeds a critical value, the lymph changes from a *laminar* to a *turbulent* state, in which the lymph flow is complicated and contains eddies. To simplify the problem, we can impose to $\mathbf{f}$ the so-called *Reynolds condition*, $\langle \mathbf{f} \cdot \mathbf{u} \rangle = \varepsilon$, where ε is the average rate of energy injection.

In mechanical Lie algebra terms (see, e.g. [II06c]), the Navier-Stokes PDE (2) can be written:

$$\dot{\omega} = -[\mathbf{u}, \omega] + \text{curl } \mathbf{f} + \nu \Delta \omega.$$

On the other hand, the *Euler equation* for ideal lymph flows,

$$\dot{\mathbf{u}} + \mathbf{u} \cdot \nabla \mathbf{u} + \nabla p/\rho = 0, \tag{3}$$

reads in Lie algebra terms,

$$\dot{\omega} = -[\mathbf{u}, \omega], \qquad \omega = \text{curl } \mathbf{u}.$$

Equation (3) is related to the Navier-Stokes PDE (2) in the same way as the classical Euler equation of a rigid body (with a fixed point),

$$\dot{\pi} = \pi \times \omega,$$

is associated with a more general equation, involving friction and external angular momentum [Arn78, AK98]

$$\dot{\pi} = \pi \times \omega + \mathbf{F} - \nu \pi, \tag{4}$$

the 'friction operator' ν is symmetric and positive definite. The distributed mass force f, which appeared in the Navier-Stokes equation (2), is similar to the external angular momentum $\mathbf{F}$, and it is the origin of the lymph flow motion. The viscous friction $\nu \Delta \mathbf{u}$ is analogous to the term $-\nu\pi$ in (4) slowing the rigid body motion. The similarity becomes especially noticeable if one rewrites the equations in components in the eigenbasis of the friction operator.

For example, for the Navier-Stokes equation with periodic boundary conditions one can expand the lymph vorticity field and the force $\mathbf{f}$ into the ordinary Fourier series. The equations in both of the cases have the form of *Galerkin approximation*:

$$\dot{x}_i = \sum_{j,k} a_{ijk} x_j x_k + \sum_i f_i - \nu_i x_i. \tag{5}$$

The first term corresponds to the Euler equation (3) and describes the inertial lymph motion. It follows from the properties of (3) that the divergence of this term is equal to zero. Furthermore, the Euler equation of an ideal lymph flow in any dimension, as well as that of a rigid body, has a quadratic positive definite first integral, the kinetic energy. Therefore, for $\mathbf{f} = \mathbf{u} = 0$, the vector field on the right-hand side of (5) is tangent to certain ellipsoids centered at the lymph origin. This implies that during the lymph evolution defined by this equation there is neither growth nor decay of solutions.

The term $-\nu_i x_i$ in (5), corresponding to the lymph friction, dominates over the constant *lymph pumping* $\mathbf{f}$ when considered sufficiently far away from the lymph origin. Hence, in that remote lymph region, the *lymph motion* is directed towards the origin, and an infinite growth of solutions is impossible. Also, since the lymph pumping $\mathbf{f}$ pushes a phase point out of any neighborhood of the origin, while the friction returns it from a distance, lymph motion in the system of a rigid body (4) approaches an intermediate regime-attractor. For instance, this lymph attractor can be a stable stagnation point or a periodic lymph motion, while for sufficiently high dimension of the phase space it can appear to be a *chaotic motion* sensitive to the initial conditions.

Here we recall that *chaos theory*, of which turbulence is the most extreme form, started in 1963, when E. Lorenz from MIT took the Navier-Stokes PDEs (2) and reduced them into three first-order coupled nonlinear ODEs, to demonstrate the idea of sensitive dependence upon initial conditions and associated *chaotic behavior*. The 3D *phase-portrait* of the Lorenz system (6) shows the celebrated '*Lorenz mask*', a special type of *fractal attractor*:

$$\dot{x} = a(y - x), \qquad \dot{y} = bx - y - xz, \qquad \dot{z} = xy - cz, \tag{6}$$

where x, y and z are dynamical variables, constituting the 3D *phase-space* of the *Lorenz system*; and a, b and c are the parameters of the system.

3.2. Quantum ECM Fluid Dynamics and Quantum Turbulence

As already stated, the general *ECM flow/behavior* obeys the laws of *quantum fluid dynamics*, of which the extreme case is *ECM quantum turbulence*. ECM quantum turbulence is the only way to prevent *ECM stagnation*, which is the main cause of many local human diseases.

3.2.1. Quantum Turbulence

Quantum turbulence was firstly discovered in superfluid helium (^{4}He) in the 1950s, to be re-discovered in the fermionic superfluid helium (^{3}He) in 1972. However, the field moved in a new direction starting around the mid 1990s (see [Tsu06, TKK10]). Briefly, quantum turbulence is comprised of quantized vortices that are definite topological defects arising from the order parameter appearing in *Bose-Einstein condensation* (BEC). Hence quantum turbulence is expected to yield a simpler model of turbulence than does classical turbulence based on the Navier-Stokes PDE (2).

Bose-Einstein condensation is often considered to be a *macroscopic quantum phenomenon* (see, e.g. [PS03]). This is because bosons occupy the same single-particle ground state below a critical temperature through Bose-Einstein condensation to form a *macroscopic wave function* (order parameter) extending over the entire system. As a direct result of the formation of a macroscopic wave function, quantized vortices appear in the Bose-condensed system. A *quantized vortex* is a vortex of inviscid superflow, and any rotational motion of a superfluid is sustained by quantized vortices. A quantized vortex is stable and well-defined topological defect, very different from classical vortices in a conventional fluid. Hydrodynamics dominated by quantized vortices is called *quantum superfluid dynamics*, and turbulence comprised of quantized vortices is known as *quantum turbulence* (QT) [Tsu06, TKK10].

Liquid ^{4}He enters a superfluid state below the λ point ($T_\lambda = 2.17$ K) with Bose-Einstein condensation of the ^{4}He atoms. The λ transition is closely related to the Bose-Einstein condensation of ^{4}He atoms, as first proposed by [Lon38]. The Bose-condensed system exhibits the macroscopic wave-function $\psi(\mathbf{x}, t) = |\psi(\mathbf{x}, t)|e^{i\theta(\mathbf{x},t)}$ as an *order parameter*. The superfluid velocity field is given by $\mathbf{v}_x = (\hbar/m)\nabla\theta$, with boson mass m, representing the potential flow. Since the macroscopic wave function should be single-valued for the space coordinate $\mathbf{x}$, the circulation $\Gamma = \oint \mathbf{v} \cdot d\ell$ for an arbitrary closed loop in the fluid is quantized by the quantum $\kappa = h/m$. A vortex with quantized circulation is called a quantized vortex. Any rotational motion of a superfluid is sustained only by quantized vortices.

The idea of quantized circulation was first proposed by L. Onsager (1949), for a series of annular rings in a rotating superfluid. Later, R. Feynman (1955) considered that a vortex in a superfluid could take the form of a vortex filament, with the quantized circulation κ and a core of atomic dimension. W.F. Vinen (1957-61) confirmed Feynman's findings experimentally, by showing that the dissipation arises from mutual friction between vortices and the normal flow. Vinen also succeeded in observing quantized circulation using vibrating wires in rotating superfluid ^{4}He. Subsequently, many experimental studies have examined *superfluid turbulence* (ST) in thermal counterflow systems, and have revealed a variety of physical phenomena. Since the dynamics of quantized vortices are nonlinear and non-local, it has not been easy to quantitatively understand these observations on the basis of vortex dynamics. K.W. Schwarz (1985-88) clarified the picture of ST based on tangled vortices by numerical simulation of the quantized vortex filament model in the thermal counterflow. However, since the thermal counterflow has no analogy in conventional fluid dynamics, this study was not helpful in clarifying the relationship between ST and *classical turbulence* (CT). Superfluid turbulence is often called quantum turbulence (QT), which emphasizes the belief that it is comprised of quantized vortices (see [Tsu06, TKK10] and references therein).

3.2.2. Bose-Einstein Superfluids in ECM Dynamics

Recall from the previous subsection, what happens if we rotate a cylindrical vessel with a classical viscous fluid inside. Even if the fluid is initially at rest, it starts to rotate and eventually reaches a steady rotation with the same rotational speed as the vessel. In that case, one can say that the system contains a vortex that mimics solid-body rotation. A rotation of arbitrary angular velocity can be sustained by a single vortex.

However, this does not occur in a *quantum superfluid*. Because of quantization of circulation, *superfluids* respond to rotation, not with a single vortex, but with a *lattice of quantized vortices*. R. Feynman noted that in uniform rotation with angular velocity Ω the curl of the superfluid velocity is the circulation per unit area, and since the curl is 2Ω, a lattice of quantized vortices with number density $n_0 = \text{curl } v_x/\kappa = 2\Omega/\kappa$ (the '*Feynman rule*') arranges itself parallel to the rotation axis [FH65].

This observation has been reproduced by a simulation of the Gross-Pitaevskii (GP) equation for the *macroscopic wave-function* $\psi(\mathbf{x},t) = |\psi(\mathbf{x},t)|e^{i\theta(\mathbf{x},t)}$ in 2D and 3D spaces (see [Tsu06, TKK10] and references therein). In a weakly interacting Bose system, the macroscopic wave-function $\psi(\mathbf{x},t)$ appears as the order parameter of the Bose-Einstein condensate (representing our quantum ECM superfluid model) obeying the GP equation, or the modified cubic NLS equation (extended by the linear term $-\mu\psi$):

$$i\hbar\partial_t\psi = -\frac{\hbar^2}{2m}\Delta\psi + V|\psi|^2\psi - \mu\psi. \tag{7}$$

Writing $\psi = |\psi|\exp(i\theta)$, the squared amplitude $|\psi|^2$ is the ***ECM superfluid density*** and the gradient of the phase θ gives the ***ECM superfluid velocity*** $\mathbf{v}_x = (\hbar/m)\nabla\theta$, corresponding to frictionless flow of the ECM. This relation causes quantized vortices to appear with quantized ECM circulation. The only characteristic scale of the GP model is the coherence length defined by $\xi = \hbar/(\sqrt{2mV}\,|\psi|)$, which determines the ECM vortex core size. The GP model can explain not only the ECM vortex dynamics but also phenomena related to vortex cores, such as ECM reconnection and nucleation. The trapping potential is

$$V_{\text{ex}} = \frac{1}{2}m\omega^2[(1+\epsilon_x)x^2 + (1+\epsilon_y)y^2],$$

where $\omega = 2\pi \times 219$ Hz, and the parameters ϵ_x and ϵ_y describe small deviations from axial symmetry corresponding to experiments.

3.2.3. Kolmogorov Energy Spectra

The Kolmogorov energy spectra were confirmed for both decaying [KT05a] and steady [KT05b] QT by the GP model. The normalized GP equation is:

$$i\partial_t\psi = -\frac{1}{2}\Delta\psi + V|\psi|^2\psi - \mu\psi, \tag{8}$$

which determines the dynamics of the macroscopic wave-function $\psi(\mathbf{x},t) = f(\mathbf{x},t)\exp[i\phi(\mathbf{x},t)]$. The ECM superfluid dynamics in the GP model are compressible. The total number of ECM agents is $N = \int |\psi|^2 d\mathbf{x}$ and the total ECM energy is:

$$E(t) = \frac{1}{N}\int^{*}\psi\left(-\Delta + \frac{V}{2}f^2\right)\psi\,d\mathbf{x},$$

as represented by the sum of the interaction energy $E_{int}(t)$, the quantum energy $E_q(t)$, and the kinetic energy $E_{kin}(t)$ [NAB97a, NAB97b],

$$E_{int}(t) = \frac{V}{2N}\int f^4 d\mathbf{x}, \qquad E_q(t) = \frac{1}{N}\int[\nabla f]^2 d\mathbf{x},$$

$$E_{kin}(t) = \frac{1}{N}\int[f\nabla\phi]^2 d\mathbf{x}.$$

The kinetic energy is further divided into a compressible part $E^c_{kin}(t)$ due to compressible excitations and an incompressible part $E^i_{kin}(t)$ due to vortices. The Kolmogorov spectrum is expected for $E^i_{kin}(t)$ [Tsu06, TKK10].

3.2.4. Shock-Waves, Solitons and Rogue Waves in Quantum ECM Dynamics

The realistic (i.e., adaptive, wave-form, nonlinear and stochastic) ECM dynamics can be modeled using (an adaptive form of) *nonlinear Schrödinger equation*(NLS), also called *Gross-Pitaevskii equation* (GP), defining the time-dependent complex-valued *macroscopic wave function* $\psi = \psi(x,t)$, whose absolute square $|\psi(x,t)|^2$ represents the *crowd density function*. In natural quantum units ($\hbar = 1,\ m = 1$), our NLS equation reads:

$$\mathrm{i}\partial_t\psi = -\frac{1}{2}\partial_{xx}\psi + V|\psi|^2\psi, \qquad (\mathrm{i} = \sqrt{-1}; \quad \text{with } \partial_z\psi = \frac{\partial\psi}{\partial z}), \tag{9}$$

where $V = V(w,x)$ denotes the adaptive heat potential.

The general ECM NLS equation (9) can be exactly solved (see [II09b]) using the power series expansion method of *Jacobi elliptic functions*. Consider the $\psi-$function describing a single plane wave, with the wave number k and circular frequency ω:

$$\psi(x,t) = \phi(\xi)\,\mathrm{e}^{\mathrm{i}(kx-\omega t)}, \qquad \text{with } \xi = x - kt \text{ and } \phi(\xi) \in \mathbb{R}. \tag{10}$$

Its substitution into the NLS equation (9) gives the nonlinear ECM oscillator ODE:

$$\phi''(\xi) + [\omega - \frac{1}{2}k^2]\,\phi(\xi) - V\phi^3(\xi) = 0. \tag{11}$$

We can seek a solution $\phi(\xi)$ for (11) as a linear function:

$$\phi(\xi) = a_0 + a_1\mathrm{sn}(\xi),$$

where $\mathrm{sn}(x) = \mathrm{sn}(x,m)$ are Jacobi elliptic sine functions with *elliptic modulus* $m \in [0,1]$, such that $\mathrm{sn}(x,0) = \sin(x)$ and $\mathrm{sn}(x,1) = \tanh(x)$. The solution of (11) has been calculated to be

$$\begin{aligned}\phi(\xi) &= \pm m\sqrt{\frac{1}{V}}\,\mathrm{sn}(\xi), \quad \text{for } m \in [0,1]; \text{ and}\\ \phi(\xi) &= \pm\sqrt{\frac{1}{V}}\,\tanh(\xi), \quad \text{for } m = 1.\end{aligned}$$

This gives the exact periodic solution of (9) as [II09b]

$$\psi_1(x,t) = \pm m\sqrt{\frac{1}{V(w)}}\,\mathrm{sn}(x-kt)\,\mathrm{e}^{\mathrm{i}[kx-\frac{1}{2}t(1+m^2+k^2)]},$$

$$\text{for } m \in [0,1); \tag{12}$$

$$\psi_2(x,t) = \pm\sqrt{\frac{1}{V(w)}}\,\tanh(x-kt)\,\mathrm{e}^{\mathrm{i}[kx-\frac{1}{2}t(2+k^2)]},$$

$$\text{for } m = 1, \tag{13}$$

where (12) defines the general solution, while (13) defines the *ECM envelope shock-wave* (or, 'dark soliton') solution of the ECM NLS equation (9).

Alternatively, if we seek a solution $\phi(\xi)$ as a linear function of Jacobi elliptic cosine functions, such that $\mathrm{cn}(x,0)=\cos(x)$ and $\mathrm{cn}(x,1)=\mathrm{sech}(x)$,

$$\phi(\xi)=a_0+a_1\mathrm{cn}(\xi),$$

then we get [II09b]

$$\begin{aligned} \psi_3(x,t) &= \mp m\sqrt{\frac{1}{V(w)}}\,\mathrm{cn}(x-kt)\,\mathrm{e}^{\mathrm{i}[kx-\frac{1}{2}t(1-2m^2+k^2)]}, \\ \text{for } m\in[0,1); & \qquad (14) \\ \psi_4(x,t) &= \mp\sqrt{\frac{1}{V(w)}}\,\mathrm{sech}(x-kt)\,\mathrm{e}^{\mathrm{i}[kx-\frac{1}{2}t(k^2-1)]}, \\ \text{for } m=1, & \qquad (15) \end{aligned}$$

where (14) defines the general solution, while (15) defines the *ECM envelope solitary-wave* (or, 'bright soliton') solution of the ECM NLS equation (9).

In all four solution expressions (12), (13), (14) and (15), the adaptive ECM potential $V(w)$ is yet to be calculated using either unsupervised Hebbian learning, or supervised Levenberg-Marquardt algorithm (see, e.g. [II07a]). In this way, the NLS equation (9) becomes a *quantum neural network* [II09b]. Any kind of numerical analysis can be easily performed using above closed-form ECM solutions $\psi_i(x,t)$ $(i=1,...,4)$ as initial conditions.

In addition, two new wave-solutions of the ECM NLS equation (9) have been recently provided in the form of rogue waves,[1] using the deformed Darboux transformation method:

1. The *one-rogon* ECM solution:

$$\psi_{1\mathrm{rogon}}(x,t)=\alpha\sqrt{\frac{1}{2V}}\left[1-\frac{4(1+\alpha^2 t)}{1+2\alpha^2(x-kt)^2+\sigma^2\alpha^4t^2}\right]\mathrm{e}^{\mathrm{i}[kx+1/2(\alpha^2-k^2)t]}, \qquad (16)$$

where $V>0$, while α and k denote the ECM scaling and gauge.

2. The *two-rogon* ECM solution:

$$\psi_{2\mathrm{rogon}}(x,t)=\alpha\sqrt{\frac{1}{2V}}\left[1+\frac{P_2(x,t)+\mathrm{i}Q_2(x,t)}{R_2(x,t)}\right]\mathrm{e}^{\mathrm{i}[kx+1/2(\alpha^2-k^2)t]}, \qquad (17)$$

where $V>0$, while P_2, Q_2, R_2 are certain polynomial functions of x and t.

This concludes our expose on classical lymphatic fluid dynamics (with lymphatic turbulence) and quantum ECM fluid dynamics (with quantum ECM turbulence).

[1] Rogue waves are also known as *freak waves*, *monster waves*, *killer waves*, *giant waves* and *extreme waves*. They are found in various media, including optical fibers.

4. Biomedical Applications

4.1. Nonstandard Biomesotherapy for Motor Neuron Diseases

Using the above-explained principles of lymphatic dynamics stimulation, an effective nonstandard treatment for motor neuron diseases can be performed using the Diamond Medisculpt machine.

The main purpose of this nonstandard treatment for motor neuron diseases is to stimulate and increase the natural motion of the *cerebrospinal fluid* (CSF), which would in turn increase the local metabolism and bust immune responses along the spinal cord and brain.

4.2. Diamond Fat-Burning Mechanism

Experienced *iron-man athletes* and other very-long-distance runners (beyond classical 42 km marathon) are able to switch on to the *fat-burning mechanism*. It happens after 3 to 4 or more hours of continued physical exercise of moderate intensity (e.g., running or cycling), at the moment when all the *glycogen* in the body has been exhausted, but only if the *heart-rate* is below its anaerobic (lactate) threshold (around 160 beats/minute) at that time. Otherwise, after the glycogen is exhausted the body starts burning protein (that is, muscle tissue), which can be disastrous for the athlete. To be able to achieve this consistently, iron-man athletes are trained to keep their heart-rate low (around 130 beats/minute) while still maintaining the high volume of work done. This way of training effectively pushes the *anaerobic threshold* (the maximum of the lactate curve) to the right.

It has been our experience that the same effect (burning fat instead of muscle tissue and shifting the lactate curve to the right) would be easily achieved with Diamond Medisculpt EMS treatment combined with taking of Diamond Oxygen drops.

Conclusion

In this paper we presented the biophysics of solitons and lymphatic turbulence occurring in human lymphatic and extracellular matrix dynamics controlled by Diamond Biomesotherapy. We started with a basic description of the extracellular matrix, followed by an explanation of Diamond ECM therapy. Then we presented modern research in lymphatic and ECM turbulent dynamics, from both a global-macroscopic perspective and local-quantum one.This was followed by a brief description of related Diamond therapies and fat-burning mechanics. We concluded with two biomedical applications: (i) a nonstandard Diamond Biomesotherapy for motor neuron diseases, and (ii) a Diamond fat-burning mechanism.

References

[Pis07] Pischinger, A.: *The Extracellular Matrix and Ground Regulation: Basis for a Holistic Biological Medicine*, North Atlantic Books, Berkeley, (2007).

[LB68] Leak, L.V., Burke. J.F.: Ultrastructural studies on the lymphatic anchoring filaments, *J. Cell Biol.* 36, 129–149, (1968).

[SSP11] Schulte-Merker, S., Sabine, A., Petrova, T.V.: Lymphatic vascular morphogenesis in development, physiology, and disease, *J Cell Biol.* 193, 607–618, (2011).

[TDT01] Tammi, M., Day, A.J., Turley, E.A.: Hyaluronan and homeostasis: a balancing act, *J. Biol. Chem.* 277, 4581–4584, (2001).

[FLL97] Fraser J.R., Laurent T.C., Laurent U.B.: Hyaluronan – Its nature, distribution, functions and turnover, *J. Intern. Med.* 242, 27–33, (1997).

[Ste04] Stern, R.: Hyaluronan catabolism – a new metabolic pathway, *Eur. J. Cell Biol.* 83, 317–325, (2004).

[Sor10] Sorokin, L.: The impact of the extracellular matrix on inflammation, *Nature Reviews Immunology* 10, 712–723 (October 2010).

[Kin12] King, M.W.: *The Medical Biochemistry Page*, (2012); http://themedicalbiochemistrypage.org/extracellularmatrix.php.

[Sim07] Simmons, K.: *Cells and cellular processes*, in Biology BIOL-1115/3, Uni. Winnipeg, (2007); http://kentsimmons.uwinnipeg.ca/cm1504/15intro.htm.

[DI12] Diamond International: Medisculpt user manual, (2012).

[FW80] Frasher, J.R.E., Laurent, T.C.,; Laurent, U.B.G.: Hyaluronan: its nature, distribution, functions and turnover, *J. Int. Med.* 242(1), 27–33, (1997).

[II09b] Ivancevic, V., Ivancevic, T.: *Quantum Neural Computation*, Springer, (2009).

[Kol41a] Kolmogorov, A.N.: The local structure of turbulence in incompressible viscous fluid for very large Reynolds number, *Dokl. Akad. Nauk SSSR* 30, 299–303, (1941) [reprinted in *Proc. Roy. Soc. A* 434, 9–13, (1991)].

[Kol41b] Kolmogorov, A.N.: On degeneration (decay) of isotropic turbulence in an incompressible viscous liquid, *Dokl. Akad. Nauk SSSR* 31, 538–540, (1941) [reprinted in *Proc. Roy. Soc. A* 434, 15–17, (1991)].

[NAB97a] Nore, C., Abid, M., Brachet, M.E.: Kolmogorov turbulence in low-temperature superflows, *Phys. Rev. Lett.* 78, 3296–3299, (1997).

[NAB97b] Nore, C., Abid, M., Brachet, M.E.: Decaying Kolmogorov turbulence in a model of superflow, *Phys. Fluids* 9, 2644–2669, (1997).

[II06c] Ivancevic, V., Ivancevic, T.: *Geometrical Dynamics of Complex Systems.* Springer, (2006).

[Arn78] Arnold, V.I.: *Mathematical Methods of Classical Mechanics*. Springer, New York, (1978).

[AK98] Arnold, V.I., Khezin, B.: *Topological Methods in Hydrodynamics,* Springer, (1998).

[KK01] Kawahara, G., Kida, S.: Periodic motion embedded in plane Couette turbulence: regeneration cycle and burst. *J. Fluid Mech.* 449, 291–300, (2001).

[Tsu06] Tsubota, M.: Quantized vortices in superfluid helium and Bose-Einstein condensates. *J. Physics: Conf. Ser.* 31, 88–94, (2006).

[TKK10] Tsubota, M., Kasamatsu, K, Kobayashi, M.: *Quantized vortices in superfluid helium and atomic Bose-Einstein condensates*. arXiv: cond-mat.quant-gas 1004.5458v2, (2010).

[PS03] Pitaevskii, L., Stringari, S.: *Bose-Einstein Condensation*. Oxford Univ. Press, Oxford, (2003).

[Kap38] Kapitza, P.: Viscosity of liquid helium below the λ point. *Nature* 141, 74, (1938).

[Lan41] Landau, L.: The theory of superfluidity of helium II. *J. Phys. U.S.S.R.* 5, 71-90, (1941).

[Tis38] Tisza, L.: Transport phenomena in helium II. *Nature*, 141, 913, (1938).

[AM38] Allen, J.F., Misener, A.D., Flow of liquid helium II. *Nature*, 141, 75, (1938).

[Lon38] London, F.: On the Bose-Einstein condensation. *Phys. Rev.* 54, 947–954, (1938).

[FH65] Feynman, R.P., Hibbs, A.R.: *Quantum Mechanics and Path Integrals*, McGraw-Hill, New York, (1965).

[KT05a] Kobayashi, M., Tsubota, M.: Kolmogorov spectrum of superfluid turbulence: Numerical analysis of the Gross-Pitaevskii equation with a small-scale dissipation, *Phys. Rev. Lett.* 94, 065302, (2005).

[KT05b] Kobayashi, M., Tsubota, M.: Kolmogorov spectrum of quantum turbulence, *J. Phys. Soc. Jpn.* 74, 3248–3258, (2005).

[II07a] Ivancevic, V., Ivancevic, T.: *Neuro–Fuzzy Associative Machinery for Comprehensive Brain and Cognition Modelling*. Springer, Berlin, (2007).

In: Solitons
Editor: Matthew C. Williams
ISBN: 978-1-62618-234-9

Chapter 7

NONLINEAR WAVES IN STRONGLY INTERACTING RELATIVISTIC FLUIDS

***D.A. Fogaça*[1] *, F.S. Navarra*[1] *and L.G. Ferreira Filho*[2]**
[1]Instituto de Física, Universidade de São Paulo, Paulo, Brazil
[2]Faculdade de Tecnologia, Universidade do Estado do Rio de Janeiro, Resende, Brazil

Abstract

During the past decades the study of strongly interacting fluids experienced a tremendous progress. In the relativistic heavy ion accelerators, specially the RHIC and LHC colliders, it became possible to study not only fluids made of hadronic matter but also fluids of quarks and gluons. Part of the physics program of these machines is the observation of waves in this strongly interacting medium. From the theoretical point of view, these waves are often treated with li-nearized hydrodynamics. In this text we review the attempts to go beyond linearization. We show how to use the Reductive Perturbation Method to expand the equations of (ideal and viscous) relativistic hydrodynamics to obtain nonlinear wave equations. These nonlinear wave equations govern the evolution of energy density perturbations (in hot quark gluon plasma) or baryon density perturbations (in cold quark gluon plasma and nuclear matter). Different nonlinear wave equations, such as the breaking wave, Korteweg-de Vries and Burgers equations, are obtained from different equations of state (EOS). In nuclear matter, the Walecka EOS may lead to a KdV equation. We explore equations of state such as those extracted from the MIT Bag Model and from QCD in the mean field theory approach. Some of these equations are integrable and have analytical solitonic solutions. We derive these equations also in spherical and cylindrical coordinates. We extend the analysis to two and three dimensions to obtain the Kadomtsev-Petviashvili (KP) equation, which is the generalization of the KdV. The KP is also integrable and presents analytical solitonic solutions. In viscous relativistic hydrodynamics we have second spatial derivatives which physically represent dissipation terms. We present numerical solutions and their corresponding algorithms for the cases where the equations are not integrable.

1. Introduction

The elementary particles and their interactions are well described by the Standard Model (SM), which is an extremely successful theory [1]. In this theory matter is composed by quarks and leptons and their interactions are due to the exchange of gauge bosons. The sector of the SM which describes the strong interactions at the fundamental level is called Quantum Chromodynamics, or QCD [2]. According to QCD there are six types, or flavors, of quarks: up, down, strange, charm, bottom and top, and they interact exchanging gluons. Quarks and gluons have a special charge called color, responsible for the strong interaction. They do not exist as individual particles but, due to the property of color confinement, quarks and gluons form clusters called hadrons, which can be grouped in baryons and mesons. The former are made of three quarks, as the proton, and the latter are made of a quark and an antiquark, as the pion, for example. The quarks carry a fraction of the elementary electric charge and a fraction of the baryon number. The electric and color charges and the baryon number are conserved quantities in QCD.

The hadrons also interact strongly and their interactions have been traditionally described by a quantum field theory called Quantum Hadrodynamics, or QHD, in its different versions. According to QHD, in nuclear matter neutrons and protons interact exchanging scalar and vector mesons. This potentially complicated theory becomes quite simple in the mean field approximation, in which the meson fields are treated as classical fields. The different versions of QHD in the mean field approximation are called relativistic mean field models (RMF). More recently, nuclear matter has been studied with Effective Field Theories, which incorporate the fundamental symmetries of QCD in hadron physics.

In the phase diagram of QCD, we can observe that under extreme conditions of very large temperatures and/or very large densities, the normal hadronic matter undergoes a phase transition to a deconfined phase, a new state of matter called the quark gluon plasma (QGP). Together with deconfinement, a second phase transition takes place: the chiral phase transition, during which chiral symmetry is restored and the light quarks (up and down) become massless. The hot QGP is produced in relativistic heavy ion collisions in the Relativistic Heavy Ion Collider (RHIC) at the Brookhaven National Laboratory (BNL) and even more in the Large Hadron Collider (LHC) at CERN. The cold QGP may exist in the core of compact stars.

According to our present understanding of the RHIC measurements, the QGP behaves as an almost perfect fluid and its space-time evolution can be very well described by relativistic hydrodynamics. After the discovery of this new fluid, more sophisticated measurements made possible to study the propagation of perturbations in the QGP. We may, for example, study the effect of a fast quark traversing the hot QGP medium. As it moves supersonically throughout the fluid, it may generate waves of energy density or baryon density. In some works it was even claimed that these waves may pile up and form Mach cones [3], which would affect the angular distribution of the produced particles, fluid fragments which are experimentally observed.

The study of waves in the quark-gluon fluid has been mostly performed with the assumption that the amplitude of the perturbations is small enough to justify the linearization of the Euler and continuity equations [4]. The analysis of perturbations with the linearized relativistic hydrodynamics leads to the standard second order wave equations and their trav-

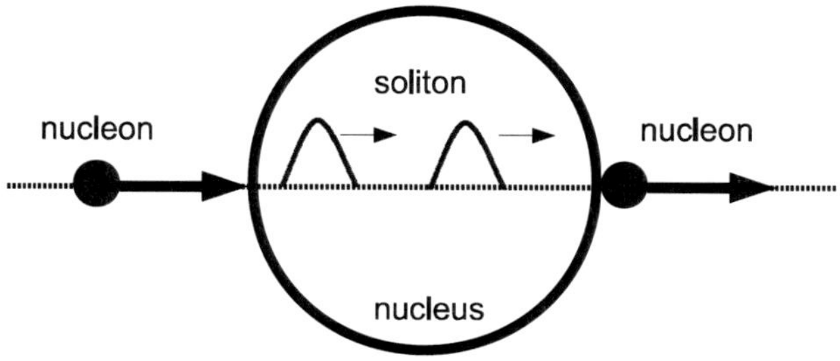

Figure 1. KdV soliton in the baryon density simulating "nuclear transparency".

eling wave solutions, such as acoustic waves in the QGP. While linearization is justified in many cases, in others it should be replaced by another technique to treat perturbations keeping the nonlinearities of the theory. Since long ago there is a technique which preserves nonlinearities in the derivation of the differential equations which govern the evolution of perturbations. This is the reductive perturbation method (RPM) [5–8].

Nonlinearities may lead, as they do in other domains of physics, to new and interesting phenomena. In a pioneering work [9], with the use of nonrelativistic ideal hydrodynamics combined with the RPM and with an appropriate equation of state of cold nuclear matter, it was shown that it is possible to derive a Korteweg- de Vries (KdV) equation for the baryon density, which has analytic solitonic solutions. This indicates that a pulse in baryon density (the KdV soliton) can propagate without dissipation through the nuclear medium. If this occurs this pulse might be responsible for an interesting phenomena: the apparent nuclear transparency in proton nucleus collisions at low energies. The incoming proton would be absorbed by the fluid (target nucleus) and turned into a density pulse. In the fluid the pulse satisfies the KdV equation and it is able to traverse the target nucleus emerging on the other side. This can be called "nuclear transparency" and it is illustrated in the Fig 1.

Perturbations in fluids with different equations of state (EOS) generate different nonlinear wave equations: the breaking wave equation, KdV, Burgers... etc. Among these equations we find the Kadomtsev-Petviashvili (KP) equation [10], which is a nonlinear wave equation in three spatial and one temporal coordinate. It is the generalization of the KdV equation to higher dimensions. This equation has been found with the application of the reductive perturbation method [5] to several different problems such as the propagation of solitons in multicomponent plasmas, dust acoustic waves in hot dust plasmas and dense electro-positron-ion plasma [11–23]. Previous studies on nonlinear waves in cold and warm nuclear matter can be found in [9, 24–29]. Works on nonlinear waves in cold QGP in the mean field approach were published in [30] and their extension to three dimensions were published in [31].

In this text we review the applications of the RPM [5, 11–22] to relativistic fluid dynamics [32, 33]. It is organized as follows. In the next section we review the basic equations of relativistic hydrodynamics. In section III we present the RPM and give the coordinate transformations to be used in the subsequent sections. In section IV we introduce the equations of state of hadronic matter and of the quark-gluon plasma, giving special attention to the ingredients which may generate solitons. In section V we list the wave equations which follow from the application of the RPM to hydrodynamics. In section VI we present the analytical solutions of some of these differential equations and in section VII we show the time evolution of several nonlinear waves obtained by numerical integration. In section VIII we make some final remarks.

2. Relativistic Hydrodynamics

2.1. Definitions and Basic Equations

Hydrodynamics [34, 35] and supersonic flow [36] became more relevant to explain recent phenomena observed at the CERN-SPS and RHIC experiments. Indeed, the hot and dense medium created in heavy ion collisions at RHIC behaves as a perfect fluid and ideal hydrodynamics was applied to describe its space-time evolution. The study of the perturbations on the fluid, such as the waves created by the passage of a supersonic parton, can be explained by the conical shock waves generated by large energy deposition in the medium, as discussed in Refs. [37–48]. Besides, the ideal fluid approach evolved to viscous hydrodynamics, as recently shown in [49–52].

In this section we present the formalism of relativistic hydrodynamics. Pedagogical texts on this subject can be found in [32, 33]. Here we give special attention to the variables which are relevant for strongly interacting fluids, such as the baryon density. We write the final equations in a more extended form, which allows the reader to make a direct comparison with the corresponding non-relativistic versions of these equations. In what follows we use $c = 1$, $\hbar = 1$ and the Boltzmann constant is taken to be one, i.e., $k_B = 1$. All the relativistic equations are written in terms of 4-vectors and the metric tensor is given by $g_{\mu\nu}$, with $g_{00} = -g_{11} = -g_{22} = -g_{33} = 1$ and $g_{\mu\nu} = 0$ if $\mu \neq \nu$. From [32, 33, 53] the fundamental equations of a relativistic ideal fluid are:

$$D\varepsilon + (\varepsilon + p)\partial_\mu u^\mu = 0 \tag{1}$$

and

$$(\varepsilon + p)Du^\alpha - \nabla^\alpha p = 0 \tag{2}$$

where:

$$D \equiv u^\mu \partial_\mu \qquad \text{and} \qquad \nabla^\alpha \equiv \Delta^{\mu\alpha}\partial_\mu \tag{3}$$

and the projection operator on the orthogonal direction to the fluid velocity u^μ is given by the tensor:

$$\Delta^{\mu\nu} = g^{\mu\nu} - u^\mu u^\nu \tag{4}$$

which has the properties $\Delta^{\mu\nu}u_\mu = \Delta^{\mu\nu}u_\nu = 0$ and $\Delta^{\mu\nu}\Delta^\alpha_\nu = \Delta^{\mu\alpha}$.

The velocity 4-vector of the fluid element is given by [32, 33, 53]: $u^\mu = (\gamma, \gamma\vec{v})$ where γ is the Lorentz factor $\gamma = (1 - v^2)^{-1/2}$ and thus $u^\mu u_\mu = 1$ and $u_\mu \partial_\nu u^\mu = (1/2)\partial_\nu(u^\mu u_\mu) = (1/2)\partial_\nu(1) = 0$.

In an ideal fluid all dissipative (viscous) effects are neglected and in order to introduce the effects of viscosity, it is necessary to add the viscous stress tensor $\Pi^{\mu\nu}$ to the energy-momentum tensor. In a simple way we may write for a viscous fluid:

$$T^{\mu\nu} = T^{\mu\nu}_{(0)} + \Pi^{\mu\nu} \tag{5}$$

where

$$T^{\mu\nu}_{(0)} = \varepsilon u^\mu u^\nu - p\Delta^{\mu\nu} \tag{6}$$

is the ideal relativistic fluid energy-momentum tensor [32, 33]. We also consider for simplicity [53] a system without conserved charges (or at zero chemical potential) and so the total momentum density is due to the flow of energy density $u_\mu T^{\mu\nu} = \varepsilon u^\nu$ and hence we must assume that:

$$u_\mu \Pi^{\mu\nu} = 0 \tag{7}$$

We take the appropriate projections of the conservation equations of the energy momentum tensor: the parallel $(u_\nu \partial_\mu T^{\mu\nu})$ and perpendicular $(\Delta^\alpha_\nu \partial_\mu T^{\mu\nu})$ to the fluid velocity. The results are:

$$u_\nu \partial_\mu T^{\mu\nu} = D\varepsilon + (\varepsilon + p)\partial_\mu u^\mu + u_\nu \partial_\mu \Pi^{\mu\nu} = 0 \tag{8}$$

and

$$\Delta^\alpha_\nu \partial_\mu T^{\mu\nu} = (\varepsilon + p)Du^\alpha - \nabla^\alpha p + \Delta^\alpha_\nu \partial_\mu \Pi^{\mu\nu} = 0 \tag{9}$$

Using the symmetrization notation:

$$A_{(\mu}B_{\nu)} = \frac{1}{2}(A_\mu B_\nu + A_\nu B_\mu) \tag{10}$$

we are able to rewrite the $u_\nu \partial_\mu \Pi^{\mu\nu}$ term in (8) as $u_\nu \partial_\mu \Pi^{\mu\nu} = \partial_\mu (u_\nu \Pi^{\mu\nu}) - \Pi^{\mu\nu}\partial_{(\mu}u_{\nu)}$. We also use the identity

$$\partial_\mu = u_\mu D + \nabla_\mu \tag{11}$$

and the choice of frame $u_\mu \Pi^{\mu\nu} = 0$. The fundamental equations of relativistic viscous fluid dynamics are finally given by:

$$D\varepsilon + (\varepsilon + p)\partial_\mu u^\mu - \Pi^{\mu\nu}\nabla_{(\mu}u_{\nu)} = 0 \tag{12}$$

and:

$$(\varepsilon + p)Du^\alpha - \nabla^\alpha p + \Delta^\alpha_\nu \partial_\mu \Pi^{\mu\nu} = 0 \tag{13}$$

2.2. The Viscous Stress Tensor

The viscous stress tensor $\Pi^{\mu\nu}$ has not been specified yet. We shall derive it with the help of the second law of thermodynamics, which states that entropy density must always increase locally. Moreover we shall consider a system in thermodynamic equilibrium $(dp = 0)$ with zero chemical potential. Based on these statements, the thermodynamic relations are given by:

$$\varepsilon + p = Ts \qquad \text{and} \qquad Tds = d\varepsilon \tag{14}$$

The second law of thermodynamics is considered in the covariant form:

$$\partial_\mu s^\mu \geq 0 \tag{15}$$

where the 4-current s^μ is given by:

$$s^\mu = su^\mu \tag{16}$$

Inserting relations (14) into the second law (15) and assuming that $Dp = 0$ we find:

$$\partial_\mu s^\mu = Ds + s\partial_\mu u^\mu = \frac{D\varepsilon}{T} + \frac{(\varepsilon + p)}{T}\partial_\mu u^\mu \tag{17}$$

Using now equation (12) we obtain:

$$\partial_\mu s^\mu = \frac{1}{T}\Pi^{\mu\nu}\nabla_{(\mu}u_{\nu)} \geq 0 \tag{18}$$

The viscous tensor may be decomposed into a traceless part $\pi^{\mu\nu}$ $(\pi^\mu_\mu = 0)$ and a non-vanishing trace part, so that:

$$\Pi^{\mu\nu} = \pi^{\mu\nu} + \Delta^{\mu\nu}\Pi \tag{19}$$

In terms of the notation introduced in [53]:

$$\nabla_{\langle\mu}u_{\nu\rangle} \equiv 2\nabla_{(\mu}u_{\nu)} - \frac{2}{3}\Delta_{\mu\nu}\nabla_\alpha u^\alpha \tag{20}$$

it can be shown that:

$$u_\mu \pi^{\mu\nu} = 0 \tag{21}$$

$$u^\mu u^\nu \nabla_{\langle\mu}u_{\nu\rangle} = 0 \tag{22}$$

$$g^{\mu\nu}\nabla_{\langle\mu}u_{\nu\rangle} = 0 \tag{23}$$

and

$$\Delta^{\mu\nu}\Delta_{\mu\nu} = 3 \tag{24}$$

Inserting (19) to (24) into (18) we find:

$$\partial_\mu s^\mu = \frac{1}{2T}\pi^{\mu\nu}\nabla_{\langle\mu}u_{\nu\rangle} + \frac{1}{T}\Pi\nabla_\alpha u^\alpha \geq 0 \tag{25}$$

and this inequality (a positive sum of squares) is satisfied if:

$$\pi^{\mu\nu} = \eta\nabla^{\langle\mu}u^{\nu\rangle}\ , \qquad \Pi = \zeta\nabla_\alpha u^\alpha\ , \qquad \eta \geq 0 \quad \text{and} \quad \zeta \geq 0 \tag{26}$$

where η is the *shear viscosity coefficient* and ζ is the *bulk viscosity coefficient*. The final expression for the viscous tensor is given by the substitution of (26) in (19):

$$\Pi^{\mu\nu} = \eta\nabla^{\langle\mu}u^{\nu\rangle} + \zeta\Delta^{\mu\nu}\nabla_\alpha u^\alpha \tag{27}$$

2.3. The Relativistic Navier-Stokes Equation

The system of equations (12), (13) and (27) is called relativistic Navier-Stokes equation. The temporal component $(\alpha = 0)$ of (13) multiplied by v^i is given by:

$$(\varepsilon + p)(D\gamma)v^i - v^i\nabla^0 p + v^i\Delta^0_\nu\partial_\mu\Pi^{\mu\nu} = 0 \tag{28}$$

and the spatial component $(\alpha = i)$ of (13):

$$(\varepsilon + p)(D\gamma)v^i = -(\varepsilon + p)\gamma(Dv^i) + \nabla^i p - \Delta^i_\nu\partial_\mu\Pi^{\mu\nu} \tag{29}$$

Inserting (29) into (28) we find:

$$(\varepsilon+p)\gamma(Dv^i)+(v^i\nabla^0-\nabla^i)p-(v^i\Delta^0_\nu-\Delta^i_\nu)\partial_\mu\Pi^{\mu\nu}=0 \tag{30}$$

which can be rewritten with the help of (3) and (4) as:

$$(\varepsilon+p)\gamma^2\left(\frac{\partial}{\partial t}+\vec{v}\cdot\vec{\nabla}\right)v^i+v^i\frac{\partial p}{\partial t}-\partial^i p-(v^i\Delta^0_\nu-\Delta^i_\nu)\partial_\mu\Pi^{\mu\nu}=0 \tag{31}$$

The ideal relativistic fluid limit $\Pi^{\mu\nu}=0$ ($\eta=\zeta=0$) in the last equation provides the relativistic version of the Euler equation [32, 33, 53]. For future use we rewrite (31) in detail. To do so, we recall (3), (4) and (27) to obtain the expression for $\partial_\mu\Pi^{\mu\nu}$ in (31):

$$\partial_\mu\Pi^{\mu\nu}=\eta\left\{\partial_\mu\partial^\mu u^\nu+\partial_\mu\partial^\nu u^\mu-\partial_\mu\left[\gamma\left(\frac{\partial}{\partial t}+\vec{v}\cdot\vec{\nabla}\right)(u^\mu u^\nu)\right]\right\}$$
$$+\left(\zeta-\frac{2}{3}\eta\right)\partial^\nu\left[\frac{\partial\gamma}{\partial t}+\vec{\nabla}\cdot(\gamma\vec{v})\right]-\left(\zeta-\frac{2}{3}\eta\right)\partial_\mu\left\{u^\mu u^\nu\left[\frac{\partial\gamma}{\partial t}+\vec{\nabla}\cdot(\gamma\vec{v})\right]\right\} \tag{32}$$

and then insert this last result into (31). We find after some algebra [54]:

$$(\varepsilon+p)\gamma^2\left(\frac{\partial}{\partial t}+\vec{v}\cdot\vec{\nabla}\right)\vec{v}+\vec{v}\frac{\partial p}{\partial t}+\vec{\nabla}p$$
$$-\eta\vec{v}\left\{\partial_\mu\partial^\mu\gamma+\partial_\mu\frac{\partial u^\mu}{\partial t}-\partial_\mu\left[\gamma\left(\frac{\partial}{\partial t}+\vec{v}\cdot\vec{\nabla}\right)(\gamma u^\mu)\right]\right\}-\vec{v}\left(\zeta-\frac{2}{3}\eta\right)\frac{\partial}{\partial t}\left[\frac{\partial\gamma}{\partial t}+\vec{\nabla}\cdot(\gamma\vec{v})\right]$$
$$+\vec{v}\left(\zeta-\frac{2}{3}\eta\right)\partial_\mu\left\{\gamma u^\mu\left[\frac{\partial\gamma}{\partial t}+\vec{\nabla}\cdot(\gamma\vec{v})\right]\right\}$$
$$+\eta\left\{\partial_\mu\partial^\mu(\gamma\vec{v})-\partial_\mu\vec{\nabla}u^\mu-\partial_\mu\left[\gamma\left(\frac{\partial}{\partial t}+\vec{v}\cdot\vec{\nabla}\right)(\gamma\vec{v}u^\mu)\right]\right\}-\left(\zeta-\frac{2}{3}\eta\right)\vec{\nabla}\left[\frac{\partial\gamma}{\partial t}+\vec{\nabla}\cdot(\gamma\vec{v})\right]$$
$$-\left(\zeta-\frac{2}{3}\eta\right)\partial_\mu\left\{\gamma\vec{v}u^\mu\left[\frac{\partial\gamma}{\partial t}+\vec{\nabla}\cdot(\gamma\vec{v})\right]\right\}=0 \tag{33}$$

which is the relativistic version of the Navier-Stokes equation. The perfect fluid, described by $\eta=\zeta=0$ in (33) gives the relativistic version of Euler equation [32, 33]:

$$\frac{\partial\vec{v}}{\partial t}+(\vec{v}\cdot\vec{\nabla})\vec{v}=-\frac{1}{(\varepsilon+p)\gamma^2}\left(\vec{\nabla}p+\vec{v}\frac{\partial p}{\partial t}\right) \tag{34}$$

2.4. Causality and The Relativistic Navier-Stokes Equation

The relativistic Navier-Stokes equation (33) does not constitute a causal theory [53]. This fact can be understood when small perturbations are considered in a system in equilibrium with energy density ε_0, pressure p_0 and at rest ($\vec{v}=\vec{0}$).

Such perturbations can be described by:

$$\varepsilon=\varepsilon_0+\delta\varepsilon(t,\vec{x})\,, \qquad p=p_0+\delta p(t,\vec{x}) \qquad \textit{and} \qquad u^\mu=(1,\vec{0})+\delta u^\mu(t,\vec{x}) \tag{35}$$

where " δ " denotes small deviation from equilibrium. For simplicity we consider perturbations which depend only on the x space coordinate. For the particular direction $\alpha = y$ we insert (35) into the (13) to find:

$$(\varepsilon_0 + \delta\varepsilon + p_0 + \delta p) D(\delta u^y) - \nabla^y (p_0 + \delta p) + \Delta^y_\nu \partial_\mu \Pi^{\mu\nu} = 0$$

Considering only the dependence on the x coordinate it becomes:

$$(\varepsilon_0 + p_0) \frac{\partial}{\partial t} \delta u^y + \frac{\partial}{\partial x} \Pi^{xy} + O(\delta^2) = 0 \tag{36}$$

Analogously we insert (35) into the viscous tensor (19) for the particular direction $\nu = y$ and considering only the x coordinate dependence:

$$\Pi^{xy} = -\eta \frac{\partial}{\partial x} \delta u^y + O(\delta^2) \tag{37}$$

Substituting (37) in (36) we find:

$$(\varepsilon_0 + p_0) \frac{\partial}{\partial t} \delta u^y - \eta \frac{\partial^2}{\partial x^2} \delta u^y = O(\delta^2) \tag{38}$$

which provides, after performing the linearization approximation (i.e., neglecting $O(\delta^2)$) the diffusion-type evolution equation:

$$\frac{\partial}{\partial t} \delta u^y - \frac{\eta}{(\varepsilon_0 + p_0)} \frac{\partial^2}{\partial x^2} \delta u^y = 0 \tag{39}$$

for the perturbation $\delta u^y = \delta u^y(t, x)$.

The acausality property of the Navier-Stokes equation can be studied from the Laplace-Fourier wave ansatz for δu^y:

$$\delta u^y(t, x) = f_{(w,k)} e^{ikx - wt} \tag{40}$$

where $f_{(w,k)}$ are the coefficient for a given frequency w and a given wave number k. When inserting (40) in (143) we obtain the dispersion-relation of the diffusion equation:

$$w = \frac{\eta}{(\varepsilon_0 + p_0)} k^2 \tag{41}$$

which gives the speed of diffusion of a mode with wavenumber k :

$$V_{diff}(k) = \frac{dw}{dk} = 2 \frac{\eta}{(\varepsilon_0 + p_0)} k \tag{42}$$

The causality violation occurs when

$$\lim_{k \to \infty} V_{diff}(k) \to \infty \tag{43}$$

which exceeds the speed of light, i.e , the speed of diffusion grows without bound for sufficiently large wavenumber due its linear dependence on the wavenumber.

A possible way to regulate the viscous fluid theory is given by the "Maxwell-Cattaneo law". In this approach a new transport coefficient called relaxation time (τ_π) is added to the equation (37) as follows:

$$\tau_\pi \frac{\partial}{\partial t}\Pi^{xy} + \Pi^{xy} = -\eta \frac{\partial}{\partial x}\delta u^y + O(\delta^2) \tag{44}$$

The derivative of above equation with respect to x provides:

$$\tau_\pi \frac{\partial}{\partial t}\frac{\partial}{\partial x}\Pi^{xy} + \frac{\partial}{\partial x}\Pi^{xy} = -\eta \frac{\partial^2}{\partial x^2}\delta u^y + O(\delta^2) \tag{45}$$

which becomes, after substituting $\frac{\partial}{\partial x}\Pi^{xy}$ from (36) and performing the linearization approximation, the following evolution equation:

$$\tau_\pi \frac{\partial^2}{\partial t^2}\delta u^y + \frac{\partial}{\partial t}\delta u^y - \frac{\eta}{(\varepsilon_0 + p_0)}\frac{\partial^2}{\partial x^2}\delta u^y = 0 \tag{46}$$

where the term with second derivative with respect to time provides the following dispersion relation when (40 is used:

$$w = \frac{1}{2\tau_\pi} \pm \sqrt{\frac{1}{4\tau_\pi{}^2} - \frac{\eta k^2}{(\varepsilon_0 + p_0)\tau_\pi}} = \frac{1}{2\tau_\pi} \pm ik\sqrt{\frac{\eta}{(\varepsilon_0 + p_0)\tau_\pi} - \frac{1}{4\tau_\pi{}^2 k^2}}$$

$$|w|^2 = w\bar{w} = k^2 \frac{\eta}{(\varepsilon_0 + p_0)\tau_\pi} \tag{47}$$

The speed of diffusion of a mode with wavenumber k is now:

$$V_{diff_{causal}}(k) = \frac{d|w|}{dk} = \sqrt{\frac{\eta}{(\varepsilon_0 + p_0)\tau_\pi}} \tag{48}$$

which does not violate causality:

$$\lim_{k\to\infty} V_{diff}(k) = \sqrt{\frac{\eta}{(\varepsilon_0 + p_0)\tau_\pi}} \tag{49}$$

where $\tau_\pi \neq 0$. Beyond the Maxwell-Cattaneo law, the more complete formulation of viscous hydrodynamics is the Müller-Israel-Stewart theory, which contains the Maxwell-Cattaneo law as a limit. A more precise study is found in [53]. We just wish to point out that there are alternative approaches, in which causality is preserved.

We present a different approximation scheme, which goes beyond the linear approximation, preserves nonlinear terms and does not violate causality: the reductive perturbation method, a technique that will be presented in the next section. When this method is applied to the relativistic Navier-Stokes equation (33) we arrive at the conclusion that, for the purpose of studying perturbations, the equation:

$$\frac{\partial \vec{v}}{\partial t} + (\vec{v}\cdot\vec{\nabla})\vec{v} = -\frac{1}{(\varepsilon + p)}\left[\vec{\nabla}p + \vec{v}\frac{\partial p}{\partial t}\right] + \frac{1}{(\varepsilon + p)}\left[\eta\vec{\nabla}^2\vec{v} + \left(\zeta + \frac{1}{3}\eta\right)\vec{\nabla}(\vec{\nabla}\cdot\vec{v})\right] \tag{50}$$

is equivalent to (33). In other words, considering (50) or (33) leads to the same wave equation.

2.5. Continuity Equations

2.5.1. Entropy

The continuity equation for the entropy density is given by (25):

$$\partial_\mu s^\mu = \frac{1}{2T}\eta\,\nabla^{\langle\mu}u^{\nu\rangle}\nabla_{\langle\mu}u_{\nu\rangle} + \frac{1}{T}\zeta(\nabla_\alpha u^\alpha)^2 \tag{51}$$

Using (3), (4), $u_\nu\partial_\mu u^\nu = 0$ and $u^\nu u_\nu = 1$ we can rewrite the above equation in the form:

$$\partial_\mu s^\mu = -\frac{\eta}{T}(\partial^\mu u^\nu)\partial_\nu u_\mu + \frac{1}{T}\left(\frac{2}{3}\eta+\zeta\right)(\partial_\mu u^\mu)^2 \tag{52}$$

Using (16) and $u^\mu = (\gamma, \gamma\vec{v})$ the last equation can also be rewritten as [54]:

$$\gamma\frac{\partial s}{\partial t} + \gamma\vec{\nabla}s\cdot\vec{v} + s\frac{\partial\gamma}{\partial t} + s\vec{\nabla}\gamma\cdot\vec{v} + \gamma s\vec{\nabla}\cdot\vec{v} = -\frac{\eta}{T}\left(\frac{\partial\gamma}{\partial t}\right)^2 - 2\frac{\eta}{T}\left[\vec{\nabla}\gamma\cdot\frac{\partial}{\partial t}(\gamma\vec{v})\right]$$

$$-\frac{\eta}{T}(\partial^i u^j)\partial_j u_i + \frac{1}{T}\left(\frac{2}{3}\eta+\zeta\right)\left[\frac{\partial\gamma}{\partial t} + \gamma\vec{\nabla}\cdot\vec{v} + \vec{\nabla}\gamma\cdot\vec{v}\right]^2 \tag{53}$$

which is the relativistic version of the continuity equation for the entropy density s. In the case of an ideal fluid ($\eta = \zeta = 0$) we recover the entropy density conservation:

$$\gamma\frac{\partial s}{\partial t} + \gamma\vec{\nabla}s\cdot\vec{v} + s\frac{\partial\gamma}{\partial t} + s\vec{\nabla}\gamma\cdot\vec{v} + \gamma s\vec{\nabla}\cdot\vec{v} = 0$$

which, with the use of:

$$\frac{\partial\gamma}{\partial t} = \gamma^3 v\frac{\partial v}{\partial t} \qquad \text{and} \qquad \vec{\nabla}\gamma = \gamma^3 v\vec{\nabla}v \tag{54}$$

becomes [32, 33]:

$$\frac{\partial s}{\partial t} + \gamma^2 vs\left(\frac{\partial v}{\partial t} + \vec{v}\cdot\vec{\nabla}v\right) + \vec{\nabla}\cdot(s\vec{v}) = 0 \tag{55}$$

2.5.2. Baryon Density

The relativistic version of the continuity equation for the baryon density ρ_B in ideal relativistic hydrodynamics is [32]:

$$\partial_\nu j_B{}^\nu = 0 \tag{56}$$

Since $j_B{}^\nu = u^\nu\rho_B$ and using (54) the last equation can be rewritten as [32]:

$$\frac{\partial\rho_B}{\partial t} + \gamma^2 v\rho_B\left(\frac{\partial v}{\partial t} + \vec{v}\cdot\vec{\nabla}v\right) + \vec{\nabla}\cdot(\rho_B\vec{v}) = 0 \tag{57}$$

2.6. Non-Relativistic Limit

In what follows we recover the non-relativistic limit of the continuity equation for the entropy density, continuity equation for the baryon density and also for the Navier-Stokes equation. The non-relativistic limit is essentially given by $v^2 << 1$ and $\gamma \cong 1$.

2.6.1. Continuity Equation

Using $v^2 << 1$ and $\gamma \cong 1$ in the equation for entropy density (53) we find:

$$\frac{\partial s}{\partial t} + \vec{\nabla} \cdot (s\vec{v}) = -\frac{\eta}{T}(\partial^i u^j)\partial_j u_i + \frac{1}{T}\left(\frac{2}{3}\eta + \zeta\right)(\vec{\nabla} \cdot \vec{v})^2 \tag{58}$$

For a perfect fluid $\eta = \zeta = 0$ we obtain the usual continuity equation [32, 33]:

$$\frac{\partial s}{\partial t} + \vec{\nabla} \cdot (s\vec{v}) = 0$$

For the baryon density continuity equation we have, after applying $v^2 << 1$ and $\gamma \cong 1$ in (57):

$$\frac{\partial \rho_B}{\partial t} + \vec{\nabla} \cdot (\rho_B \vec{v}) = 0 \tag{59}$$

2.6.2. Navier-Stokes Equation

The energy density of the fluid element of mass M and momentum P is:

$$\varepsilon = \frac{E}{Vol} = \frac{\sqrt{M^2 + P^2}}{Vol}$$

In the non-relativistic limit: $M >> P$ we have:

$$\varepsilon \cong \frac{\sqrt{M^2}}{Vol} = \frac{M}{Vol}$$

The volumetric density of fluid matter is given by $\frac{M}{Vol} = \rho$ and hence:

$$\varepsilon \cong \rho \tag{60}$$

The pressure is defined as the ratio of the force per area of the fluid element A. The force is given by the time derivative of the momentum P. The pressure is then:

$$p = \frac{\text{force}}{A} = \frac{1}{A}\frac{dP}{dt} \tag{61}$$

and so

$$\vec{\nabla} p + \vec{v}\frac{\partial p}{\partial t} = \frac{1}{A}\vec{\nabla}\frac{dP}{dt} + \frac{1}{A}\vec{v}\frac{d^2 P}{dt^2}$$

For $v^2 << 1$ we find:

$$|\vec{v}|\frac{d^2 P}{dt^2} << \left|\vec{\nabla}\frac{dP}{dt}\right|$$

and consequently:

$$\vec{\nabla} p + \vec{v}\frac{\partial p}{\partial t} \cong \frac{1}{A}\vec{\nabla}\frac{dP}{dt}$$

With the use of (61) we finally obtain:

$$\vec{\nabla} p + \vec{v}\frac{\partial p}{\partial t} \cong \vec{\nabla} p \tag{62}$$

According to these last approximations and using (60) we conclude that $\varepsilon + p$ takes the form:

$$\varepsilon + p \cong \rho + \frac{1}{A}\frac{dP}{dt}$$

and with $\rho >> \frac{1}{A}\frac{dP}{dt}$ (because $M >> P$) we find:

$$\varepsilon + p \cong \rho \tag{63}$$

Inserting $\gamma \cong 1$, (62) and (63) in (33) we find:

$$\begin{aligned}
&\rho\left[\frac{\partial\vec{v}}{\partial t} + (\vec{v}\cdot\vec{\nabla})\vec{v}\right] + \vec{\nabla}p \\
&\eta\Bigg\{-\vec{\nabla}^2\vec{v} - \vec{\nabla}(\vec{\nabla}\cdot\vec{v}) - \left(\frac{\partial\vec{v}}{\partial t}\cdot\vec{\nabla}\right)\vec{v} - (\vec{v}\cdot\vec{\nabla})\frac{\partial\vec{v}}{\partial t} - \left(\vec{\nabla}\cdot\frac{\partial\vec{v}}{\partial t}\right)\vec{v} \\
&-\vec{\nabla}\cdot\vec{v}\,\frac{\partial\vec{v}}{\partial t} - \vec{v}\cdot\vec{\nabla}\left[(\vec{v}\cdot\vec{\nabla})\vec{v}\right]\Bigg\} \\
&+\left(\zeta - \frac{2}{3}\eta\right)\Bigg\{-\vec{\nabla}(\vec{\nabla}\cdot\vec{v}) - \frac{\partial\vec{v}}{\partial t}\vec{\nabla}\cdot\vec{v} - \vec{v}\,\vec{\nabla}\cdot\frac{\partial\vec{v}}{\partial t} - \left[(\vec{v}\cdot\vec{\nabla})\vec{v}\right]\vec{\nabla}\cdot\vec{v}\Bigg\} = 0
\end{aligned} \tag{64}$$

Neglecting the terms of order v^3, which are $\vec{v}\cdot\vec{\nabla}\left[(\vec{v}\cdot\vec{\nabla})\vec{v}\right]$ and $\left[(\vec{v}\cdot\vec{\nabla})\vec{v}\right]\vec{\nabla}\cdot\vec{v}$ we have:

$$\begin{aligned}
&\rho\left[\frac{\partial\vec{v}}{\partial t} + (\vec{v}\cdot\vec{\nabla})\vec{v}\right] + \vec{\nabla}p - \eta\vec{\nabla}^2\vec{v} - \left(\zeta + \frac{1}{3}\eta\right)\vec{\nabla}(\vec{\nabla}\cdot\vec{v}) \\
&-\left(\zeta + \frac{1}{3}\eta\right)\frac{\partial}{\partial t}\left[\vec{v}(\vec{\nabla}\cdot\vec{v})\right] - \eta\frac{\partial}{\partial t}\left[(\vec{v}\cdot\vec{\nabla})\vec{v}\right] = 0
\end{aligned} \tag{65}$$

We still need to estimate the relative size of the terms in last equation. This is made by the comparison of terms without viscosity to others with viscosity. From (65):

$$\rho(\vec{v}\cdot\vec{\nabla})\vec{v} - \eta\frac{\partial}{\partial t}\left[(\vec{v}\cdot\vec{\nabla})\vec{v}\right] = \frac{M}{Vol}(\vec{v}\cdot\vec{\nabla})\vec{v} - \frac{\eta}{M^2}\frac{\partial}{\partial t}\left[(\vec{P}\cdot\vec{\nabla})\vec{P}\right]$$

where we have used the volumetric density of the fluid matter $\frac{M}{Vol} = \rho$ and the non-relativistic momentum $\vec{P} = M\vec{v}$ for the element with mass M for the fluid matter. In the non-relativistic limit we have, as usual: $M >> P$ and the last expression becomes:

$$\rho(\vec{v}\cdot\vec{\nabla})\vec{v} - \eta\frac{\partial}{\partial t}\left[(\vec{v}\cdot\vec{\nabla})\vec{v}\right] \cong \rho(\vec{v}\cdot\vec{\nabla})\vec{v} \tag{66}$$

since

$$\frac{M}{Vol}(\vec{v}\cdot\vec{\nabla})\vec{v} >> -\frac{\eta}{M^2}\frac{\partial}{\partial t}\left[(\vec{P}\cdot\vec{\nabla})\vec{P}\right]$$

Analogously we also have from (65):

$$\rho\frac{\partial\vec{v}}{\partial t} - \left(\zeta + \frac{1}{3}\eta\right)\frac{\partial}{\partial t}\left[\vec{v}(\vec{\nabla}\cdot\vec{v})\right] \cong \rho\frac{\partial\vec{v}}{\partial t} \tag{67}$$

since

$$\frac{M}{Vol}\frac{\partial \vec{v}}{\partial t} >> -\left(\zeta + \frac{1}{3}\eta\right)\frac{1}{M^2}\frac{\partial}{\partial t}\left[\vec{P}(\vec{\nabla}\cdot\vec{P})\right]$$

The non-relativistic Navier-Stokes equation is then given by using (66) and (67) in (65):

$$\frac{\partial \vec{v}}{\partial t} + (\vec{v}\cdot\vec{\nabla})\vec{v} = -\frac{1}{\rho}\vec{\nabla}p + \frac{\eta}{\rho}\vec{\nabla}^2\vec{v} + \frac{1}{\rho}\left(\zeta + \frac{1}{3}\eta\right)\vec{\nabla}(\vec{\nabla}\cdot\vec{v}) \tag{68}$$

Again, the perfect fluid is described by $\eta = \zeta = 0$ in (68) and gives the non-relativistic version of Euler equation [32, 33]:

$$\frac{\partial \vec{v}}{\partial t} + (\vec{v}\cdot\vec{\nabla})\vec{v} = -\frac{1}{\rho}\vec{\nabla}p \tag{69}$$

3. The Reductive Perturbation Method

3.1. Linearization

We start from the equations of ideal relativistic hydrodynamics and, using the linearization approximation, we derive a wave equation for perturbations in the pressure. This equation has traveling wave solutions which represent acoustic waves. In the derivation presented here we follow closely the reference [55]. In the presence of perturbations the energy density and pressure for the relativistic fluid are written as (35):

$$\varepsilon(\vec{r},t) = \varepsilon_0 + \delta\varepsilon(\vec{r},t) \tag{70}$$

and

$$p(\vec{r},t) = p_0 + \delta p(\vec{r},t) \tag{71}$$

respectively. The uniform relativistic fluid is defined by ε_0 and p_0, while $\delta\varepsilon$ and δp correspond to perturbations in this fluid. Energy-momentum conservation implies that for an ideal fluid $\Pi^{\mu\nu} = 0$ and from (5) we have:

$$\partial_\mu T^{\mu\nu} = 0 \tag{72}$$

where $T^{\mu\nu}$ is the energy-momentum tensor given by:

$$T^{\mu\nu} = (\varepsilon + p)u^\mu u^\nu - p g^{\mu\nu} \tag{73}$$

Linearization consists in keeping only first order terms such as $\delta\varepsilon$, δP and $\vec{v}$ and neglect terms proportional to:

$$v^2,\ \ v\delta\varepsilon,\ \ v\delta P,\ \ \vec{v}\cdot\vec{\nabla}v,\ \ (\vec{v}\cdot\vec{\nabla})\vec{v} \tag{74}$$

and also neglect higher powers of these products or other combinations of them. Naturally we have $\gamma \sim 1$. From (72) we have:

$$u^\mu\partial_\nu[(\varepsilon + p)u^\nu] + (\varepsilon + p)u^\nu\partial_\nu u^\mu - \partial_\nu(p g^{\nu\mu}) = 0 \tag{75}$$

The temporal component ($\mu = 0$) of the above equation is given by:

$$\gamma\partial_0[(\varepsilon + p)\gamma] + \gamma\partial_i[(\varepsilon + p)u^i] + (\varepsilon + p)u^0\partial_0\gamma + (\varepsilon + p)u^i\partial_i\gamma - \partial_0 p = 0 \tag{76}$$

which, after using (74) and $\gamma \sim 1$, becomes:

$$\partial_0(\varepsilon+p)+\partial_i[(\varepsilon+p)v^i]-\partial_0 p=0$$

or

$$\frac{\partial\varepsilon}{\partial t}+\vec{\nabla}\cdot[(\varepsilon+p)\vec{v}]=0 \tag{77}$$

For the j-th spatial component ($\mu=j$) in (75) we have:

$$u^j\partial_0[(\varepsilon+p)u^0]+u^j\partial_i[(\varepsilon+p)u^i]+(\varepsilon+p)u^0\partial_0 u^j+(\varepsilon+p)u^i\partial_i u^j-\partial^j p=0$$

which, with the use of (74) and $\gamma \sim 1$, becomes:

$$\frac{\partial}{\partial t}[(\varepsilon+p)\vec{v}]+\vec{\nabla}p=0 \tag{78}$$

Substituting the expansions (70) and (71) in (77) and (78) we find:

$$\frac{\partial}{\partial t}[\varepsilon_0+\delta\varepsilon]+\vec{\nabla}\cdot[(\varepsilon_0+\delta\varepsilon+p_0+\delta p)\vec{v}]=0 \tag{79}$$

and

$$\frac{\partial}{\partial t}[(\varepsilon_0+\delta\varepsilon+p_0+\delta p)\vec{v}]+\vec{\nabla}[p_0+\delta p]=0 \tag{80}$$

Neglecting the terms listed in (74) in (79) and (80) these equations become:

$$\frac{\partial(\delta\varepsilon)}{\partial t}+(\varepsilon_0+p_0)\vec{\nabla}\cdot\vec{v}=0 \tag{81}$$

and

$$(\varepsilon_0+p_0)\frac{\partial\vec{v}}{\partial t}+\vec{\nabla}(\delta p)=0 \tag{82}$$

Equation (81) expresses energy conservation and equation (82) is Newton's second law. Integrating (82) with respect to the time and setting the integration constant to zero we find:

$$\vec{v}=-\frac{1}{(\varepsilon_0+p_0)}\int\vec{\nabla}(\delta p)dt \tag{83}$$

which inserted in (81) yields:

$$\frac{\partial(\delta\varepsilon)}{\partial t}-\int\vec{\nabla}^2(\delta p)dt=0 \tag{84}$$

Performing the time derivative we obtain:

$$\frac{\partial^2(\delta\varepsilon)}{\partial t^2}-\vec{\nabla}^2(\delta p)=0 \tag{85}$$

Assuming that

$$\delta\varepsilon=\frac{\partial\varepsilon}{\partial p}\delta p \tag{86}$$

with $\partial\varepsilon/\partial p$ being a constant, we have (85) rewritten as:

$$\frac{\partial \varepsilon}{\partial p}\frac{\partial^2(\delta p)}{\partial t^2} - \vec{\nabla}^2(\delta p) = 0 \tag{87}$$

The above expression is a wave equation from where we can identify the velocity of propagation as:

$$c_s = \left(\frac{\partial p}{\partial \varepsilon}\right)^{1/2} \tag{88}$$

where c_s is the speed of sound. Equation (87) can then be finally written as:

$$\vec{\nabla}^2(\delta p) - \frac{1}{c_s{}^2}\frac{\partial^2(\delta p)}{\partial t^2} = 0 \tag{89}$$

which describes the propagation of a pressure wave in the fluid. This study ensures that the analysis of perturbations with the linearized relativistic hydrodynamics leads to the standard second order linear wave equations and their traveling wave solutions, such as acoustic waves in the hadronic medium.

3.2. Beyond Linearization

While linearization is justified in many cases, in others, where perturbations are not so small, it should be replaced by another technique to treat perturbations keeping the nonlinearities of the theory. This is where a physical theory, in our case relativistic hydrodynamics, may benefit from developments in applied mathematics. Indeed, since long ago there is a technique which preserves nonlinearities in the derivation of the differential equations which govern the evolution of perturbations. This is the reductive perturbation method (RPM) [5–8].

We start the RPM description by considering the simple linear wave equation:

$$\frac{\partial F}{\partial t} + \alpha\frac{\partial F}{\partial x} = 0 \tag{90}$$

which describes one-dimensional waves in an ideal fluid. In the above equation α is a constant. Let us expand of $F(x,t)$ around the constant value F_0 in terms of the small expansion parameter σ $(0 < \sigma < 1)$:

$$F(x,t) = F_0 + \sigma F_1(x,t) + \sigma^2 F_2(x,t) + \sigma^3 F_3(x,t) + \dots \tag{91}$$

Inserting (91) into (90) we obtain a series:

$$\sum_j \sigma^j \left\{ \frac{\partial F_j}{\partial t} + \alpha\frac{\partial F_j}{\partial x} \right\} = 0 \qquad \text{for} \quad j = 1,2,3,\dots \tag{92}$$

The coefficients of each power of σ must vanish independently, i.e., each bracket must vanish. This condition yields a set of differential equations for the F_j:

$$\frac{\partial F_j}{\partial t} + \alpha\frac{\partial F_j}{\partial x} = 0 \tag{93}$$

Now we consider the simplest nonlinear wave equation, the so called breaking wave equation:

$$\frac{\partial F}{\partial t} + \alpha F \frac{\partial F}{\partial x} = 0 \tag{94}$$

where α is a real coefficient. Performing the expansion (91) the above equation becomes:

$$\sigma \frac{\partial F_1}{\partial t} + \sigma^2 \frac{\partial F_2}{\partial t} + \sigma \alpha F_0 \frac{\partial F_1}{\partial x} + \sigma^2 \alpha F_0 \frac{\partial F_2}{\partial x} + \sigma^2 \alpha F_1 \frac{\partial F_1}{\partial x} + \cdots = 0 \tag{95}$$

Let us look for simple nonlinear differential equations for the perturbations F_1, F_2, ... We expect to find algebraic structures similar to (94) in equation (95) in order σ and σ^2. To do so, we introduce the new coordinates τ and ξ, which are connected to t and x. This coordinate transformation is such that:

$$\frac{\partial}{\partial t} = \sigma^n \frac{\partial}{\partial \tau} \qquad \text{and} \qquad \frac{\partial}{\partial x} = \sigma^m \frac{\partial}{\partial \xi} \tag{96}$$

where n and m will be determined. We note that a highly nontrivial aspect of this transformation is that it contains the same small parameter σ used in the expansion (95). Inserting (96) into (95) we find the following equation:

$$\sigma^{n+1} \frac{\partial F_1}{\partial \tau} + \sigma^{n+2} \frac{\partial F_2}{\partial \tau} + \sigma^{m+2} \alpha F_1 \frac{\partial F_1}{\partial \xi} + \sigma \alpha F_0 \left[\sigma^m \frac{\partial F_1}{\partial \xi} + \sigma^{m+1} \frac{\partial F_2}{\partial \xi} \right] = 0 \tag{97}$$

where, for the sake of simplicity, we have chosen $F_0 = 0$ and we have truncated the sum keeping only the lowest order terms in σ. We observe that if we choose:

$$n + 1 = m + 2 \qquad \text{and} \qquad n + 2 \geq 3 \tag{98}$$

the first and third terms can be grouped together and, for the particular choice $n = 3/2$ and $m = 1/2$, equation (97) becomes:

$$\sigma^2 \left[\frac{\partial F_1}{\partial \tau} + \alpha F_1 \frac{\partial F_1}{\partial \xi} \right] + \sigma^3 \left[\frac{\partial F_2}{\partial \tau} \right] = 0 \tag{99}$$

where, just before arriving at the above expression, we have divided the whole equation by $\sigma^{1/2}$. From the term proportional to σ^2 we have:

$$\frac{\partial F_1}{\partial \tau} + \alpha F_1 \frac{\partial F_1}{\partial \xi} = 0 \tag{100}$$

as expected. Including all the higher order terms, F_2, F_3, ... etc, we would obtain a set of differential equations. Solving them we would be able to write the complete series (91). In practice, since σ is small, this series is dominated by the first terms and it is often sufficient to compute only F_1.

From this simple exercise we conclude that for a nonlinear wave equation such as (94) it is necessary to introduce the "stretching" operators (96):

$$\frac{\partial}{\partial t} = \sigma^{3/2} \frac{\partial}{\partial \tau} \qquad \text{and also} \qquad \frac{\partial}{\partial x} = \sigma^{1/2} \frac{\partial}{\partial \xi} \tag{101}$$

which are related to the "stretched" coordinates:

$$\xi = \sigma^{1/2} x \qquad \text{and} \qquad \tau = \sigma^{3/2} t \tag{102}$$

If there was a dissipative term in (94), the wave equation would have the following form:

$$\frac{\partial F}{\partial t} + \alpha F \frac{\partial F}{\partial x} = \nu \frac{\partial^2 F}{\partial x^2} \tag{103}$$

which is called Burgers equation with a dissipative (viscous) coefficient ν. It becomes, after performing the expansion (91) and keeping only terms up to order σ^2:

$$\sigma \frac{\partial F_1}{\partial t} + \sigma^2 \frac{\partial F_2}{\partial t} + \sigma \alpha F_0 \frac{\partial F_1}{\partial x} + \sigma^2 \alpha F_0 \frac{\partial F_2}{\partial x} + \sigma^2 \alpha F_1 \frac{\partial F_1}{\partial x} = \nu \sigma \frac{\partial^2 F_1}{\partial x^2} + \nu \sigma^2 \frac{\partial^2 F_2}{\partial x^2} \tag{104}$$

As before, we try to find in equation (104) a structure similar to (103). Keeping the previous choice $n = 3/2$, $m = 1/2$ and applying (101) to (104) we find:

$$\sigma^{5/2} \left[\frac{\partial F_1}{\partial \tau} + \alpha F_1 \frac{\partial F_1}{\partial \xi} \right] + \sigma^{7/2} \left[\frac{\partial F_2}{\partial \tau} \right] = \nu \sigma^2 \frac{\partial^2 F_1}{\partial \xi^2} + \nu \sigma^3 \frac{\partial^2 F_2}{\partial \xi^2} \tag{105}$$

When dissipative effects are included we perform the following transformation in the dissipation coefficient [56, 57]:

$$\nu = \sigma^{1/2} \tilde{\nu} \tag{106}$$

in (105), resulting in:

$$\sigma^{5/2} \left[\frac{\partial F_1}{\partial \tau} + \alpha F_1 \frac{\partial F_1}{\partial \xi} \right] + \sigma^{7/2} \left[\frac{\partial F_2}{\partial \tau} \right] = \tilde{\nu} \sigma^{5/2} \frac{\partial^2 F_1}{\partial \xi^2} + \tilde{\nu} \sigma^{7/2} \frac{\partial^2 F_2}{\partial \xi^2} \tag{107}$$

As before, we divide the last equation by $\sigma^{1/2}$, obtaining:

$$\sigma^2 \left[\frac{\partial F_1}{\partial \tau} + \alpha F_1 \frac{\partial F_1}{\partial \xi} \right] + \sigma^3 \left[\frac{\partial F_2}{\partial \tau} \right] = \tilde{\nu} \sigma^2 \frac{\partial^2 F_1}{\partial \xi^2} + \tilde{\nu} \sigma^3 \frac{\partial^2 F_2}{\partial \xi^2} \tag{108}$$

The terms proportional to σ^3 in (108) are neglected and from the σ^2 terms we have:

$$\frac{\partial F_1}{\partial \tau} + \alpha F_1 \frac{\partial F_1}{\partial \xi} = \tilde{\nu} \frac{\partial^2 F_1}{\partial \xi^2} \tag{109}$$

which is a Burgers equation like (103). In addition, we conclude that when a dissipative term is present in a nonlinear wave equation, apart from (101) and (102), we also need the transformation (106) for the dissipation coefficient.

If a more complete wave equation is considered, i.e., when nonlinear, dissipative and dispersive terms are present, we have the Korteweg-de Vries Burgers (KdV-B) equation:

$$\frac{\partial F}{\partial t} + \alpha F \frac{\partial F}{\partial x} + \beta \frac{\partial^3 F}{\partial x^3} = \nu \frac{\partial^2 F}{\partial x^2} \tag{110}$$

Now, β is the dispersive coefficient. Performing the same calculation we naturally find:

$$\sigma^2\left[\frac{\partial F_1}{\partial\tau}+\alpha F_1\frac{\partial F_1}{\partial\xi}+\beta\frac{\partial^3 F_1}{\partial\xi^3}\right]=\sigma^2\left[\tilde{\nu}\frac{\partial^2 F_1}{\partial\xi^2}\right] \tag{111}$$

which gives the KdV-B:

$$\frac{\partial F_1}{\partial\tau}+\alpha F_1\frac{\partial F_1}{\partial\xi}+\beta\frac{\partial^3 F_1}{\partial\xi^3}=\tilde{\nu}\frac{\partial^2 F_1}{\partial\xi^2} \tag{112}$$

When more dimensions and different coordinate systems are considered, the procedure can be systematically improved.

3.3. Some Special Cases

In the framework of the RPM we transport the equations of hydrodynamics from the space of cartesian, spherical or cylindrical coordinates to the space of the "stretched coordinates". Some well known equations and the coordinate transformations and expansions required to obtain them are given below.

3.3.1. One Dimensional KdV Equation

We write this equation in cartesian coordinates (x), in radial cylindrical coordinate and in radial spherical coordinate (r). The corresponding "stretched coordinates" are given by ξ for space and τ for time as: [9, 24–29, 58]:

$$\xi=\frac{\sigma^{1/2}}{L}(X-c_s t) \tag{113}$$

and

$$\tau=\frac{\sigma^{3/2}}{L}c_s t \tag{114}$$

where $X=x$ or $X=r$. The above transformation of coordinates must be made simultaneously with the following expansions of the baryon density, energy density and fluid velocity around their equilibrium values:

$$\hat{\rho}=\frac{\rho_B}{\rho_0}=1+\sigma\rho_1+\sigma^2\rho_2+\dots \tag{115}$$

$$\hat{\varepsilon}=\frac{\varepsilon}{\varepsilon_0}=1+\sigma\varepsilon_1+\sigma^2\varepsilon_2+\dots \tag{116}$$

$$\hat{v}=\frac{v}{c_s}=\sigma v_1+\sigma^2 v_2+\dots \tag{117}$$

For two and three dimensional nonlinear wave equations we will use the "stretched coordinates" presented in [31, 54] and references therein.

3.3.2. Two Dimensional Cylindrical Breaking Wave Equation

In this case the coordinate transformation is given by:

$$R = \frac{\sigma^{1/2}}{L}(r - c_s t) \ , \quad Z = \frac{\sigma}{L} z \ , \quad T = \frac{\sigma^{3/2}}{L} c_s t \tag{118}$$

The energy density and fluid velocity components are expanded around their equilibrium values:

$$\hat{\varepsilon} = \frac{\varepsilon}{\varepsilon_0} = 1 + \sigma\varepsilon_1 + \sigma^2\varepsilon_2 + \sigma^3\varepsilon_3 + \dots \tag{119}$$

$$\hat{v_r} = \frac{v_r}{c_s} = \sigma v_{r_1} + \sigma^2 v_{r_2} + \sigma^3 v_{r_3} + \dots \tag{120}$$

$$\hat{v_z} = \frac{v_z}{c_s} = \sigma^{3/2} v_{z_1} + \sigma^{5/2} v_{z_2} + \sigma^{7/2} v_{z_3} + \dots \tag{121}$$

3.3.3. Three Dimensional Cylindrical KP Equation

In this case we have:

$$R = \frac{\sigma^{1/2}}{L}(r - c_s t) \ , \quad \Phi = \sigma^{-1/2}\varphi \ , \quad Z = \frac{\sigma}{L} z \ , \quad T = \frac{\sigma^{3/2}}{L} c_s t \tag{122}$$

and for the baryonic density and fluid velocity components expansions are:

$$\hat{\rho} = \frac{\rho_B}{\rho_0} = 1 + \sigma\rho_1 + \sigma^2\rho_2 + \sigma^3\rho_3 + \dots \tag{123}$$

$$\hat{v_r} = \frac{v_r}{c_s} = \sigma v_{r_1} + \sigma^2 v_{r_2} + \sigma^3 v_{r_3} + \dots \tag{124}$$

$$\hat{v_\varphi} = \frac{v_\varphi}{c_s} = \sigma^{3/2} v_{\varphi_1} + \sigma^{5/2} v_{\varphi_2} + \sigma^{7/2} v_{\varphi_3} + \dots \tag{125}$$

$$\hat{v_z} = \frac{v_z}{c_s} = \sigma^{3/2} v_{z_1} + \sigma^{5/2} v_{z_2} + \sigma^{7/2} v_{z_3} + \dots \tag{126}$$

$$\hat{\rho}^{\,4/3} = \left[1 + (\sigma\rho_1 + \sigma^2\rho_2 + \dots)\right]^{4/3} \cong 1 + \frac{4}{3}\sigma\rho_1 + \frac{4}{3}\sigma^2\rho_2 + \dots \tag{127}$$

$$\hat{\rho}^{\,1/3} = \left[1 + (\sigma\rho_1 + \sigma^2\rho_2 + \dots)\right]^{1/3} \cong 1 + \frac{1}{3}\sigma\rho_1 + \frac{1}{3}\sigma^2\rho_2 + \dots \tag{128}$$

3.3.4. Three Dimensional Cartesian KP Equation

Here the stretched coordinates are:

$$X = \frac{\sigma^{1/2}}{L}(x - c_s t) \ , \quad Y = \frac{\sigma}{L} y \ , \quad Z = \frac{\sigma}{L} z \ , \quad T = \frac{\sigma^{3/2}}{L} c_s t \tag{129}$$

and the expansions are:

$$\hat{\rho} = \frac{\rho_B}{\rho_0} = 1 + \sigma\rho_1 + \sigma^2\rho_2 + \sigma^3\rho_3 + \dots \tag{130}$$

$$\hat{v_x} = \frac{v_x}{c_s} = \sigma v_{x_1} + \sigma^2 v_{x_2} + \sigma^3 v_{x_3} + \dots \tag{131}$$

$$\hat{v_y} = \frac{v_y}{c_s} = \sigma^{3/2} v_{y_1} + \sigma^2 v_{y_2} + \sigma^{5/2} v_{y_3} + \dots \tag{132}$$

$$\hat{v_z} = \frac{v_z}{c_s} = \sigma^{3/2} v_{z_1} + \sigma^2 v_{z_2} + \sigma^{5/2} v_{z_3} + \dots \tag{133}$$

$$\hat{\rho}^{4/3} \cong 1 + \frac{4}{3}\sigma\rho_1 + \frac{4}{3}\sigma^2\rho_2 + \dots \tag{134}$$

$$\hat{\rho}^{1/3} \cong 1 + \frac{1}{3}\sigma\rho_1 + \frac{1}{3}\sigma^2\rho_2 + \dots \tag{135}$$

In all cases L is a characteristic length scale of the problem, ε_0 is the equilibrium (or reference) energy density, ρ_0 is the equilibrium (or reference) baryon density and c_s is the speed of sound. When viscosity is included, we perform the transformation (106) as in [56,57]:

$$\zeta = \sigma^{1/2}\tilde{\zeta} \qquad \text{and also} \qquad \eta = \sigma^{1/2}\tilde{\eta} \tag{136}$$

Once the equations are in the "stretched spaces" and expanded, we neglect terms proportional to σ^n for $n > 2$ and organize the equations as series in powers of σ, $\sigma^{3/2}$ and σ^2. These equations form a system of differential equations which are combined to yield the final nonlinear equation for the relevant perturbation. Finally we transform the nonlinear equation back to the cartesian (cylindrical or spherical) space and solve it.

4. The Equation of State

The equations of hydrodynamics discussed in the previous sections must be supplemented with an equation of state (EOS), i.e., a relation between pressure (p) and energy density (ε) or matter density (ρ). In relativistic hydrodynamics, we can write the EOS as:

$$p = {c_s}^2 \varepsilon \tag{137}$$

where the speed of sound c_s must be smaller than one. The equation of state is derived from microscopic theories of the strongly interacting system. As mentioned above, the fundamental theory of strong interactions (QCD) predicts that cold and/or dilute systems are in the hadronic phase, where the degrees of freedom are baryons (proton, neutron, Δ, ...) and mesons (π, ρ, ...). At higher densities and/or temperatures, there is a phase transition and the formation of the quark gluon plasma (QGP), a phase where these particles are free and all hadrons are dissolved. In all phases we may have thermal equilibrium and hydrodynamical behavior. In both hadron and quark-gluon fluids we may have excitations and the nonlinear propagation of perturbations. In what follows we shall study the formation of these nonlinear waves and the role played by the equation of state.

In the pioneering study of Refs. [9, 29] the authors studied the equations of non-relativistic hydrodynamics and, with the help of the RPM, they were able to derive a KdV (Korteweg de-Vries) equation for a perturbation in the nuclear density. A key ingredient in that work was the equation of state, which established the following relation between pressure (p) and density (ρ):

$$\frac{\vec{\nabla} p}{\rho} = \vec{\nabla}\phi \tag{138}$$

where ϕ is a potential, playing the role of heat function, given by:

$$\phi = \frac{1}{\rho_0}\left[c_1\rho' + c_2\vec{\nabla}^2\rho' + ...\right] \tag{139}$$

where c_1 and c_2 are constants and ρ' is the deviation of the density from its equilibrium value $\rho' = \rho - \rho_0$. The Laplacian in (139), combined with the gradient in (138) gives origin to the cubic derivative of the KdV equation. As a result the authors arrived at the conclusion that a KdV soliton may exist in cold nuclear matter, when, for example, a light nucleus is impinged on a heavy nucleus. This conclusion relied strongly on (139) and (138), which come from an oversimplified description of the nuclear interactions. In [24–27] the nuclear interactions were described in terms of a relativistic mean field model, which is a variant of the Walecka model. Within this framework we may try to answer the question: what is the microscopic origin of higher order derivative terms appearing in the equation of state? This will be discussed in the following sections.

4.1. Hadronic Matter

In this subsection we present an equation of state derived from a successful relativistic mean field model: Quantum Hadrodynamics (QHD) or Walecka Model. For a modern approach, using chiral power counting in an effective field theory for nuclear matter with nucleons and pions as degrees of freedom, see Ref. [59].

This model is well established and is the subject of textbooks as Refs. [60, 61]. The Lagrangian density of nonlinear QHD is given by:

$$\mathcal{L} = \bar{\psi}[\gamma_\mu(i\partial^\mu - g_V V^\mu) - (M - g_S\phi)]\psi + \frac{1}{2}\left(\partial_\mu\phi\partial^\mu\phi - {m_S}^2\phi^2\right) - \frac{1}{4}F_{\mu\nu}F^{\mu\nu} +$$

$$+\frac{1}{2}{m_V}^2 V_\mu V^\mu - \frac{b}{3}\phi^3 - \frac{c}{4}\phi^4 + \mathcal{L}_d \tag{140}$$

where $F_{\mu\nu} = \partial_\mu V_\nu - \partial_\nu V_\mu$ and:

$$\mathcal{L}_d = d\frac{g_V}{{m_V}^2}\bar{\psi}(\partial_\nu\partial^\nu V_\mu)\gamma^\mu\psi \tag{141}$$

Except for the last term, this is the standard nonlinear QHD Lagrangian which is able to reproduce all the main features of nuclear matter and finite nuclei. This last term was added only in [24–27] and illustrates how to include higher order derivative terms in the equation of state.

The Lagrangian (141) is the modern version of (139) and (138). In (140) the degrees of freedom are the baryon field ψ, the neutral scalar meson field ϕ and the neutral vector meson field V_μ, with the respective couplings and masses. The last and new term (141) is designed to be small in comparison with the main baryon-vector meson interaction term $g_v\bar{\psi}\gamma_\mu V^\mu\psi$. Because of the derivatives, it is of the order of:

$$\frac{p^2}{m_V^2} \sim \frac{k_F^2}{m_V^2} \sim 0.12 \tag{142}$$

where the Fermi momentum is $k_F \simeq 0.28$ GeV and $m_V \simeq 0.8$ GeV. The form chosen for the new interaction term is not dictated by any symmetry argument, has no other deep justification and is just one possible interaction term among many others. It is used here as a prototype to study the effects of higher derivative terms, which, as it will be seen, may generate more complex wave equations, such as KdV. The parameter d is free and plays the role of a "marker". Setting d equal to zero switches off the new term and we recover the usual QHD. On the other hand $d = 1$ means that the coupling g_V is the standard one. Other values imply a correction in this coupling. As mentioned in the beginning of this section, interesting phenomena, such as KdV solitons, may appear as a consequence of the use of equations of state with higher order derivative terms.

Baryon number propagation in nuclear matter satisfies the diffusion equation:

$$\frac{\partial \rho_B}{\partial t} = D\nabla^2 \rho_B \tag{143}$$

where the diffusion constant D has been numerically calculated and studied as a function of density and temperature. For example, in [62] it was found that $D \simeq 0.35\, fm$ at densities comparable to the equilibrium nuclear density and temperatures of the order of 80 MeV. This number is small compared to any nuclear size scale and can be interpreted as indicating that

$$\frac{\partial \rho_B}{\partial t} << \nabla^2 \rho_B \tag{144}$$

and therefore the density gradients do not disappear very rapidly in nuclear matter. Because of the above inequality we can neglect the time derivatives in (141).

The usual mean field theory (MFT) approximation is based on two assumptions: a) the baryonic sources are intense and their coupling to the meson fields is strong and b) infinite nuclear matter is static, homogeneous in space and isotropic.

The first assumption above is implemented performing the substitutions:

$$V_\mu \to < V_\mu > \equiv \delta_{\mu 0} V_0 \tag{145}$$

and

$$\phi \to < \phi > \equiv \phi_0 \tag{146}$$

in (140) and obtaining $\mathcal{L}^* = \mathcal{L}\Big(V_\mu \to < V_\mu > \equiv \delta_{\mu 0} V_0 \quad \phi \to < \phi > \equiv \phi_0\Big)$:

$$\mathcal{L}^* = \bar{\psi}[(i\gamma_\mu \partial^\mu - g_V \gamma_0 V_0) - (M - g_S \phi_0)]\psi + \frac{1}{2}\left(\partial_\mu \phi_0 \partial^\mu \phi_0 - {m_S}^2 {\phi_0}^2\right) +$$

$$+\frac{1}{2}(\vec{\nabla} V_0)^2 + \frac{1}{2}{m_V}^2 {V_0}^2 - \frac{b}{3}{\phi_0}^3 - \frac{c}{4}{\phi_0}^4 \tag{147}$$

The equations of motion are given by [27, 63–65]:

$$\frac{\partial \mathcal{L}^*}{\partial \eta_i} - \partial_\mu \frac{\partial \mathcal{L}^*}{\partial(\partial_\mu \eta_i)} + \partial_\nu \partial_\mu \left[\frac{\partial \mathcal{L}^*}{\partial(\partial_\mu \partial_\nu \eta_i)}\right] = 0 \tag{148}$$

where $\eta_i = \Psi, \phi_0, V_0$ and read:

$$-\vec{\nabla}^2 V_0 + {m_V}^2 V_0 = g_V \bar{\psi}\gamma^0 \psi \tag{149}$$

$$(\partial_\mu \partial^\mu + m_S{}^2)\phi_0 = g_S \bar{\psi}\psi - b\phi_0{}^2 - c\phi_0{}^3 \tag{150}$$

$$\left[i\gamma_\mu \partial^\mu - g_V \gamma_0 V_0 - (M - g_S \phi_0) \right] \Psi = 0 \tag{151}$$

The effective mass of the nucleon is given by $M^* = M - g_S\,\phi_0$. In the case of pure QHD (147), because of the interaction between the nucleon and the vector meson, we can anticipate that the second order derivative term in the field V^μ, which comes from its equation of motion, may be transferred to the term $\bar{\psi}\gamma_\mu\psi$ and hence to the baryon density ρ_B. In the mean field approximation, only the time component of the field, V^0, contributes, yielding the term $\vec{\nabla}^2 V_0$. It is possible to estimate $\vec{\nabla}^2 V_0$ from the equation of motion (149). To do so, we first rewrite the equation (149) as function of the baryon density:

$$-\vec{\nabla}^2 V_0 + m_V{}^2 V_0 = g_V \rho_B \tag{152}$$

Next we assume that $\vec{\nabla}^2 V_0 / m_V^2 << V_0$, neglecting the Laplacian in (152) to find the first order estimate of V_0:

$$V_0 = \frac{g_V}{m_V{}^2}\rho_B \tag{153}$$

An improvement of this estimate of V_0 is obtained taking the Laplacian of (153) and substituting it in the Laplacian present in (152). After this simple algebraic procedure, we solve the resulting equation for V_0 and obtain:

$$V_0 = \frac{g_V}{m_V{}^2}\rho_B + \frac{g_V}{m_V{}^4}\vec{\nabla}^2\rho_B \tag{154}$$

The energy-momentum tensor is given by [27,63–65]:

$$T^{\mu\nu} = \frac{\partial \mathcal{L}^*}{\partial(\partial_\mu \eta_i)}(\partial^\nu \eta_i) - g^{\mu\nu}\mathcal{L}^* - \left[\partial_\beta \frac{\partial \mathcal{L}^*}{\partial(\partial_\mu \partial_\beta \eta_i)} \right](\partial^\nu \eta_i) + \frac{\partial \mathcal{L}^*}{\partial(\partial_\mu \partial_\beta \eta_i)}(\partial_\beta \partial^\nu \eta_i) \tag{155}$$

The energy density is:

$$\varepsilon = < T_{00} > \tag{156}$$

which becomes [27] :

$$\varepsilon = \frac{1}{2}(\partial_0 \phi_0)^2 + \frac{1}{2}(\vec{\nabla}\phi_0)^2 + \frac{1}{2} g_V \rho_B V_0 + \frac{1}{2} m_S{}^2 \phi_0{}^2 + \frac{b}{3}\phi_0{}^3 + \frac{c}{4}\phi_0{}^4$$
$$+ \frac{\gamma_s}{(2\pi)^3}\int_0^{k_F} d^3k (\vec{k}^2 + M^{*2})^{1/2} \tag{157}$$

Inserting (154) in the expression of the energy density (157) and also writing ϕ_0 in terms of M^* we find the following expression:

$$\varepsilon = \frac{g_V{}^2}{2m_V{}^2}\rho_B{}^2 + \frac{g_V{}^2}{2m_V{}^4}\rho_B \vec{\nabla}^2 \rho_B + \frac{m_S{}^2}{2g_S{}^2}(M - M^*)^2$$
$$+ b\frac{(M - M^*)^3}{3g_S{}^3} + c\frac{(M - M^*)^4}{4g_S{}^4} + \frac{\gamma_s}{(2\pi)^3}\int_0^{k_F} d^3k (\vec{k}^2 + M^{*2})^{1/2} \tag{158}$$

The variables ρ_B, $\vec{\nabla}^2\rho_B$, M^*, $\vec{\nabla}M^*$ and $\partial M^*/\partial t$ are independent from each other and therefore any derivative of one with respect to another will vanish. Taking the derivative of the above expression with respect to ρ_B we have:

$$\frac{\partial \varepsilon}{\partial \rho_B} = \frac{{g_V}^2}{{m_V}^2}\rho_B + \frac{{g_V}^2}{2{m_V}^4}\vec{\nabla}^2\rho_B \tag{159}$$

We will add the two sources of inhomogeneities in ρ_B (which are responsible for a non-vanishing $\vec{\nabla}^2\rho_B$). For cold nuclear matter we have then [27]:

$$\varepsilon = \frac{1}{2}\left\{\frac{\partial}{\partial t}\left[\frac{(M-M^*)}{g_S}\right]\right\}^2 + \frac{1}{2}\left\{\vec{\nabla}\left[\frac{(M-M^*)}{g_S}\right]\right\}^2 + \frac{{m_S}^2}{2{g_S}^2}(M-M^*)^2 +$$
$$b\frac{(M-M^*)^3}{3{g_S}^3} + c\frac{(M-M^*)^4}{4{g_S}^4} + \frac{{g_V}^2}{2{m_V}^2}{\rho_B}^2 + \left(d+\frac{1}{2}\right)\frac{{g_V}^2}{{m_V}^4}\rho_B\vec{\nabla}^2\rho_B +$$
$$+\frac{\gamma_s}{(2\pi)^3}\int_0^{k_F} d^3k(\vec{k}^2 + M^{*2})^{1/2} \tag{160}$$

where $\gamma_s = 4$ is the nucleon degeneracy factor. For hot nuclear matter, the energy density as described in [27] is given by:

$$\varepsilon = \frac{1}{2}\left\{\frac{\partial}{\partial t}\left[\frac{(M-M^*)}{g_S}\right]\right\}^2 + \frac{1}{2}\left\{\vec{\nabla}\left[\frac{(M-M^*)}{g_S}\right]\right\}^2 + \frac{{m_S}^2}{2{g_S}^2}(M-M^*)^2 +$$
$$b\frac{(M-M^*)^3}{3{g_S}^3} + c\frac{(M-M^*)^4}{4{g_S}^4} + \frac{{g_V}^2}{2{m_V}^2}{\rho_B}^2 + \left(d+\frac{1}{2}\right)\frac{{g_V}^2}{{m_V}^4}\rho_B\vec{\nabla}^2\rho_B +$$
$$+\frac{\gamma_s}{(2\pi)^3}\int d^3k\ h_+\ [n_{\vec{k}}(T,\nu) + \bar{n}_{\vec{k}}(T,\nu)] \tag{161}$$

The baryon density is:

$$\rho_B = \frac{\gamma_s}{(2\pi)^3}\int d^3k\ [n_{\vec{k}}(T,\nu) - \bar{n}_{\vec{k}}(T,\nu)] \tag{162}$$

with

$$n_{\vec{k}}(T,\nu) \equiv \frac{1}{1+e^{(h_+-\nu)/T}} \tag{163}$$

$$\bar{n}_{\vec{k}}(T,\nu) \equiv \frac{1}{1+e^{(h_++\nu)/T}} \tag{164}$$

$$\nu \equiv \mu_B - g_V V_0 + d\frac{g_V}{{m_V}^2}(\partial^\mu\partial_\nu V_0) \tag{165}$$

and

$$h_+ \equiv (\vec{k}^2 + M^{*2})^{1/2} \tag{166}$$

It is important to note that the term $\mathcal{L}_d$ provides the "$(d+1/2)$-term" in the energy densities (161) and (160) that generates the KdV equation. The nucleon effective mass (M^*) at zero temperature is obtained through the minimization of ε with respect to M^*:

$$\frac{\partial \varepsilon}{\partial M^*} = 0$$

which, with the help of (160) yields:

$$M^* = M - \frac{g_S{}^2}{m_S{}^2}\frac{\gamma_s}{(2\pi)^3}\int_0^{k_F} d^3k \frac{M^*}{(\vec{k}^2+M^{*2})^{1/2}}$$

$$+\frac{g_S{}^2}{m_S{}^2}\left[\frac{b}{g_S{}^3}(M-M^*)^2+\frac{c}{g_S{}^4}(M-M^*)^3\right] \tag{167}$$

Analogously, at finite temperature:

$$M^* = M - \frac{g_S{}^2}{m_S{}^2}\frac{\gamma_s}{(2\pi)^3}\int d^3k \frac{M^*}{h_+}\left[n_{\vec{k}}(T,\nu)+\bar{n}_{\vec{k}}(T,\nu)\right]$$

$$+\frac{g_S{}^2}{m_S{}^2}\left[\frac{b}{g_S{}^3}(M-M^*)^2+\frac{c}{g_S{}^4}(M-M^*)^3\right] \tag{168}$$

and we conclude by these two last expressions that "d-term" does not affect the nucleon effective mass.

4.2. Quark Gluon Plasma

The idea that quarks and gluons may exist as free particles in a deconfined phase was advanced long ago [66] in the context of compact stars. In these objects gravitation is strong enough to compress the nucleons and make them overlap with each other. The distance between the quarks can be so small that, due to asymptotic freedom, they almost do not interact. In this picture, we may have large regions of space populated with free quarks. This state is called the (cold) quark gluon plasma. This naive description of QCD in extreme conditions of density was later revisited with much more sophisticated models. Quarks and gluons may also form a hot quark gluon plasma. The hot QGP has been extensively studied in lattice simulations and also in heavy ion experiments at CERN and LHC. Today we know that a hot and deconfined phase is formed in the existing accelerators, but it is far more complicated than previously imagined. In particular the quarks and gluons are not really free. Instead, they still interact strongly with each other forming a state called the strongly interacting QGP, or sQGP. Also the non-trivial vacuum structure persists until relatively large temperatures.

A starting point to study in a unified way both the cold and hot QGP is the MIT bag model. According to this model in its simplest version, massless and non-interacting quarks live in a spherical cavity (the "bag") in the physical QCD vacuum, which is a medium. The confining property is represented by a constant term, "B", called the bag constant. In the next subsection we shall briefly show how to calculate the QGP equation of state with the help of this model.

4.2.1. The MIT Bag Model

In what follows we consider only quarks u and d. Each quark has three color states and they are massless. We also have eight massless gluons and we neglect the interactions at QGP. The confinement property is included in the model through the introduction of a constant, positive energy per unit volume in the vacuum:

$$\mathcal{B} = \left(\frac{E}{V}\right)_{vac} \tag{169}$$

that can be interpreted as the energy needed to create a bubble or bag in the vacuum, in which the noninteracting quarks and gluons are confined. $\mathcal{B}$ is known as "bag constant".

The parameter $\mathcal{B}$ can be extracted from a phenomenological analysis of hadron spectroscopy or from lattice QCD calculations. There is a relationship between $\mathcal{B}$ and the critical temperature of quark-hadron transition T_c which is determined by considering that during the phase transition the pressure is zero.

The quarks have baryon number $1/3$ and the chemical potential for gluons is zero. The baryon density, energy density and pressure for the QGP are given by:

$$\rho_B = \frac{1}{3}\frac{\gamma_Q}{(2\pi)^3}\int d^3k\ [n_{\vec{k}} - \bar{n}_{\vec{k}}] \tag{170}$$

where

$$n_{\vec{k}} \equiv n_{\vec{k}}(T) = \frac{1}{1+e^{(k-\frac{1}{3}\mu_B)/T}} \tag{171}$$

and

$$\bar{n}_{\vec{k}} \equiv \bar{n}_{\vec{k}}(T) = \frac{1}{1+e^{(k+\frac{1}{3}\mu_B)/T}} \tag{172}$$

where μ_B is the baryon chemical potential. The energy density is given by:

$$\varepsilon = \mathcal{B} + \frac{\gamma_G}{(2\pi)^3}\int d^3k\ k\ (e^{k/T}-1)^{-1} + \frac{\gamma_Q}{(2\pi)^3}\int d^3k\ k\ [n_{\vec{k}} + \bar{n}_{\vec{k}}] \tag{173}$$

where the first term of the expression above is the gluon contribution [67,68]. The pressure is given by:

$$p = -\mathcal{B} + \frac{1}{3}\left\{\frac{\gamma_G}{(2\pi)^3}\int d^3k\ k\ (e^{k/T}-1)^{-1} + \frac{\gamma_Q}{(2\pi)^3}\int d^3k\, k\left[n_{\vec{k}} + \bar{n}_{\vec{k}}\right]\right\} \tag{174}$$

The degeneracy factors are:

$$\gamma_G = 2(\text{polarizations}) \times 8(\text{colors}) = 16 \qquad \text{for gluons} \tag{175}$$

and

$$\gamma_Q = 2(\text{spins}) \times 2(\text{flavours}) \times 3(\text{colors}) = 12 \qquad \text{for quarks} \tag{176}$$

The integral of the gluon distribution function can be calculated analytically [68] and the thermodynamics of QGP can be summarized in the following expressions:

$$\rho_B = \frac{2}{\pi^2}\int_0^\infty dk\ k^2[n_{\vec{k}} - \bar{n}_{\vec{k}}] \tag{177}$$

and

$$3(p+\mathcal{B}) = \varepsilon - \mathcal{B} = \frac{8\pi^2}{15}\,T^4 + \frac{6}{\pi^2}\int_0^\infty dk\ k^3[n_{\vec{k}} + \bar{n}_{\vec{k}}] \tag{178}$$

which provide us with the EOS of QGP with all baryon densities and at all temperatures:

$$p = \frac{1}{3}\varepsilon - \frac{4}{3}\mathcal{B} \tag{179}$$

The sound speed c_s is given by:

$$c_s{}^2 = \frac{\partial p}{\partial \varepsilon} = \frac{1}{3} \tag{180}$$

4.2.2. The Cold QGP

In several systems, such as, for example, the core of a neutron star, the temperature is zero and baryon density is considerable, giving rise to a rich scenario to study baryon density perturbations. The quark distribution function becomes the step function with $\mu_B = 3k_F$ and (177) becomes:

$$\rho_B = \frac{\gamma_s}{6\pi^2} k_F{}^3 \tag{181}$$

Expression (178) becomes:

$$3(p+\mathcal{B}) = \varepsilon - \mathcal{B} = \frac{6}{\pi^2}\left(\frac{k^4}{4}\Big|_0^{k_F}\right) = \frac{3}{2\pi^2} k_F{}^4 \tag{182}$$

Using (181) we find:

$$3(p+\mathcal{B}) = \varepsilon - \mathcal{B} = \left(\frac{3}{2}\right)^{7/3} \pi^{2/3} \rho_B{}^{4/3}$$

and so

$$\varepsilon(\rho_B) = \left(\frac{3}{2}\right)^{7/3} \pi^{2/3} \rho_B{}^{4/3} + \mathcal{B} \tag{183}$$

Inserting (183) into (179) we find:

$$p(\rho_B) = \frac{1}{3}\left(\frac{3}{2}\right)^{7/3} \pi^{2/3} \rho_B{}^{4/3} - \mathcal{B} \tag{184}$$

From (183) and (184) is possible to rewrite the sum of energy density and pressure as:

$$\varepsilon + p = \frac{4}{3}\left(\frac{3}{2}\right)^{7/3} \pi^{2/3} \rho_B{}^{4/3} \tag{185}$$

From (179) we have:

$$\vec{\nabla} p = \frac{1}{3}\vec{\nabla}\varepsilon \qquad \text{and also} \qquad \frac{\partial p}{\partial t} = \frac{1}{3}\frac{\partial \varepsilon}{\partial t} \tag{186}$$

Using (183) in (186) we find:

$$\vec{\nabla} p = \frac{4}{9}\left(\frac{3}{2}\right)^{7/3} \pi^{2/3} \rho_B{}^{1/3}\, \vec{\nabla}\rho_B \tag{187}$$

and

$$\frac{\partial p}{\partial t} = \frac{4}{9}\left(\frac{3}{2}\right)^{7/3} \pi^{2/3} \rho_B{}^{1/3} \frac{\partial \rho_B}{\partial t} \tag{188}$$

4.2.3. The Hot QGP

The hot QGP is formed in heavy ion collisions in the central rapidity region where experimentally we have $\rho_B = 0$. For our purposes the most relevant physical quantity is the energy density ε. Since $\rho_B = 0$, the baryon chemical potential is zero ($\mu_B = 0$) and so the distribution functions given by (171) and (172) are the same:

$$n_{\vec{k}} = \bar{n}_{\vec{k}} = \frac{1}{1+e^{k/T}}$$

and then (178) takes the form:

$$3(p+\mathcal{B}) = \varepsilon - \mathcal{B} = \frac{8\pi^2}{15}\, T^4 + \frac{12}{\pi^2}\int_0^\infty dk \,\frac{k^3}{[1+e^{k/T}]} \tag{189}$$

We can performing the above integral analytically [68] to find:

$$3(p+\mathcal{B}) = \varepsilon - \mathcal{B} = \frac{37}{30}\pi^2 T^4 \tag{190}$$

From basic thermodynamics [67] we know that

$$s = \left(\frac{\partial p}{\partial T}\right)_V \tag{191}$$

which leads to the specific form for s, when using (190):

$$s = \frac{\partial}{\partial T}\left(-\mathcal{B} + \frac{37}{90}\pi^2 T^4\right) = 4\frac{37}{90}\pi^2 T^3 \tag{192}$$

The parameter $\mathcal{B}$, the "bag constant" can be defined at the temperature T_B. In the bag surface, (190) is given by:

$$\mathcal{B} = \frac{37}{90}\pi^2 (T_B)^4 \tag{193}$$

Choosing $\mathcal{B}^{1/4} = 170 MeV$ corresponds to $T_B = 91\, MeV$. Rewriting (190) as:

$$T = \left[\frac{30}{37\pi^2}(\varepsilon - \mathcal{B})\right]^{1/4} \tag{194}$$

and inserting it into (192) we find:

$$s = s(\varepsilon) = 4\frac{37}{90}\pi^2 \left[\frac{30}{37\pi^2}(\varepsilon - \mathcal{B})\right]^{3/4}$$

In a compact notation:

$$s(\varepsilon) = A(\varepsilon - \mathcal{B})^{3/4} \tag{195}$$

where

$$A \equiv 4\frac{37}{90}\pi^2 \left[\frac{30}{37\pi^2}\right]^{3/4} \tag{196}$$

4.3. Mean Field Theory for Quantum Chromodynamics

In spite of its phenomenological success the MIT bag model gives a poor representation of the quark gluon plasma. From heavy ion collisions there is convincing evidence that quarks and gluons interact strongly forming rather a "prefect fluid" than an ideal gas. Therefore, the picture of free partons must be modified. There are several ways to do that. Here we discuss the approach proposed in [30], which can be called mean field QCD (MFQCD). It allows us to start from QCD Lagrangian and derive an equation of state, which incorporates the effects of a sizeable strong coupling constant and also non -perturbative effects from the QCD vacuum. As an interesting by-product this equation of state supports the existence of KdV solitons in a cold QGP. In the following section we summarize the basic ideas of this approach and derive the expressions for the pressure and energy density, which are relevant for the hydrodynamical study.

We first introduce the mean field approximation for QCD, extending previous works along the same line [69, 70]. We consider a system of quarks and gluons which are represented by the QCD Lagrangian density:

$$\mathcal{L}_{QCD} = -\frac{1}{4}F^a_{\mu\nu}F^{a\mu\nu} + \sum_{q=1}^{N_f} \bar{\psi}^q_i \left[i\gamma^\mu(\delta_{ij}\partial_\mu - igT^a_{ij}G^a_\mu) - \delta_{ij}m_q \right] \psi^q_j \tag{197}$$

with

$$F^{a\mu\nu} = \partial^\mu G^{a\nu} - \partial^\nu G^{a\mu} + gf^{abc}G^{b\mu}G^{c\nu} \tag{198}$$

where ψ^q_i and G^a_μ represent the quark and gluon fields respectively. The summation on q runs over all quark flavors, m_q is the mass of the quark of flavor q, i and j are the color indices of the quarks, T^a are the SU(3) generators and f^{abc} are the SU(3) antisymmetric structure constants. For simplicity we consider only light quarks with the same mass m. Moreover, we drop the summation and consider only one flavor. At the end of our calculation the number of flavors will be recovered. Following [69, 70], we shall start writing the gluon field as:

$$G^{a\mu} = A^{a\mu} + \alpha^{a\mu} \tag{199}$$

where $A^{a\mu}$ and $\alpha^{a\mu}$ are the low ("soft") and high ("hard") momentum components of the gluon field respectively. We will assume that $A^{a\mu}$ represents the soft modes which populate the vacuum and $\alpha^{a\mu}$ represents the modes for which the running coupling constant is small. In a cold quark gluon plasma the density is much larger than the ordinary nuclear matter density. These high densities imply a very large number of sources of the gluon field. Assu-ming that the coupling constant is not very small, the existence of intense sources implies that the bosonic fields tend to have large occupation numbers at all energy levels, and therefore they can be treated as classical fields. This is the famous approximation for bosonic fields used in relativistic mean field models of nuclear matter [60]. It has been applied to QCD in the past and amounts to assume that the "hard" gluon field, α^a_μ, is simply a function of the coordinates [60]:

$$\alpha^a_\mu(\vec{x},t) = \delta_{\mu 0}\,\alpha^a_0(\vec{x},t) \tag{200}$$

with $\partial_\nu \alpha^a_\mu \neq 0$. This space and time dependence goes beyond the standard mean field approximation, where α^a_μ is constant in space and time [60] and consequently $\partial_\nu \alpha^a_\mu = 0$.

We keep assuming, as in [30], that the soft gluon field $A^{a\mu}$ is independent of position and time and thus $\partial^{\nu}A^{a\mu} = 0$. Following the same steps of [30] we obtain the following effective Lagrangian:

$$\mathcal{L} = -\frac{1}{2}\alpha_0^a(\vec{\nabla}^2\alpha_0^a) + \frac{m_G{}^2}{2}\alpha_0^a\alpha_0^a - \mathcal{B}_{QCD} + \bar{\psi}_i\Big(i\delta_{ij}\gamma^{\mu}\partial_{\mu} + g\gamma^0 T^a_{ij}\alpha_0^a - \delta_{ij}m\Big)\psi_j \tag{201}$$

where the constant $\mathcal{B}_{QCD}$ is the same bag constant for QCD as defined in [30]. In fact, the effective Lagrangian (201) is quite similar to the effective Lagrangian obtained in [30]. The new feature of (201) is the term $-\frac{1}{2}\alpha_0^a(\vec{\nabla}^2\alpha_0^a)$. For simplicity, we take the quarks to be massless. From the above Lagrangian, using (155), it is straightforward to derive the energy-momentum tensor $T_{\mu\nu}$, which gives us the energy density ε and pressure p:

$$\varepsilon = \left(\frac{27g^2}{16m_G{}^2}\right)\rho_B{}^2 + \left(\frac{27g^2}{16m_G{}^4}\right)\rho_B\vec{\nabla}^2\rho_B + \left(\frac{27g^2}{16m_G{}^6}\right)\rho_B\vec{\nabla}^2(\vec{\nabla}^2\rho_B)$$
$$+\left(\frac{27g^2}{16m_G{}^8}\right)\vec{\nabla}^2\rho_B\vec{\nabla}^2(\vec{\nabla}^2\rho_B) + \mathcal{B}_{QCD} + 3\frac{\gamma_Q}{2\pi^2}\frac{k_F{}^4}{4} \tag{202}$$

and the pressure is:

$$p = \left(\frac{27g^2}{16m_G{}^2}\right)\rho_B{}^2 + \left(\frac{9g^2}{4m_G{}^4}\right)\rho_B\,\vec{\nabla}^2\rho_B - \left(\frac{9g^2}{8m_G{}^6}\right)\rho_B\,\vec{\nabla}^2(\vec{\nabla}^2\rho_B)$$
$$-\left(\frac{9g^2}{16m_G{}^4}\right)\vec{\nabla}\rho_B\cdot\vec{\nabla}\rho_B + \left(\frac{9g^2}{16m_G{}^6}\right)\vec{\nabla}^2\rho_B\,\vec{\nabla}^2\rho_B - \left(\frac{9g^2}{8m_G{}^8}\right)\vec{\nabla}^2\rho_B\,\vec{\nabla}^2(\vec{\nabla}^2\rho_B)$$
$$-\left(\frac{9g^2}{16m_G{}^8}\right)\vec{\nabla}(\vec{\nabla}^2\rho_B)\cdot\vec{\nabla}(\vec{\nabla}^2\rho_B) - \left(\frac{9g^2}{8m_G{}^6}\right)\vec{\nabla}\rho_B\cdot\vec{\nabla}(\vec{\nabla}^2\rho_B)$$
$$-\mathcal{B}_{QCD} + \frac{\gamma_Q}{2\pi^2}\frac{k_F{}^4}{4} \tag{203}$$

where γ_Q is the quark degeneracy factor $\gamma_Q = 2(\text{spin})\times 3(\text{flavor}) = 6$ and k_F is the Fermi momentum defined by the baryon number density by $\rho_B = k_F{}^3/\pi^2$. The other parameters g, m_G and $\mathcal{B}_{QCD}$ are the coupling of the hard gluons, the dynamical gluon mass and the bag constant in terms of the gluon condensate, respectively. An improved version of the EOS of (202) and (203) was used in the study of three dimensional solitons in cold QGP. These solitons are solutions of the Kadomtsev-Petviashvili (KP) equation, which is the three dimensional generalization of the KdV equation. The complete description of the calculation may be found in [31].

5. Nonlinear Wave Equations

In the previous sections we presented a review of the equations of hydrodynamics and we introduced the equations of state, which represent the microscopic dynamics of the corresponding fluids. Furthermore, we have presented a mathematical prescription (the RPM) to study perturbations in these fluids preserving the nonlinearities of the differential equations. The application of the RPM to several systems of interest leads to the nonlinear differential equations which we discuss in this section. Most of these wave equations were developed in [24–28,31,54,58].

5.1. KdV Equation in Nuclear Matter

5.1.1. Cold Nuclear Matter

We insert (160) in the Euler equation given by (34). We then combine this Euler equation with (57) and follow the RPM procedure to obtain the following KdV equation:

$$\frac{\partial \hat{\rho}_1}{\partial t} + c_s \frac{\partial \hat{\rho}_1}{\partial X} + (3 - {c_s}^2) c_s \hat{\rho}_1 \frac{\partial \hat{\rho}_1}{\partial X} + \left(d + \frac{1}{2}\right)\left(\frac{{g_V}^2 \rho_0}{2M{m_V}^4 c_s}\right)\frac{\partial^3 \hat{\rho}_1}{\partial X^3} + \mathcal{G}\frac{\hat{\rho}_1}{t} = 0 \tag{204}$$

where $\mathcal{G}$ is a "geometrical factor":

$$\mathcal{G} \equiv \begin{cases} 0 & \text{for cartesian coordinates: } X = x \\ 1 & \text{for spherical coordinates: } X = r \end{cases} \tag{205}$$

The non-relativistic limit is obtained by the approximation $3 - {c_s}^2 \cong 3$ and so (204) becomes:

$$\frac{\partial \hat{\rho}_1}{\partial t} + c_s \frac{\partial \hat{\rho}_1}{\partial X} + 3 c_s \hat{\rho}_1 \frac{\partial \hat{\rho}_1}{\partial X} + \left(d + \frac{1}{2}\right)\left(\frac{{g_V}^2 \rho_0}{2M{m_V}^4 c_s}\right)\frac{\partial^3 \hat{\rho}_1}{\partial X^3} + \mathcal{G}\frac{\hat{\rho}_1}{t} = 0 \tag{206}$$

which could have been obtained with the use of the same procedure applied to the nonrelativistic Euler equation (69) with $\rho = M\rho_B$.

5.1.2. Hot Nuclear Matter

We follow the same steps listed above, but changing the (160) by the EOS (161) to find a KdV given by:

$$\frac{\partial \hat{\rho_1}}{\partial t} + c_s \frac{\partial \hat{\rho_1}}{\partial X} + \left(2 - {c_s}^2 - \frac{\mu_B {m_V}^2 {c_s}^2}{2{g_V}^2 \rho_0}\right) c_s \hat{\rho_1} \frac{\partial \hat{\rho_1}}{\partial X}$$
$$+ \left(d + \frac{1}{2}\right)\left(\frac{c_s}{2{m_V}^2}\right)\frac{\partial^3 \hat{\rho_1}}{\partial X^3} + \mathcal{G}\frac{\hat{\rho_1}}{t} = 0 \tag{207}$$

In the last two wave equations $\hat{\rho}_1 \equiv \sigma \rho_1$ as defined in (115) and $\mathcal{G}$ is defined by (205). Choosing $d = -\frac{1}{2}$ in the above equations we eliminate the cubic derivative term and obtain a breaking wave equation.

5.2. Breaking Wave Equation in QGP

5.2.1. Cold QGP

Inserting the cold QGP MIT EOS, (182) to (188), into the ideal hydrodynamical equations (34) and (57) and following the RPM procedure we obtain the following breaking wave equation:

$$\frac{\partial \hat{\rho}_1}{\partial t} + c_s \frac{\partial \hat{\rho}_1}{\partial x} + \frac{2}{3} c_s \hat{\rho}_1 \frac{\partial \hat{\rho}_1}{\partial x} = 0 \tag{208}$$

where again $\hat{\rho}_1 \equiv \sigma \rho_1$.

5.2.2. Hot QGP

Now we insert the hot QGP MIT EOS relations, (189) to (196), into the ideal hydrodynamical equations (34) and (55) and following the RPM procedure we obtain the following breaking wave equation:

$$\frac{\partial \hat{\varepsilon}_1}{\partial t} + c_s \frac{\partial \hat{\varepsilon}_1}{\partial x} + \left[1 + \left(\frac{T_B}{T_0} \right)^4 \right] \frac{c_s}{2} \hat{\varepsilon}_1 \frac{\partial \hat{\varepsilon}_1}{\partial x} = 0 \tag{209}$$

where $\hat{\varepsilon}_1 \equiv \sigma \varepsilon_1$.

5.3. KP and KdV Equations From MFQCD

The energy density and pressure are given respectively by (202) and (203) and the wave equations are for perturbations in the baryon density given by $\hat{\rho}_1 \equiv \sigma \rho_1$ as defined by (123) and (130). We use the ideal hydrodynamical equations (34) and (57) following the RPM procedure. In the calculations we find the following relation:

$$\left(\frac{27 g^2 \rho_0{}^2}{8 m_G{}^2} \right) c_s{}^2 + 3\pi^{2/3} \rho_0{}^{4/3} c_s{}^2 = \left(\frac{27 g^2 \rho_0{}^2}{8 m_G{}^2} \right) + \pi^{2/3} \rho_0{}^{4/3} = A \tag{210}$$

which relates the speed of sound to the background density ρ_0:

$$c_s{}^2 = \frac{\left(\frac{27 g^2 \rho_0{}^2}{8 m_G{}^2} \right) + \pi^{2/3} \rho_0{}^{4/3}}{\left(\frac{27 g^2 \rho_0{}^2}{8 m_G{}^2} \right) + 3\pi^{2/3} \rho_0{}^{4/3}} \tag{211}$$

The wave equations are the following.

5.3.1. Cylindrical KP Equation

$$\frac{\partial}{\partial r} \left\{ \frac{\partial \hat{\rho}_1}{\partial t} + c_s \frac{\partial \hat{\rho}_1}{\partial r} + \left[\frac{3}{2}(1 - c_s{}^2) - \frac{\pi^{2/3} \rho_0{}^{4/3}}{3A} \right] c_s \hat{\rho}_1 \frac{\partial \hat{\rho}_1}{\partial r} + \left[\frac{9 g^2 \rho_0{}^2 c_s}{8 m_G{}^4 A} \right] \frac{\partial^3 \hat{\rho}_1}{\partial r^3} + \frac{\hat{\rho}_1}{2t} \right\}$$
$$+ \frac{1}{2 c_s t^2} \frac{\partial^2 \hat{\rho}_1}{\partial \varphi^2} + \frac{c_s}{2} \frac{\partial^2 \hat{\rho}_1}{\partial z^2} = 0 \tag{212}$$

5.3.2. KP Equation

$$\frac{\partial}{\partial x} \left\{ \frac{\partial \hat{\rho}_1}{\partial t} + c_s \frac{\partial \hat{\rho}_1}{\partial x} + \left[\frac{3}{2}(1 - c_s{}^2) - \frac{\pi^{2/3} \rho_0{}^{4/3}}{3A} \right] c_s \hat{\rho}_1 \frac{\partial \hat{\rho}_1}{\partial x} + \left[\frac{9 g^2 \rho_0{}^2 c_s}{8 m_G{}^4 A} \right] \frac{\partial^3 \hat{\rho}_1}{\partial x^3} \right\}$$
$$+ \frac{c_s}{2} \frac{\partial^2 \hat{\rho}_1}{\partial y^2} + \frac{c_s}{2} \frac{\partial^2 \hat{\rho}_1}{\partial z^2} = 0 \tag{213}$$

5.3.3. KdV Equation

The one dimensional cartesian particular case of (213) is obtained by neglecting the y and z dependence, so that (213) becomes the KdV :

$$\frac{\partial\hat{\rho}_1}{\partial t}+c_s\frac{\partial\hat{\rho}_1}{\partial x}$$
$$+\left[\frac{(2-c_s{}^2)}{2}-\left(\frac{27g^2\rho_0{}^2}{8m_G{}^2}\right)\frac{(2c_s{}^2-1)}{2A}-\frac{\pi^{2/3}\rho_0{}^{4/3}}{A}\left(c_s{}^2-\frac{1}{6}\right)\right]c_s\hat{\rho}_1\frac{\partial\hat{\rho}_1}{\partial x}$$
$$+\left[\frac{9g^2\rho_0{}^2c_s}{8m_G{}^4A}\right]\frac{\partial^3\hat{\rho}_1}{\partial x^3}=0 \tag{214}$$

Taking the limit $m_G \to \infty$ we obtain from (210) and (211):

$$A=\pi^{2/3}\rho_0{}^{4/3}\ , \qquad\qquad c_s{}^2=\frac{1}{3}$$

and (214) becomes:

$$\frac{\partial\hat{\rho}_1}{\partial t}+c_s\frac{\partial\hat{\rho}_1}{\partial x}+\frac{2}{3}c_s\hat{\rho}_1\frac{\partial\hat{\rho}_1}{\partial x}=0 \tag{215}$$

and we recover exactly the result (208), the breaking wave equation for $\hat{\rho}_1$ at zero temperature in the QGP derived from the MIT equation of state.

5.3.4. Breaking Wave Equation

Neglecting the spatial derivatives in (202) and (203), the equation (214) reduces to:

$$\frac{\partial\hat{\rho}_1}{\partial t}+c_s\frac{\partial\hat{\rho}_1}{\partial x}+$$
$$+\left[\frac{(2-c_s{}^2)}{2}-\left(\frac{27g^2\rho_0{}^2}{8m_G{}^2}\right)\frac{(2c_s{}^2-1)}{2A}-\frac{\pi^{2/3}\rho_0{}^{4/3}}{A}\left(c_s{}^2-\frac{1}{6}\right)\right]c_s\hat{\rho}_1\frac{\partial\hat{\rho}_1}{\partial x}=0 \tag{216}$$

which is also a breaking wave equation for $\hat{\rho}_1$ with the ρ_0, m_G and g dependence in its coefficients.

5.4. KP-Burgers Equation in Hot QGP

The relations (189) to (196) for the hot QGP are now inserted in the relativistic viscous hydrodynamical equations, i.e., the Navier-Stokes (33) and the continuity for the entropy density (53). The ideal case is recovered when the viscous coefficients are set to zero. Following the RPM procedure we obtain the following wave equations:

5.4.1. Cylindrical KP-Burgers

$$\frac{\partial}{\partial r}\left\{\frac{\partial\hat{\varepsilon}_1}{\partial t}+c_s\frac{\partial\hat{\varepsilon}_1}{\partial r}+\frac{c_s}{2}\left[1+\left(\frac{T_B}{T_0}\right)^4\right]\hat{\varepsilon}_1\frac{\partial\hat{\varepsilon}_1}{\partial r}+\frac{\hat{\varepsilon}_1}{2t}-\frac{1}{2T_0}\left(\frac{\zeta}{s}+\frac{4}{3}\frac{\eta}{s}\right)\frac{\partial^2\hat{\varepsilon}_1}{\partial r^2}\right\}$$
$$+\frac{1}{2c_st^2}\frac{\partial^2\hat{\varepsilon}_1}{\partial\varphi^2}+\frac{c_s}{2}\frac{\partial^2\hat{\varepsilon}_1}{\partial z^2}=0 \tag{217}$$

5.4.2. Cylindrical Burgers

Neglecting the φ and z dependence, the equation (217) becomes:

$$\frac{\partial\hat{\varepsilon}_1}{\partial t}+c_s\frac{\partial\hat{\varepsilon}_1}{\partial r}+\frac{c_s}{2}\left[1+\left(\frac{T_B}{T_0}\right)^4\right]\hat{\varepsilon}_1\frac{\partial\hat{\varepsilon}_1}{\partial r}+\frac{\hat{\varepsilon}_1}{2t}=\frac{1}{2T_0}\left(\frac{\zeta}{s}+\frac{4}{3}\frac{\eta}{s}\right)\frac{\partial^2\hat{\varepsilon}_1}{\partial r^2} \tag{218}$$

Setting $\eta=\zeta=0$ (ideal fluid) the (218) becomes:

$$\frac{\partial\hat{\varepsilon}_1}{\partial t}+c_s\frac{\partial\hat{\varepsilon}_1}{\partial r}+\frac{c_s}{2}\left[1+\left(\frac{T_B}{T_0}\right)^4\right]\hat{\varepsilon}_1\frac{\partial\hat{\varepsilon}_1}{\partial r}+\frac{\hat{\varepsilon}_1}{2t}=0 \tag{219}$$

6. Analytical Solutions of Nonlinear Wave Equations

We present some cases where particular analytical solutions exist. For the KdV equation we have the soliton solution. Soliton or solitary wave is a localized pulse which propagates without change in shape. For a detailed study of the KdV solitons we recommend the reading of [71]. The soliton will be also used as an initial condition in the study of the numerical solution of the spherical KdV and the breaking wave equations. For the Burgers, cKP, KP and cylindrical KP-Burgers we present the exact solutions as developed in [13, 14, 17, 19, 72–75] where several techniques to solve nonlinear wave equations are presented.

6.1. KdV Equation in Nuclear Matter

We first show the soliton solution of the KdV equations (204) and (206) in cold and hot nuclear matter, respectively [24–27]. We have only soliton solutions in the cartesian case $\mathcal{G}=0$, so $X=x$. We also choose $d=1/2$ as in [27]. The equation (204) can be integrated and solved exactly and its soliton solution is given by:

$$\hat{\rho}_1(x,t)=\frac{3(u-c_s)}{c_s}(3-{c_s}^2)^{-1}sech^2\left[\frac{{m_V}^2}{g_V}\sqrt{\frac{(u-c_s)c_sM}{2\rho_0}}(x-ut)\right] \tag{220}$$

and the solution of (206) is found by setting $3-{c_s}^2\cong 3$ in (220):

$$\hat{\rho}_1(x,t)=\frac{(u-c_s)}{c_s}sech^2\left[\frac{{m_V}^2}{g_V}\sqrt{\frac{(u-c_s)c_sM}{2\rho_0}}(x-ut)\right] \tag{221}$$

Finally, the solution of (207) is:

$$\hat{\rho}_1(x,t)=\frac{3(u-c_s)}{c_s}\left(2-{c_s}^2-\frac{\mu_B{m_V}^2{c_s}^2}{2{g_V}^2\rho_0}\right)^{-1}sech^2\left[\sqrt{\frac{(u-c_s){m_V}^2}{2c_s}}(x-ut)\right] \tag{222}$$

6.2. KP Equations in Cold QGP

The cKP equation (212) has the exact analytical soliton solution [31]:

$$\hat{\rho}_1(r,\varphi,z,t) = \frac{h_1}{h_2} sech^2 \left\{ \frac{\sqrt{h_1}}{2} \left[ar + bz - \left(u + a\frac{c_s \varphi^2}{2} \right) t \right] \right\} \tag{223}$$

where u is a parameter which satisfies $u > ac_s + b^2 c_s / 2a$ and the phase velocity given by $u + a\frac{c_s\varphi^2}{2}$. The constants appearing in the above expression are:

$$h_1 = \frac{u - ac_s - b^2 c_s/2a}{a^3 c_2} \qquad \text{and} \qquad h_2 = \frac{c_1}{3a^2 c_2} \tag{224}$$

where

$$c_1 \equiv \left[\frac{3}{2}(1 - c_s{}^2) - \frac{\pi^{2/3} \rho_0{}^{4/3}}{3A} \right] c_s \tag{225}$$

and

$$c_2 \equiv \left[\frac{9 g^2 \rho_0{}^2 c_s}{8 m_G{}^4 A} \right] \tag{226}$$

For the KP equation (213) we have the following soliton solution [31]:

$$\hat{\rho}_1(x,y,z,t) = \frac{3(U-w)}{\mathcal{A} c_1} sech^2 \left[\sqrt{\frac{(U-w)}{4\mathcal{A}^3 c_2}} \left(\mathcal{A}x + \mathcal{B}y + \mathcal{C}z - Ut \right) \right] \tag{227}$$

where $\mathcal{A}$, $\mathcal{B}$, $\mathcal{C}$ are real constants and w is given by:

$$w = \mathcal{A} c_s + \frac{\mathcal{B}^2 c_s}{2} + \frac{\mathcal{C}^2 c_s}{2} \tag{228}$$

We shall consider $\mathcal{A} > 0$ for simplicity and the parameter U such that $U > w$. For the KdV equation (214) we have:

$$\hat{\rho}_1(x,t) = \frac{3(u - c_s)}{c_3} sech^2 \left[\sqrt{\frac{(u - c_s)}{4c_4}} (x - ut) \right] \tag{229}$$

where u is an arbitrary supersonic velocity and the constants c_3 and c_4 are given by:

$$c_3 \equiv \left[\frac{(2 - c_s{}^2)}{2} - \left(\frac{27 g^2 \rho_0{}^2}{m_G{}^2} \right) \frac{(2c_s{}^2 - 1)}{2A} - \frac{\pi^{2/3} \rho_0{}^{4/3}}{A} \left(c_s{}^2 - \frac{1}{6} \right) \right] c_s \tag{230}$$

and

$$c_4 \equiv \left[\frac{9 g^2 \rho_0{}^2 c_s}{m_G{}^4 A} \right] \tag{231}$$

6.3. KP-Burgers Equation in Hot QGP

The cKP-B (217) has the solutions:

$$\hat{\varepsilon}_1(r,z,\varphi,t) = \frac{2DA}{c_s T_0}\left(\frac{\zeta}{s}+\frac{4}{3}\frac{\eta}{s}\right)\left[1+\left(\frac{T_B}{T_0}\right)^4\right]^{-1}$$
$$-\frac{2DA}{c_s T_0}\left(\frac{\zeta}{s}+\frac{4}{3}\frac{\eta}{s}\right)\left[1+\left(\frac{T_B}{T_0}\right)^4\right]^{-1}\times$$
$$\times\, tanh\left\{D\left[Ar+Bz-A\frac{c_s\varphi^2 t}{2}-\left(Ac_s+\frac{B^2c_s}{2A}+\frac{DA^2}{T_0}\left(\frac{\zeta}{s}+\frac{4}{3}\frac{\eta}{s}\right)\right)t\right]\right\} \tag{232}$$

$$\hat{\varepsilon}_1(r,z,\varphi,t) = -\frac{2DA}{c_s T_0}\left(\frac{\zeta}{s}+\frac{4}{3}\frac{\eta}{s}\right)\left[1+\left(\frac{T_B}{T_0}\right)^4\right]^{-1}$$
$$-\frac{2DA}{c_s T_0}\left(\frac{\zeta}{s}+\frac{4}{3}\frac{\eta}{s}\right)\left[1+\left(\frac{T_B}{T_0}\right)^4\right]^{-1}\times$$
$$\times\, tanh\left\{D\left[Ar+Bz-A\frac{c_s\varphi^2 t}{2}-\left(Ac_s+\frac{B^2c_s}{2A}-\frac{DA^2}{T_0}\left(\frac{\zeta}{s}+\frac{4}{3}\frac{\eta}{s}\right)\right)t\right]\right\} \tag{233}$$

or the following solutions:

$$\hat{\varepsilon}_1(r,z,\varphi,t) = \frac{2DA}{c_s T_0}\left(\frac{\zeta}{s}+\frac{4}{3}\frac{\eta}{s}\right)\left[1+\left(\frac{T_B}{T_0}\right)^4\right]^{-1}$$
$$-\frac{2DA}{c_s T_0}\left(\frac{\zeta}{s}+\frac{4}{3}\frac{\eta}{s}\right)\left[1+\left(\frac{T_B}{T_0}\right)^4\right]^{-1}\times$$
$$\times\, coth\left\{D\left[Ar+Bz-A\frac{c_s\varphi^2 t}{2}-\left(Ac_s+\frac{B^2c_s}{2A}+\frac{DA^2}{T_0}\left(\frac{\zeta}{s}+\frac{4}{3}\frac{\eta}{s}\right)\right)t\right]\right\} \tag{234}$$

$$\hat{\varepsilon}_1(r,z,\varphi,t) = -\frac{2DA}{c_s T_0}\left(\frac{\zeta}{s}+\frac{4}{3}\frac{\eta}{s}\right)\left[1+\left(\frac{T_B}{T_0}\right)^4\right]^{-1}$$
$$-\frac{2DA}{c_s T_0}\left(\frac{\zeta}{s}+\frac{4}{3}\frac{\eta}{s}\right)\left[1+\left(\frac{T_B}{T_0}\right)^4\right]^{-1}\times$$
$$\times\, coth\left\{D\left[Ar+Bz-A\frac{c_s\varphi^2 t}{2}-\left(Ac_s+\frac{B^2c_s}{2A}-\frac{DA^2}{T_0}\left(\frac{\zeta}{s}+\frac{4}{3}\frac{\eta}{s}\right)\right)t\right]\right\} \tag{235}$$

where the real constants to be chosen are A, B and D.

7. Numerical Solutions

In this section we present numerical results. The solutions of the differential equations can be grouped in those which are smooth and those which exhibit some non-smooth behavior, such as rapid oscillations or the formation of "walls", specially at later times. This kind of behavior appears when there is a lack of balance between the different terms of the equations. Therefore, before presenting numbers and plots, we discuss, in the next subsection, the conditions for finding stable solutions.

7.1. Soliton Stability

As we have seen, perturbations in fluids with different equations of state generate different nonlinear wave equations. Some examples are the Kadomtsev-Petviashvili (KP) equation:

$$\frac{\partial}{\partial x}\left\{\frac{\partial F}{\partial t}+\alpha_1 F\frac{\partial F}{\partial x}+\alpha_2\frac{\partial^3 F}{\partial x^3}\right\}+\alpha_3\frac{\partial^2 F}{\partial y^2}+\alpha_4\frac{\partial^2 F}{\partial z^2}=0. \tag{236}$$

and its particular cases, such as the KdV:

$$\frac{\partial F}{\partial t}+\alpha_1 F\frac{\partial F}{\partial x}+\alpha_2\frac{\partial^3 F}{\partial x^3}=0 \tag{237}$$

and the breaking wave equation:

$$\frac{\partial F}{\partial t}+\alpha_1 F\frac{\partial F}{\partial x}=0 \tag{238}$$

Another example of a nonlinear wave in a dissipative system is the Burgers equation:

$$\frac{\partial F}{\partial t}+\alpha_1 F\frac{\partial F}{\partial x}+\alpha_2\frac{\partial^2 F}{\partial x^2}=0 \tag{239}$$

where α_1 to α_4 are real constants. Having derived a particular differential equation, we can check whether the obtained equation is consistent with the physical picture of a small amplitude and long wave length perturbation propagating over large distances. We shall follow the analysis performed in Ref. [7]. Let us assume that the nonlinear equations listed above for a generic function F have a solitary wave solution with a typical large length $L \simeq 1/\sigma$ $(0 < \sigma << 1)$. Considering the general case, the KP equation has a dispersion $\frac{\partial^4 F}{\partial x^4}$ term that is about $\frac{\partial^4 F}{\partial x^4} \simeq \sigma^4 F$. It must arise at a propagation distance (or equivalently propagation time T) D, accounted for in the equation by the term $\frac{\partial^2 F}{\partial x \partial t} \simeq \sigma \frac{F}{T}$. If both the dispersion and propagation terms have the same size, then $T \simeq 1/\sigma^3$. Regarding the nonlinear term, if it has the form $\frac{\partial}{\partial x}(F\frac{\partial F}{\partial x})$ its order of magnitude is $\sigma^2 F^2$. The formation of the soliton requires that the nonlinear effect balances the dispersion. Hence it must have the same order of magnitude and $F^2\sigma^2 = F\sigma^4$. Hence $F \simeq \sigma^2$. We can then conclude that $F << L << D$ and the above equation describes the propagation of a wave with small amplitude (F) and large wave length (L) which travels large distances (D). In the case of KP we have terms that describe the transverse evolution of the wave. We can estimate their sizes only if we make assumptions about the transverse length scales. In most cases the resulting flow is one-dimensional along the x direction with some "leakage" to the transverse directions.

7.2. Numerical Analysis

In what follows we apply the numerical tools developed in the appendix to several cases.

7.2.1. Nuclear Soliton

We start our numerical analysis showing in Fig. 2 the solution of the linear KdV equation at $T = 0$, ($\mathcal{G} = 0$ and $\mathcal{X} = x$) (204) with $d = \frac{1}{2}$. In Fig. 2(a), we use the analytical solution (220) as initial condition. As expected this pulse propagates without dissipation nor dispersion: it is a soliton wave. This is the situation illustrated in Fig. 1. Any change in the initial condition has noticeable consequences as it can be seen in Fig. 2(b), where we follow the evolution of the numerical solution of (204) for an initial pulse given by (220) multiplied by a factor 20. As it can be seen, the amplitude grows, the width decreases and secondary bumps appear propagating behind the first.

In Fig. 3, we show the equivalent plot for the spherical case: $\mathcal{G} = 1$ and $\mathcal{X} = r$. In contrast to the linear case there is a strong damping of the pulse. The dependence on the initial conditions is also strong.

In Fig. 4 we show for the linear case and for the "optimal" initial condition (220) the evolution of the pulse with time for different temperatures. We can see that, increasing the temperature the pulses move faster and go farther. The same feature can be observed in the spherical case, as shown in Fig. 5 .

Setting $d = -\frac{1}{2}$ in the wave equations (204) and (207) we eliminate the third order derivative terms. The corresponding wave equations are breaking wave equations. Out of smooth initial perturbations, given by (220), these equations create shock waves. We can see this process in one dimensional Cartesian coordinates in Fig. 6 . We observe a steepening of the profile until the formation of the shock, followed by the dispersion of the wave. We see that the higher is the initial amplitude, the sooner the wave breaking and dispersion occurs.

In Fig. 7 we fix one initial profile and study its time evolution for two different temperatures. 7(a) and 7(b) show the development of a shock wave at $T = 20\ MeV$ and $70\ MeV$ respectively. As it can be seen, with increasing temperatures the pulse moves faster and the shock formation and the subsequent dispersive breaking occurs later. For the radial case a similar behavior is observed.

7.2.2. Breaking Wave Equation in QGP

The KdV equation can be written as:

$$\frac{\partial u}{\partial t} + c_s \frac{\partial u}{\partial t} + \beta \frac{\partial}{\partial x}\left(\frac{u^2}{2}\right) + \mu \frac{\partial^3 u}{\partial x^3} = 0 \tag{240}$$

which has the following analytical soliton solution:

$$f(x,t) = \frac{3(u - c_s)}{\beta}\, sech^2\left[\sqrt{\frac{(u - c_s)}{4\mu}}(x - ut)\right] \tag{241}$$

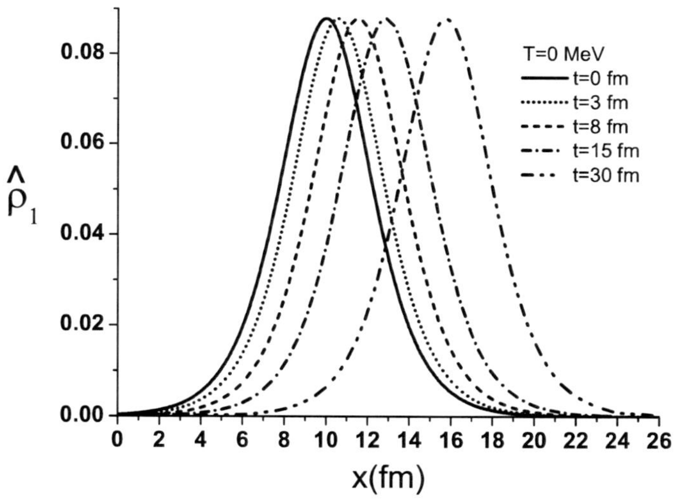

(a)

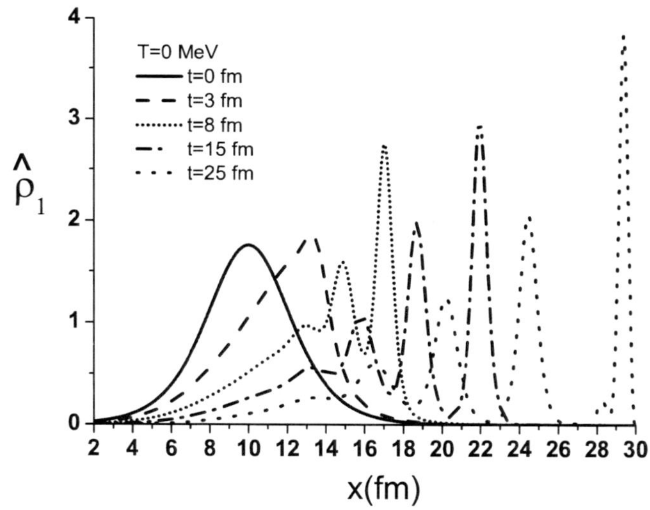

(b)

Figure 2. a) The evolution of the analytic solution of the KdV equation. b) The evolution of the analytic solution multiplied by a factor 20.

In the numerical study of (208) and (209) we use the following soliton-like profile:

$$\hat{\rho}_1(x,t_0) = A\ sech^2\left[\frac{x}{B}\right] \tag{242}$$

In this equation A and B represent the amplitude and width of the initial baryon density pulse, respectively.

We present in Fig. 8 the numerical solution of (208) for different times. In Fig. 8(a) we show for $A = 0.075$ and $B = 1$ fm and in 8(b) for $A = 0.35$ and $B = 1$ fm. It is possible to

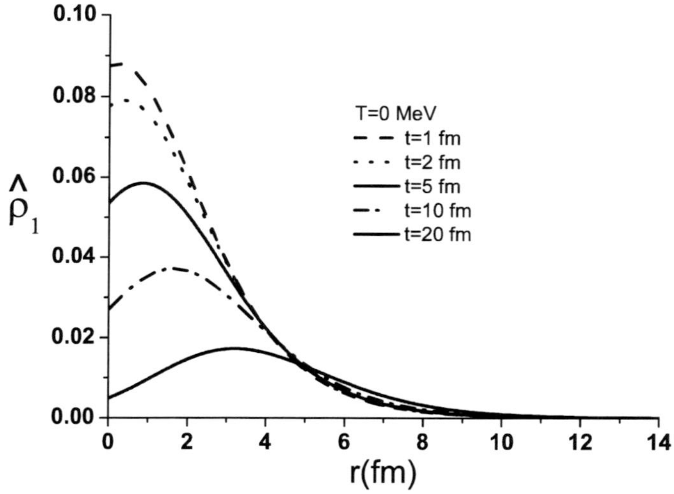

(a)

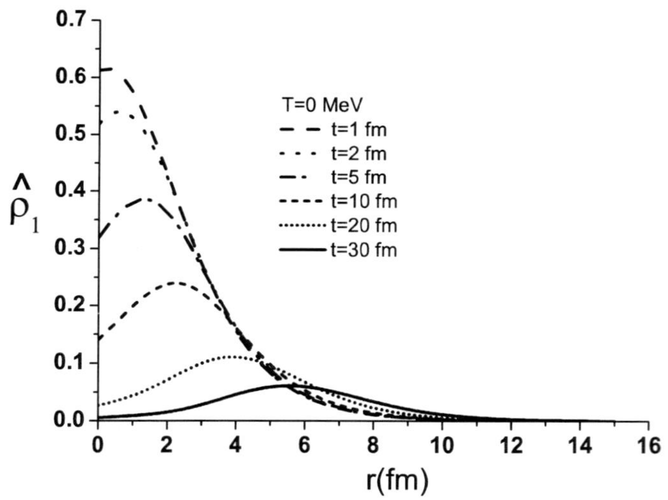

(b)

Figure 3. Analogous to Fig. 2 for spherical coordinates.

observe the evolution of the initial gaussian-like pulse with the formation of a "wall" on the right side. In 8(b) the "wall" formation and dispersion occurs much earlier than in 8(a) due higher initial amplitude.

In Fig. 9(a) we show the solution of (209) with the initial condition given by (242) with $A = 0.01$, $B = 1$ fm and $T = 300\,MeV$. Fig. 9(b) shows the same but with $A = 0.1$ and $B = 1$ fm. As in the zero temperature case, increasing the initial amplitude the breaking process and dispersion develops earlier. From Fig. 9(a) we can conclude that it is possible to find an approximate solitonic behavior even when the differential equation is not the KdV one.

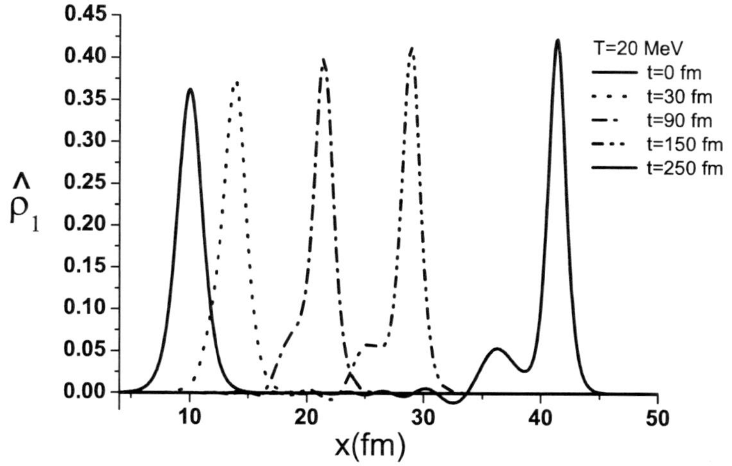

(a)

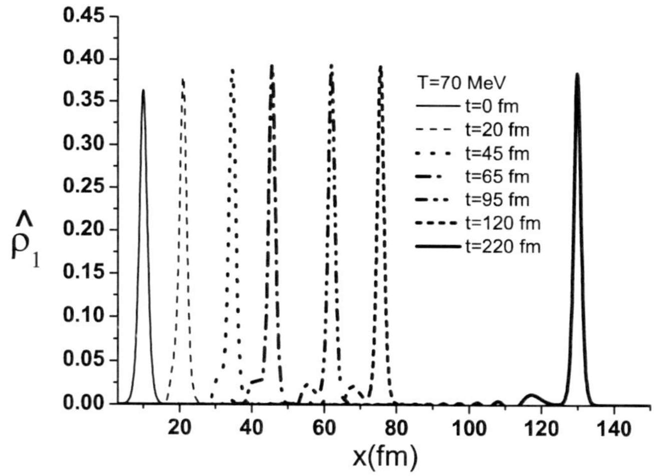

(b)

Figure 4. The panels show calculations with temperatures a) $T = 20$ and b) $T = 70\ MeV$.

7.2.3. KP Equation in Cold QGP

We now study the conditions in which the solution (223) must be real and therefore the constant h_1 must be positive. Moreover, following Refs. [22,23] we assume that $a^2 + b^2 = 1$ and consider $\hat{\rho_1}$ a normalized perturbation [31], within the region (in the $u - a$ plane) Eq. (223) is well defined and we can have solitons. The parameters are chosen to be: $\rho_0 = 1$ fm^{-3} , $g = 1.15$ and $m_G = 460\ MeV$, which imply $c_s \simeq 0.64$ [31]. The stability analysis can be made more rigorous with the introduction of the Sagdeev potential [22,23] by using $\eta = \xi - ut = ar + bz - d\frac{c_s\varphi^2 t}{2} - ut$ to rewrite equation (212) as an energy balance equation. For our present purposes the requirements in [31] are sufficient.

The plot of the soliton evolution is presented in Fig. 10 and in Fig. 11. We show

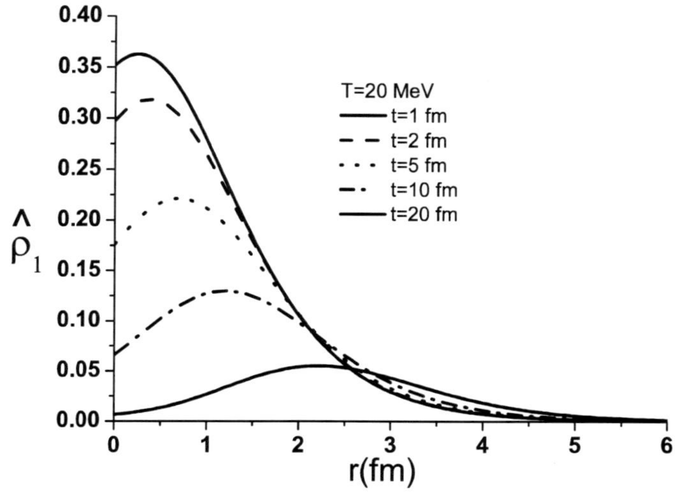

(a)

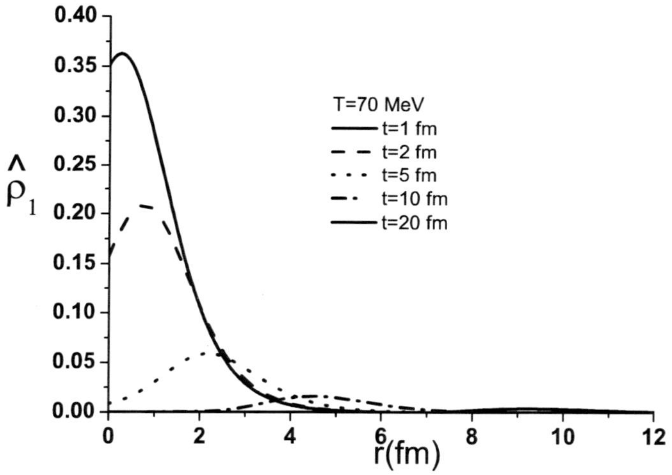

(b)

Figure 5. Analogous to Fig. 4 for spherical coordinates.

a plot of (223) with fixed $\varphi = 0^o$, $a = 0.6$, $b = 0.8$, $u = 0.73$ and z varying in the range $0\,\text{fm} \leq z \leq 30\,\text{fm}$ which satisfies the soliton conditions. In Fig. 10(a) the pulse is observed at $t = 18$ fm whereas in Fig. 10(b) at $t = 28$ fm. From the Fig. 10(a) we can see that the cylindrical pulse expands outwards in the radial direction. The regions with larger z expand with a delay with respect to the central ($z = 0$) region.

Keeping $z = 1$ fm fixed, we show the time evolution of (223) from $t = 10$ fm (Fig. 11(a)) to $t = 22$ fm (Fig. 11(b)). The azimuthal angle varies in the range $20^o \leq \varphi \leq 150^o$. From the parenthesis in (223) we can see that the expansion velocity grows with the angle. This asymmetry can be clearly seen in the figure, where the large angle "backward" region moves faster the small angle "forward" region. The breaking of z invariance and azimuthal

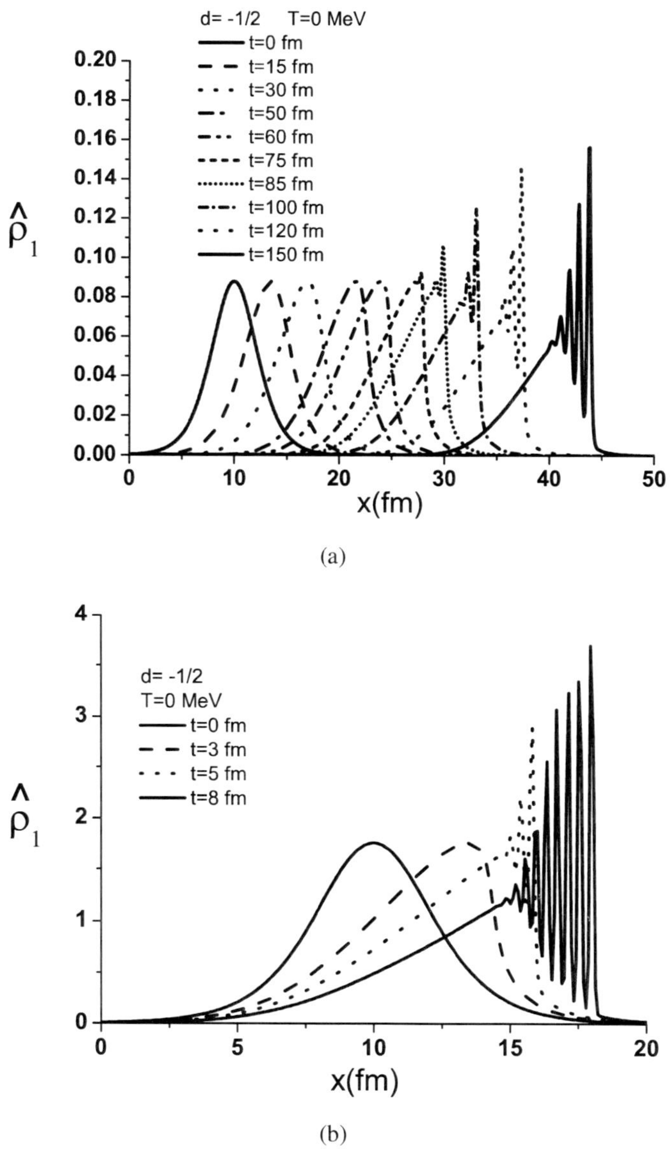

Figure 6. a) Shock wave formation in one dimensional Cartesian coordinates. b) Initial profile multiplied by a factor 20.

symmetry is entangled with the soliton stability [31] and with the physical properties of the system (contained in the parameters h_1, h_2 and c_s).

We perform the study of the existence condition for the solution (227), which must be real and therefore the constant $U - w$ must be positive. The parameters are the same: $\rho_0 = 1\ fm^{-3}$, $g = 1.15$ and $m_G = 460$ MeV, which imply $c_s \simeq 0.64$ [31]. We also set $\mathcal{C} = 0.5$ and extend the condition in Refs. [22, 23] to $\mathcal{A}^2 + \mathcal{B}^2 + \mathcal{C}^2 = 1$. As mentioned before, $U > w$ and again, $\hat{\rho_1}$ is a normalized perturbation [31] and within the region (in the $U - \mathcal{A}$ plane), (227) is well defined and we can have solitons [31]. The stability analysis

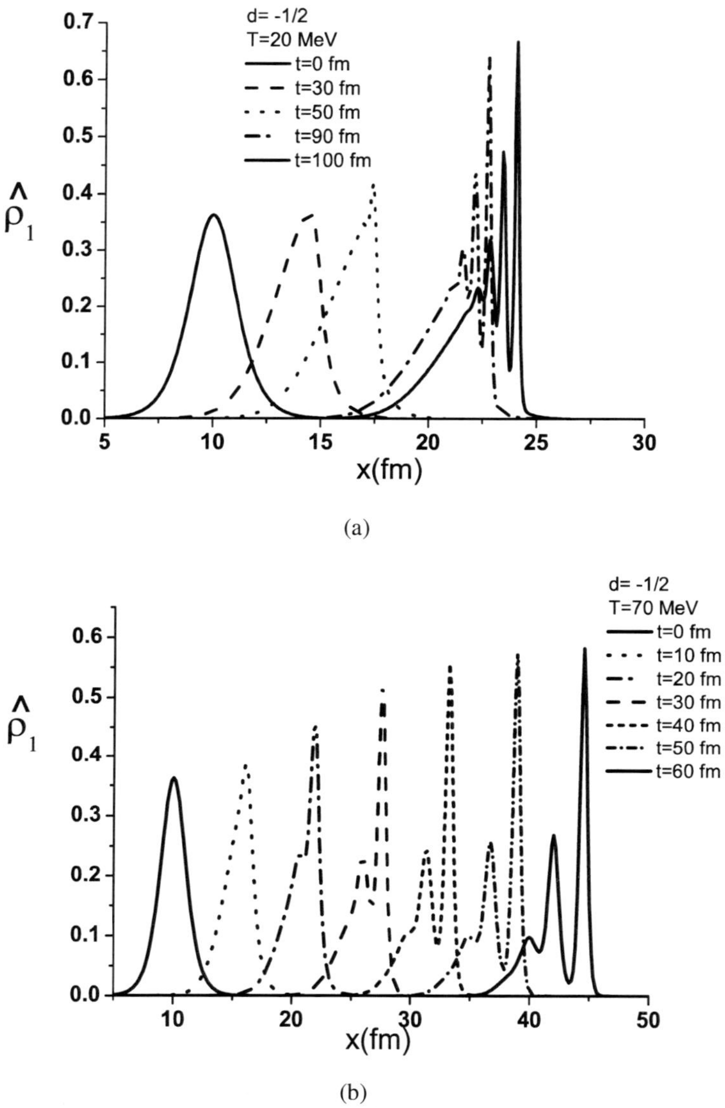

Figure 7. Shock wave formation in one dimensional Cartesian coordinates for different temperatures: a) $T = 20$ and b) $T = 70\ MeV$.

can be performed more rigorously with the introduction of the Sagdeev potential [22, 23] by using $\mathcal{A}x + \mathcal{B}y + \mathcal{C}\,z - Ut$, to rewrite equation (213) as an energy balance equation. A simple example of soliton evolution is presented in Fig. 12. The plot of (227) with fixed $z = 1$ fm, $\mathcal{A} = 0.6$, $\mathcal{B} \cong 0.62$, $U = 0.66$ and y varying in the range $0\,\text{fm} \leq y \leq 50\,\text{fm}$. The pulse is observed at two times: $t = 30$ fm (Fig. 12(a)) and $t = 120$ fm (Fig. 12(b)). From the figure we can see that the cartesian pulse expands outwards in the x direction keeping its shape and form as expected.

Analogously to the nuclear soliton, the particular case of KP (213): the KdV (214), has the exact soliton solution (229). In Fig. 13 we choose $u = 0.8$, $\rho_0 = 2\,fm^{-3}$ and $c_s{}^2 = 0.5$.

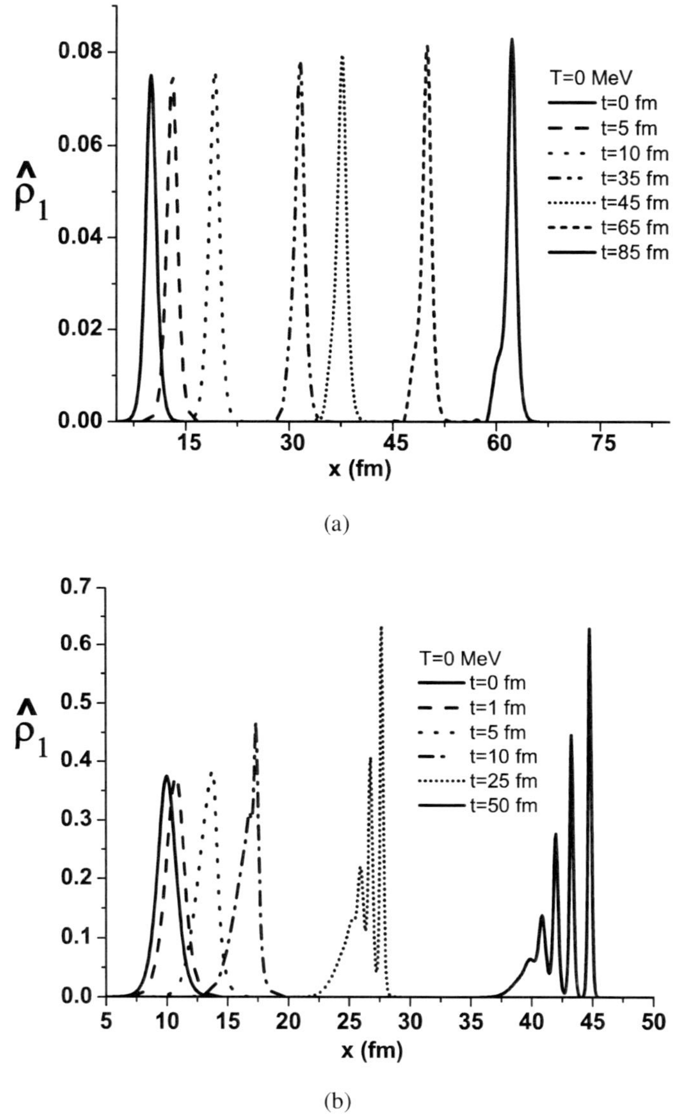

Figure 8. Time evolution of a baryon density pulse for cold QGP.

In Fig. 13(a), the numerical solution of (214) for (229) as initial condition is studied for different times. As expected, the evolution of the initial gaussian-like pulse as a well defined soliton, keeping its shape and form. In Fig. 13(b), we show again the numerical solution of (214) for (229) multiplied by a factor 10. Now the initial soliton posses amplitude 6.0 and starts to develop secondary peaks, which are called "radiation" in the literature. Further time evolution would increase the strength of these peaks until the complete loss of localization.

For the other particular case (215) of the KdV (214), which we show in Fig. 14, where again we use (229) with the same parameters listed above. Fig. 14(a), we have the breaking following by its dispersion of the initial pulse. And in Fig. 14(b), we use (229) multiplied by a factor 10, which we observe the anticipation of the breaking following by its dispersion

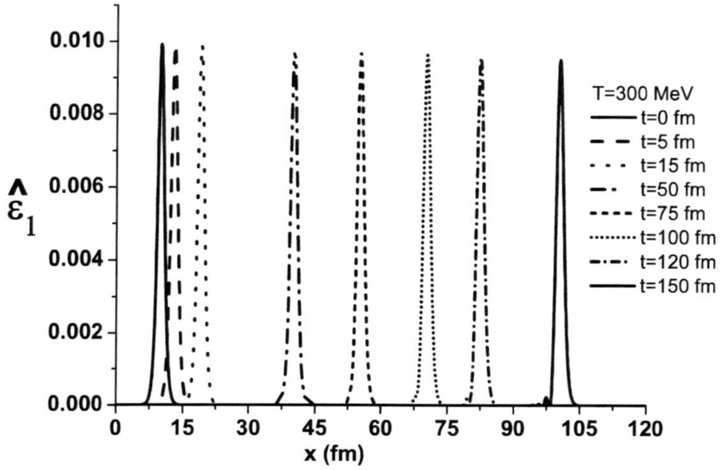

(a)

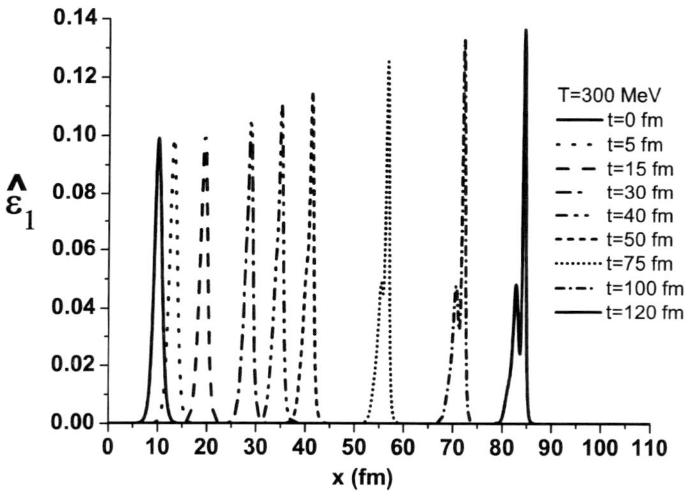

(b)

Figure 9. Time evolution of an energy density pulse at $T = 300\,MeV$.

of the initial pulse much earlier in comparison to Fig. 14(a).

7.2.4. Burgers Equation in Hot QGP

We apply the perturbation concept to study the radial expansion of cylindrical flux tubes in a hot QGP. These tubes are treated as perturbations in the energy density of the system which is formed in heavy ion collisions at RHIC and LHC as well explained in [54]. There is a "competition" between the viscosity and the tube expansion. In Fig. 15(a) the QGP expands and the viscosity dissipates the tube evolution. In Fig. 15(b) the opposite occurs: the tube expands together with the QGP expansion but the viscosity is not sufficient to suppress the tube. In in Fig. 15 we have two extreme situations and something between

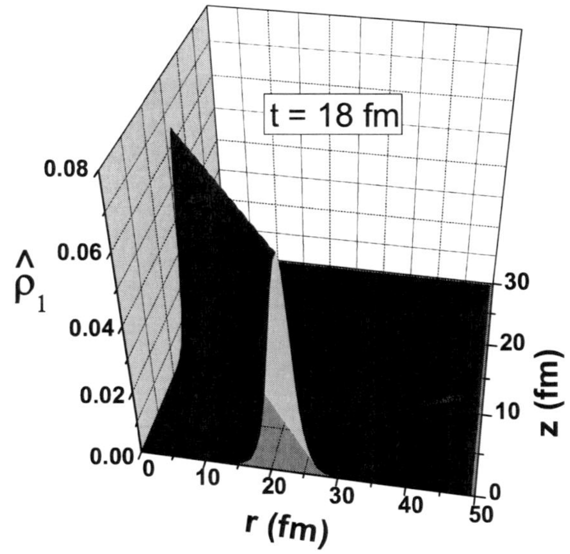

(a)

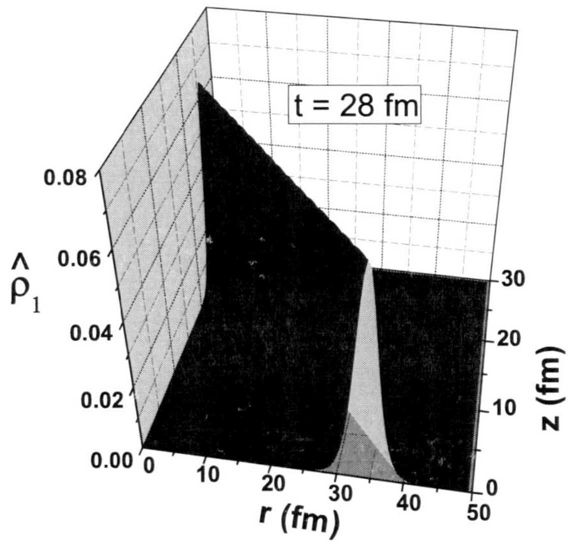

(b)

Figure 10. Cylindrical soliton time evolution in $r-z$ plane.

may occur. We want to study such situations through the behavior of the initial condition, which will describe the existence of such flux tubes.

We perform the numerical analysis for equations (218) and (219) with the initial condition given by a gaussian pulse in $\hat{\varepsilon}_1$:

$$\hat{\varepsilon}_1(r) = A\, e^{-r^2/r_0^2} \tag{243}$$

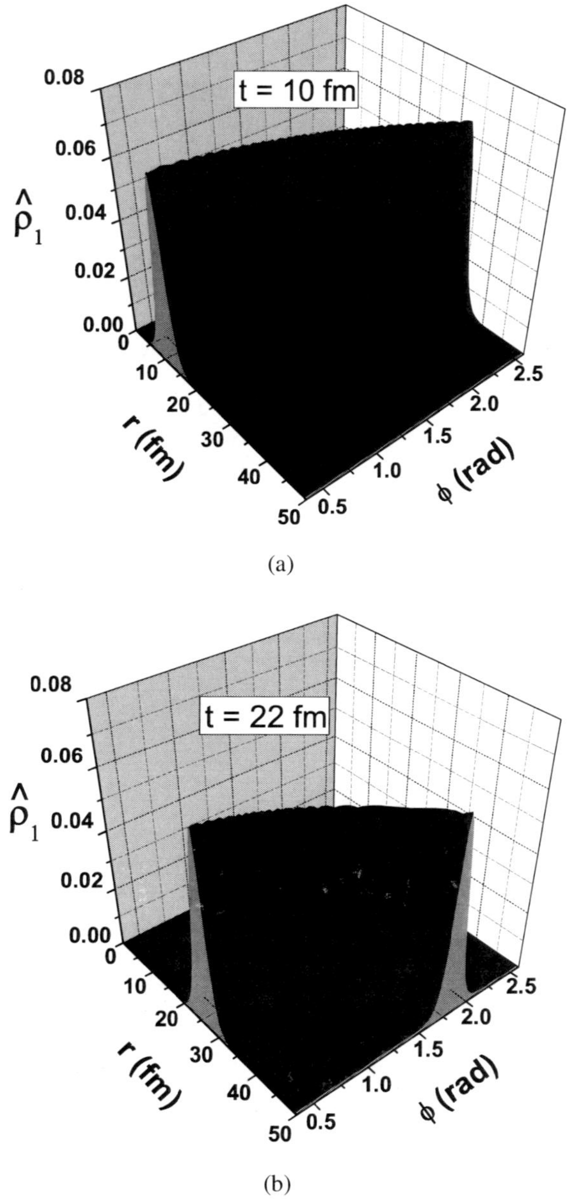

Figure 11. Cylindrical soliton time evolution in $r - \varphi$ plane.

where the amplitude A and the approximate width r_0 are parameters which depend on the dynamics of flux tube formation. We shall refer to r_0 as the initial "radius" of the tube. The tubes are perturbations, so we expect $A < 1$. According to [54] and references therein, the transverse size of the tubes is of the order of 1 fm and in our calculations $r_0 = 0.8\, fm$. We consider hot QGP at temperatures $T_0 = 150\, MeV$ and $T_0 = 500\, MeV$ treated as an ideal fluid ($\eta/s = \zeta/s = 0$) described by (219) and as a viscous fluid ($\eta/s = 0.16$ and $\zeta/s = 0$) described by (218) [54].

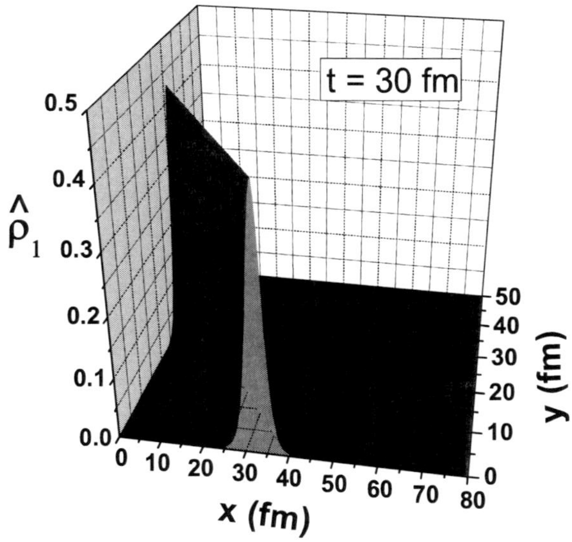

(a)

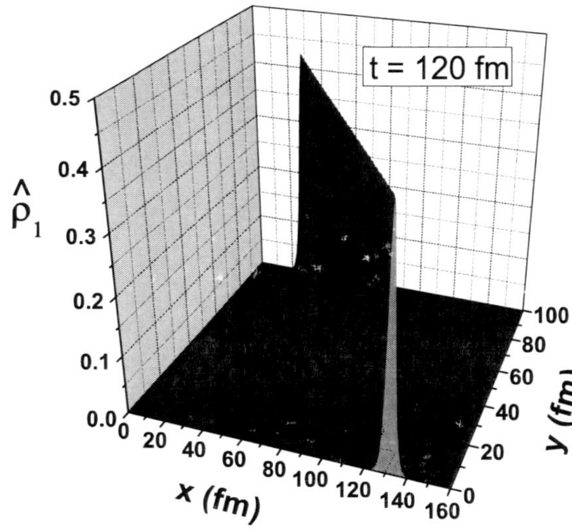

(b)

Figure 12. Cartesian soliton time evolution in $x-y$ plane.

In Fig. 16 we show numerical solutions of (218) for a viscous fluid using (243) and we can observe the increasing temperature favors the tubular structure survival.

In Fig. 17 we perform the same study for the ideal fluid (219) with (243) and we show that breaking with dispersion occurs. By comparison between Fig. 16 and Fig. 17 we conclude that viscosity dissipates the breaking followed by dispersion of the pulse. The tube expands radially with a supersonic velocity and in less than 4 fm/c it becomes a "ring", with a hole in the middle. Moreover, by this time the amplitude is already reduced

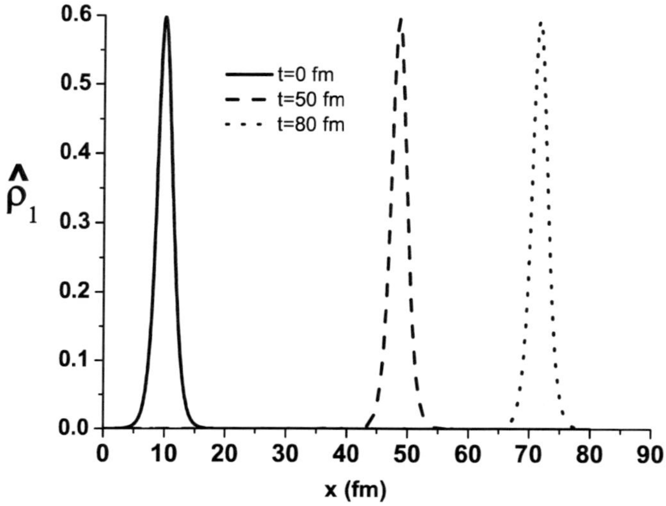

(a)

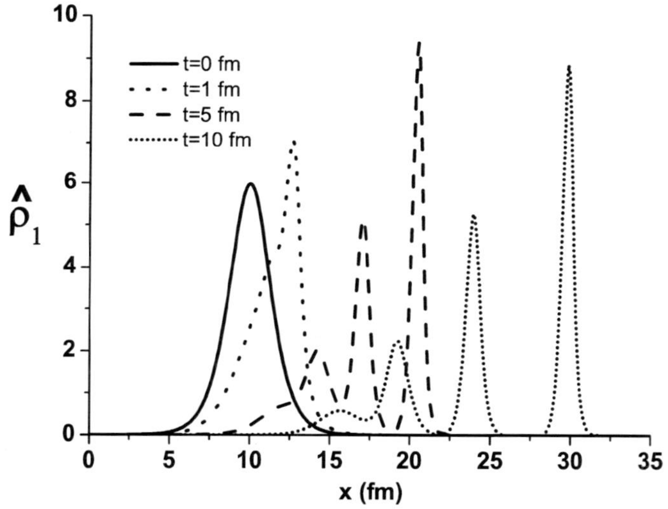

(b)

Figure 13. a) Soliton propagation in one dimensional Cartesian coordinate and b) Initial profile multiplied by a factor 10 providing peaks plus "radiation".

by a factor two and the tube (or ring) looses the strength to "push away" the surrounding matter [54].

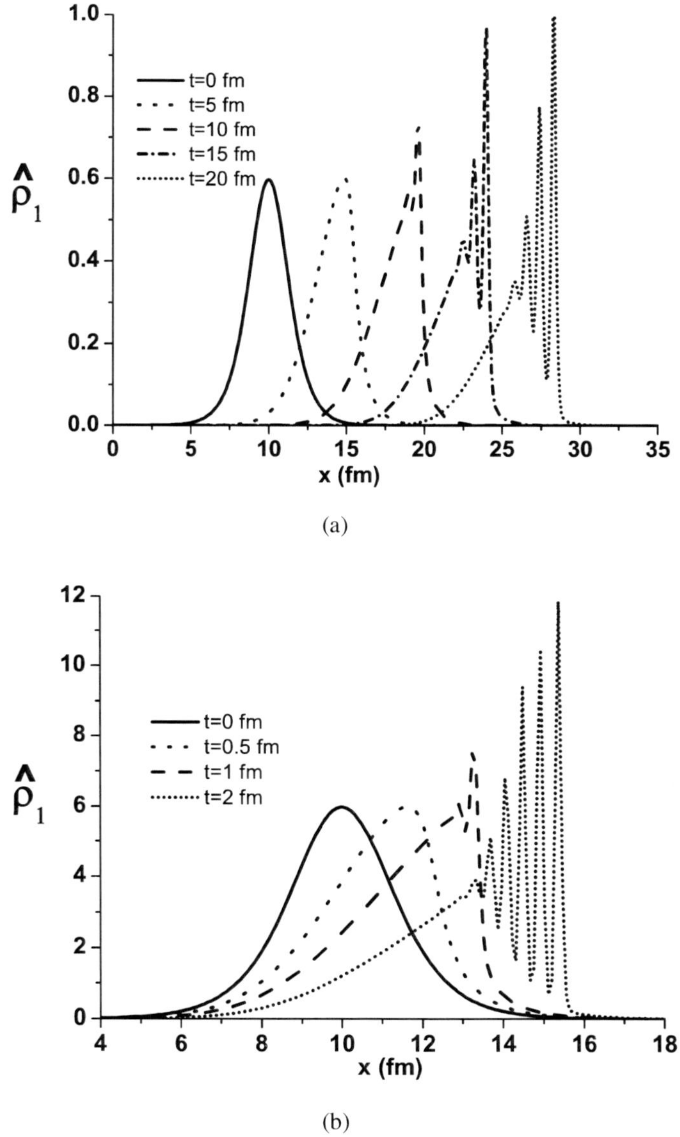

Figure 14. a) Breaking wave and b) breaking wave anticipation.

7.2.5. KP-Burgers in Hot QGP

As an example of time evolution for the analytical solution of the cKP-B equation (217) we plot the time evolution of (232) with $\varphi = 0\,rad$, $D = 1$, $A = B = 0.5$, $T_0 = 300\,MeV$ and the viscous fluid with $\eta/s = 0.16$ and $\zeta/s = 0$. The Fig. 18 shows the time evolution of the analytical traveling wave.

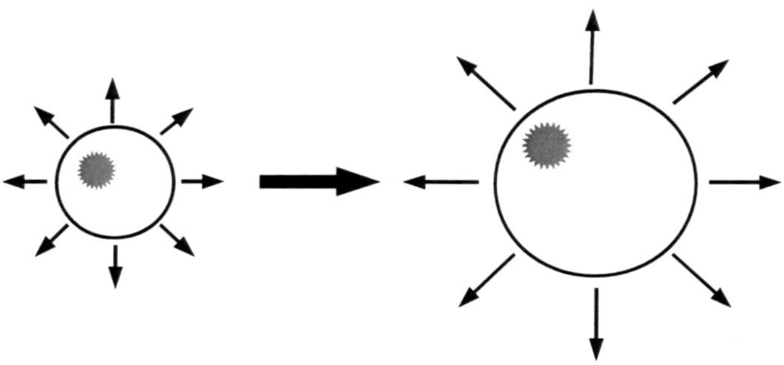

(a)

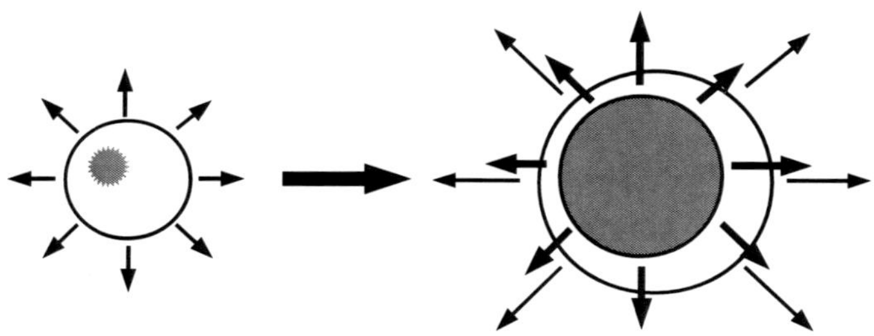

(b)

Figure 15. The large circle represents the front view of a cylindrical portion of QGP fluid. The small and dark circle represents a flux tube, i.e., a cylindrical perturbation with a higher energy density than the background. a) The tube expands slowly and the background faster. b) The tube expands much faster than the background.

As it can be seen, (232) obtained depends directly of the dissipative (viscous) coefficient term of (217), i.e., it is a dissipation dominated solution. The Fig. 18 shows a shock wave propagation in a time interval of 50 fm.

Conclusion

The discovery of the quark gluon plasma in the high energy heavy ion colliders brought relativistic hydrodynamics to the main stage of hadron physics. Encouraged by the vigorous experimental program at CERN theoreticians of hydrodynamics embarked in an ambitious project: the calculation of observables quantities with relativistic viscous hydrodynamics. The measurement of two and three particle correlations may be useful to study the propagation of waves in the QGP. Among the sources of waves we have fast partons crossing the medium and also flux tubes formed in the initial stage of heavy ion collisions. In this work we have emphasized that these waves are most likely nonlinear and should be studied with the appropriate formalism, which, in our opinion, is the Reductive Perturbation Method. After making a survey of relativistic hydrodynamics we have presented the RPM in a simple and pedagogical way. The equation of state of the two most relevant strongly interacting fluids, i.e., nuclear matter and the quark gluon plasma, was discussed. In both

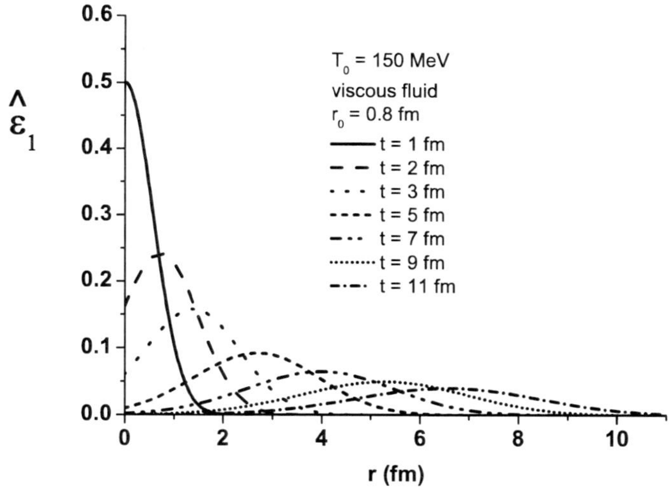

(a)

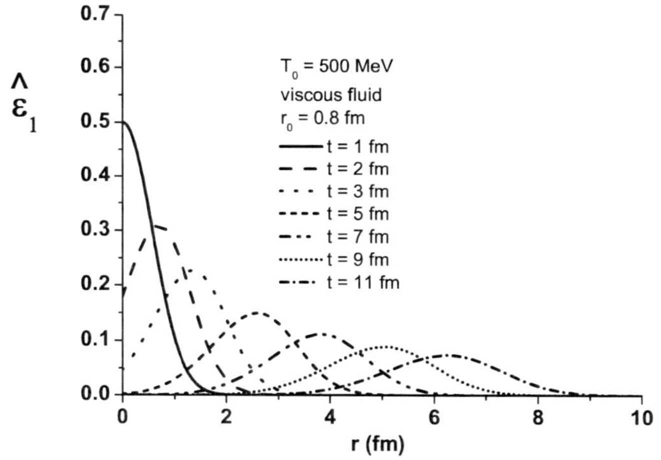

(b)

Figure 16. Solutions for several times: a) $T_0 = 150\ MeV$ and b) $T_0 = 500\ MeV$.

cases we have made an effort to give a pedagogical introduction for non experts. In both cases we have shown how to obtain a KdV soliton. The main responsible for the appearance of these solitons are higher order derivative terms in the vector fields appearing in the Lagrangian of the system. In QHD it is enough to relax the strict mean field approximation, in which all gradients vanish, and allow for spatial inhomogeneities in the vector field. A slightly more careful treatment of the vector field equation of motion yields the desired term, derived from the Laplacian $\nabla^2 V_0$. In QCD, the situation is more complicated because the gluon is massless. This makes impossible a simple estimate of the corresponding quantity $\nabla^2 A_0^a$. However a careful treatment of the non-vanishing vacuum condensates leads to a dynamically generated gluon mass, m_G, which introduces a mass and a size scale and

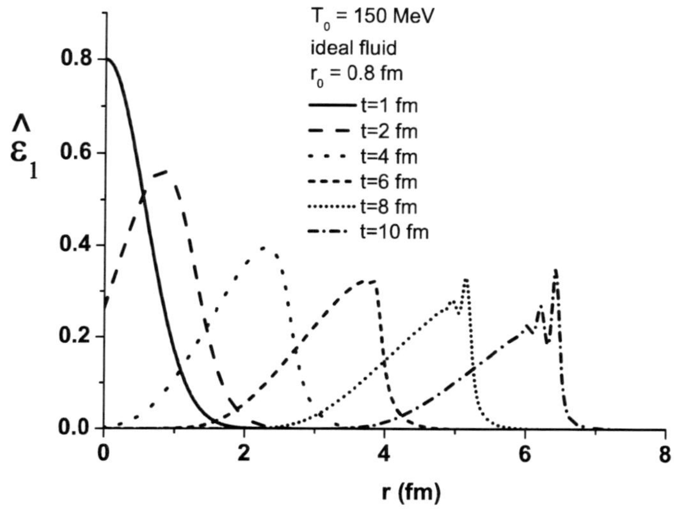

(a)

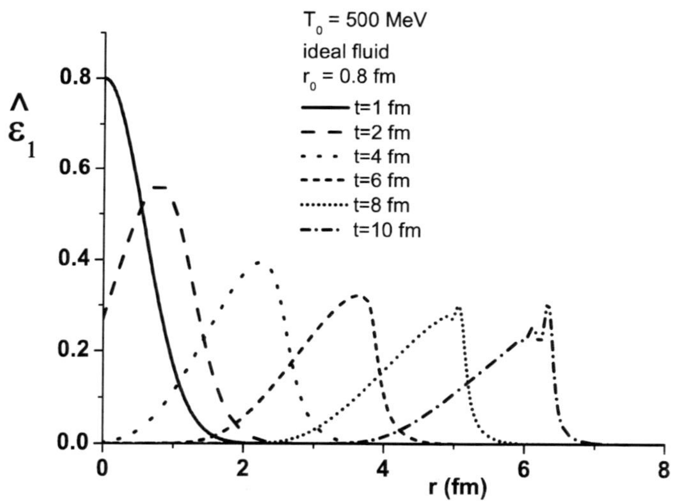

(b)

Figure 17. Analogous to Fig. 16 but for the ideal fluid. a) $T_0 = 150\, MeV$ and b) $T_0 = 500\, MeV$.

renders possible the estimate of the desired Laplacian and the existence of solitons in the QGP.

Combining the equations of hydrodynamics with the equation of state and applying the RPM we have derived several differential equations for the perturbations in energy and baryon density. These equations connect properties of the waves, such as width and speed, with the microscopic dynamical quantities of the fluids, such as particle masses and couplings. Several of them have analytical solutions, which were presented and discussed. Some others must be solved numerically. As expected for nonlinear equations, the results

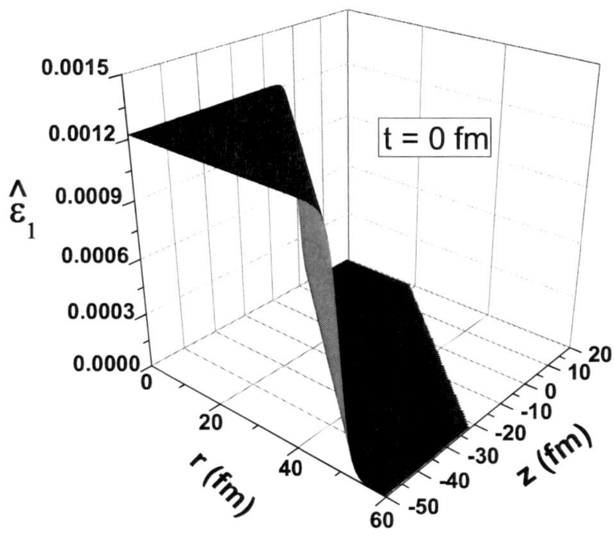

(a)

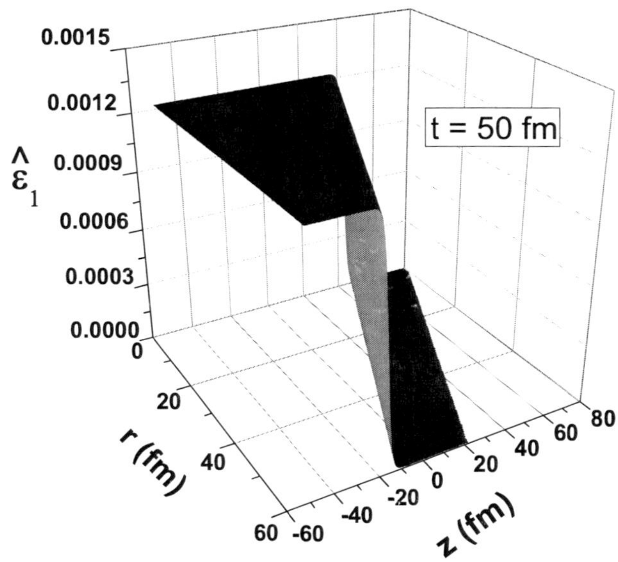

(b)

Figure 18. Shock wave evolution: a) $t = 0\ fm$ and b) $t = 50\ fm$.

depend very strongly on the initial conditions. An interesting finding is that, even when we do not have a KdV equation, in many cases the breaking wave equation has very long living localized solutions, which resemble to solitons. In some other cases, the initial pulses loose their localization and/or start to present rapid oscillations. All these features may manifest themselves directly or indirectly in the experimental data. The analysis made here is still qualitative and a closer contact with phenomenology is still to be made. For now, the obtained results suggest that viscosity strongly affects the propagation of perturbations in the quark gluon plasma. In order to confirm this statement the next step is to apply the RPM to the Müller-Israel-Stewart theory.

We hope to have convinced the reader that the study of nonlinear waves in hadron physics is an interesting and fast moving field. This study will help to interpret and understand the data from the LHC.

Appendix: Method of Finite Differences

The most general form of a one-dimensional nonlinear wave equation with a second order dissipative term and a third order dispersive terms is given by:

$$\frac{\partial u}{\partial t}+\beta\frac{\partial}{\partial x}\left(\frac{u^2}{2}\right)+\alpha\frac{\partial^2 u}{\partial x^2}+\mu\frac{\partial^3 u}{\partial x^3}=0 \tag{244}$$

For $\alpha=0$, β and $\mu\neq 0$, it is the usual third order nonlinear Kortweg-de Vries equation (KdV). For $\mu=0$, β and $\alpha\neq 0$, it represents the Burgers equations. Finally, when only $\beta\neq 0$, we have the breaking wave equation. In order to solve this equation numerically, we divide the integration region ($0\leq x\leq(n+1)h$ and $0\leq t\leq m\Delta t$) with a space step h and a time step Δt. Then, the wave function $u(x,t)$, solution of equation (244), assume the discrete values $u_{i,0},u_{i,1},\ldots,u_{i,j-2},u_{i,j-1},u_{i,j},u_{i,j+1},u_{i,j+2},\ldots,u_{i,n},u_{i,n+1}$ in a given time t_i. The expansion of this solution in Taylor's series leads to:

$$u_{i,j+1}=u_{i,j}+\left(\frac{\partial u}{\partial x}\right)_{i,j}h+\frac{1}{2!}\left(\frac{\partial^2 u}{\partial x^2}\right)_{i,j}h^2+\frac{1}{3!}\left(\frac{\partial^3 u}{\partial x^3}\right)_{i,j}h^3+O(h^4) \tag{245}$$

or, alternatively, to:

$$u_{i,j-1}=u_{i,j}-\left(\frac{\partial u}{\partial x}\right)_{i,j}h+\frac{1}{2!}\left(\frac{\partial^2 u}{\partial x^2}\right)_{i,j}h^2-\frac{1}{3!}\left(\frac{\partial^3 u}{\partial x^3}\right)_{i,j}h^3+O(h^4) \tag{246}$$

We combine these expressions conveniently to obtain finite difference expressions for first, second and third order centered partial space derivatives [77]:

$$\left(\frac{\partial u}{\partial x}\right)_{i,j}\approx\frac{u_{i,j+1}-u_{i,j-1}}{2h}=O_xU_i,\ j=2,\cdots,n-1 \tag{247}$$

$$\left(\frac{\partial^2 u}{\partial x^2}\right)_{i,j}\approx\frac{u_{i,j+1}-2u_{i,j}+u_{i,j-1}}{h^2}=O_{xx}U_i\ \ j=2,\cdots,n-1 \tag{248}$$

and

$$\left(\frac{\partial^3 u}{\partial x^3}\right)_{i,j}\approx\frac{-u_{i,j+2}+2u_{i,j+1}-2u_{i,j-1}+u_{i,j-2}}{2h^3}=O_{xxx}U_i\ \ j=2,\cdots,n-1 \tag{249}$$

in which we define

$$U_i=\begin{pmatrix}u_{i,2}\\ \vdots\\ u_{i,j+1}\\ u_{i,j}\\ \vdots\\ u_{i,n-1}\end{pmatrix} \tag{250}$$

($u_{i,0} = u_{i,1} = u_{i,n} = u_{i,n+1} = 0$ are the boundary values on x axe) and the operators:

$$O_x = \begin{pmatrix} 0 & 1/2h & 0 & \cdots & 0 \\ -1/2h & 0 & 1/2h & \cdots & 0 \\ \vdots & \ddots & \ddots & \ddots & \vdots \\ 0 & \cdots & -1/2h & 0 & 1/2h \\ 0 & \cdots & 0 & -1/2h & 0 \end{pmatrix} \tag{251}$$

$$O_{xx} = \begin{pmatrix} -2/h^2 & 1/h^2 & 0 & \cdots & 0 \\ 1/h^2 & -2/h^2 & 1/h^2 & \cdots & 0 \\ \vdots & \ddots & \ddots & \ddots & \vdots \\ 0 & \cdots & 1/h^2 & -2/h^2 & 1/h^2 \\ 0 & \cdots & 0 & 1/h^2 & -2/h^2 \end{pmatrix} \tag{252}$$

and

$$O_{xxx} = \begin{pmatrix} 0 & -1/h^3 & 1/2h^3 & 0 & 0 & \cdots & 0 \\ 1/h^3 & 0 & -1/h^3 & 1/2h^3 & 0 & \cdots & 0 \\ -1/2h^3 & 1/h^3 & 0 & -1/h^3 & 1/2h^3 & \cdots & 0 \\ \vdots & \ddots & \ddots & \ddots & \ddots & \ddots & \vdots \\ 0 & \cdots & -1/2h^3 & 1/h^3 & 0 & -1/h^3 & 1/2h^3 \\ 0 & \cdots & 0 & -1/2h^3 & 1/h^3 & 0 & -1/h^3 \\ 0 & \cdots & 0 & 0 & -1/2h^3 & 1/h^3 & 0 \end{pmatrix} \tag{253}$$

Analogously, we can expand the solution of equation (244) around a time t_i:

$$U_{i+1} = U_i + \left(\frac{\partial u}{\partial t}\right)_{i,j} \Delta t + \frac{1}{2}\left(\frac{\partial^2 u}{\partial t^2}\right)_{i,j} (\Delta t)^2 + O(h^3) \tag{254}$$

to obtain

$$\left(\frac{\partial u}{\partial t}\right)_{i,j} \approx \frac{U_{i+1} - U_i}{\Delta t} \tag{255}$$

When we apply (255) and the operators (247), (248) and (249) in equation (244), it can be represented equally in times t_i or t_{i+1}, i.e., the following:

$$\frac{U_{i+1} - U_i}{\Delta t} = -\beta O_x[(1/2)U_i^2] - \alpha O_{xx} U_i - \mu O_{xxx} U_i \tag{256}$$

or

$$\frac{U_{i+1} - U_i}{\Delta t} = -\beta O_x[(1/2)U_{i+1}^2] - \alpha O_{xx} U_{i+1} - \mu O_{xxx} U_{i+1} \tag{257}$$

As long as the error one makes in both situations is the same, the Crank-Nicolson scheme [78] prescribes to take the mean value of these two possibilities, therefore

$$U_{i+1} = U_i - \frac{1}{2}\Delta t \beta O_x[(1/2)U_i^2 + (1/2)U_{i+1}^2] - \frac{1}{2}\Delta t \alpha O_{xx}(U_{i+1} + U_i)$$
$$- \frac{1}{2}\Delta t \mu O_{xxx}(U_{i+1} + U_i) \tag{258}$$

Given a initial condition U_0, U_{i+1} is the solution of (258) at any posterior time. However, this is a set of nonlinear algebraic equations. This problem becomes much more simple if we linearize it. We expand the nonlinear term in a Taylor's series, around the $i-th$ time t_i [79]:

$$\frac{1}{2}U_{i+1}^2 = \frac{1}{2}U_i^2 + \Delta t\left(\frac{\partial(1/2)u^2}{\partial u}\right)_i\left(\frac{\partial u}{\partial t}\right)_i + O(\Delta t^2) \tag{259}$$

which leads to

$$\frac{1}{2}U_{i+1}^2 = \frac{1}{2}U_i^2 + U_i(U_{i+1} - U_i) \tag{260}$$

where we have used (255).

Replacing this result in (258) we get:

$$U_{i+1} = U_i - \frac{1}{2}\Delta t\beta O_x(U_iU_{i+1}) - \frac{1}{2}\Delta t\alpha O_{xx}(U_{i+1} + U_i)$$

$$-\frac{1}{2}\Delta t\mu O_{xxx}(U_{i+1} + U_i) \tag{261}$$

Using the matrix format of the operators O_x (251), O_{xx} (252) and O_{xxx} (253), equation (261) becomes a quin-diagonal algorithm:

$$\begin{pmatrix} 1-2s & -q_{i,j+1}+s & -p & 0 & 0 & \cdots & 0 \\ q_{i,j-1}+s & 1-2s & -q_{i,j+1}+s & -p & 0 & \cdots & 0 \\ p & q_{i,j-1}+s & 1-2s & -q_{i,j+1}+s & -p & \cdots & 0 \\ \vdots & \ddots & \ddots & \ddots & \ddots & \ddots & \vdots \\ 0 & \cdots & p & q_{i,j-1}+s & 1-2s & -q_{i,j+1}+s & -p \\ 0 & \cdots & 0 & p & q_{i,j-1}+s & 1-2s & -q_{i,j+1}+s \\ 0 & \cdots & 0 & 0 & p & q_{i,j-1}+s & 1-2s \end{pmatrix} \times$$

$$\times \begin{pmatrix} u_{i+1,2} \\ \vdots \\ u_{i+1,j-1} \\ u_{i+1,j} \\ u_{i+1,j+1} \\ \vdots \\ u_{i+1,n-1} \end{pmatrix} = \begin{pmatrix} r_{i,2} \\ \vdots \\ r_{i,j-1} \\ r_{i,j} \\ r_{i,j+1} \\ \vdots \\ r_{i,n-1} \end{pmatrix} \tag{262}$$

in which

$$p = -\frac{1}{4}\mu\frac{\Delta t}{h^3} \tag{263}$$

$$q_{i,j} = -\frac{1}{4}\beta\frac{\Delta t}{h}u_{i,j} - 2p \tag{264}$$

$$s = \frac{1}{2}\alpha\frac{\Delta t}{h^2} \tag{265}$$

and

$$r_{i,j} = (1+2s)u_{i,j} + pu_{i,j+2} - (2p+s)u_{i,j+1} + (2p-s)u_{i,j-1} - pu_{i,j-2} \tag{266}$$

Therefore, from a given initial condition:

$$U_0 = \begin{pmatrix} u_{0,2} \\ \vdots \\ u_{0,j-1} \\ u_{0,j} \\ u_{0,j+1} \\ \vdots \\ u_{0,n-1} \end{pmatrix} \tag{267}$$

the set of linear algebraic equations (262) can be iteratively solved, to obtain the solution $u(x,t)$ of equation (244) in any posterior time t_{i+1} is given by:

$$U_{i+1} = \begin{pmatrix} u_{i+1,2} \\ \vdots \\ u_{i+1,j-1} \\ u_{i+1,j} \\ u_{i+1,j+1} \\ \vdots \\ u_{i+1,n-1} \end{pmatrix} \tag{268}$$

We turn now to the two dimensional extension of the KdV equation (237), the so called Kadomtsev-Petviashvilli (KP) equation (236):

$$\frac{\partial}{\partial x}\left\{\frac{\partial u}{\partial t} + \frac{\alpha_1}{2}\frac{\partial u^2}{\partial x} + \alpha_2 \frac{\partial^3 u}{\partial x^3}\right\} + \alpha_3 \frac{\partial^2 u}{\partial y^2} = 0 \tag{269}$$

Repeating all the preceding procedure for this equation, we obtain:

$$O_x U_{i+1} = O_x U_i - \frac{1}{2}\Delta t \alpha_1 O_{xx}(U_i U_{i+1}) - \frac{1}{2}\Delta t \alpha_2 O_{xxxx}(U_{i+1} + U_i)$$
$$- \frac{1}{2}\Delta t \alpha_3 O_{yy}(U_{i+1} + U_i) \tag{270}$$

in which we define,

$$U_i = \begin{pmatrix} u_{i,2,k} \\ \vdots \\ u_{i,j+1,k} \\ u_{i,j,k} \\ \vdots \\ u_{i,n-1,k} \end{pmatrix} \quad k = 1,2,...,l \tag{271}$$

($u_{i,j,0} = u_{i,j,l+1} = 0$ are the boundary values in the y direction) and the operators [77]:

$$O_{xxxx} = \begin{pmatrix} 6/h^4 & -4/h^4 & 1/h^4 & 0 & 0 & \cdots & 0 \\ -4/h^4 & 6/h^4 & -4/h^4 & 1/h^4 & 0 & \cdots & 0 \\ 1/h^4 & -4/h^4 & 6/h^4 & -4/h^4 & 1/h^4 & \cdots & 0 \\ \vdots & \ddots & \ddots & \ddots & \ddots & \ddots & \vdots \\ 0 & \cdots & 1/h^4 & -4/h^4 & 6/h^4 & -4/h^4 & 1/h^4 \\ 0 & \cdots & 0 & 1/h^4 & -4/h^4 & 6/h^4 & -4/h^4 \\ 0 & \cdots & 0 & 0 & 1/h^4 & -4/h^4 & 6/h^4 \end{pmatrix} \tag{272}$$

is the finite differences fourth order centered partial space derivative operator, and

$$O_{yy} = \begin{pmatrix} -2/h_y^2 & 1/h_y^2 & 0 & \cdots & 0 \\ 1/h_y^2 & -2/h_y^2 & 1/h_y^2 & \cdots & 0 \\ \vdots & \ddots & \ddots & \ddots & \vdots \\ 0 & \cdots & 1/h_y^2 & -2/h_y^2 & 1/h_y^2 \\ 0 & \cdots & 0 & 1/h_y^2 & -2/h_y^2 \end{pmatrix} \tag{273}$$

in which h_y is the space step in the y direction.

Replacing the matrix representation of the operators in equation (270) we get:

$$\begin{pmatrix} c_{i,j,k} & d_{i,j,k} & a & 0 & 0 & \cdots & 0 \\ b_{i,j,k} & c_{i,j,k} & d_{i,j,k} & a & 0 & \cdots & 0 \\ a & b_{i,j,k} & c_{i,j,k} & d_{i,j,k} & a & \cdots & 0 \\ \vdots & \ddots & \ddots & \ddots & \ddots & \ddots & \vdots \\ 0 & \cdots & a & b_{i,j,k} & c_{i,j,k} & d_{i,j,k} & a \\ 0 & \cdots & 0 & a & b_{i,j,k} & c_{i,j,k} & d_{i,j,k} \\ 0 & \cdots & 0 & 0 & a & b_{i,j,k} & c_{i,j,k} \end{pmatrix} \begin{pmatrix} u_{i+1,2,k} \\ \vdots \\ u_{i+1,j-1,k} \\ u_{i+1,j,k} \\ u_{i+1,j+1,k} \\ \vdots \\ u_{i+1,n-1,k} \end{pmatrix}$$

$$= \begin{pmatrix} e_{i,2,k} \\ \vdots \\ e_{i,j-1,k} \\ e_{i,j,k} \\ e_{i,j+1,k} \\ \vdots \\ e_{i,n-1,k} \end{pmatrix} \quad k = 1,2,...,l \tag{274}$$

in which

$$a = \alpha_2 \frac{\Delta t}{h^3} \tag{275}$$

$$b_{i,j,k} = 1 - 4a + \frac{\alpha_1 \Delta t}{h} u_{i,j+1,k} \tag{276}$$

$$c_{i,j,k} = 6a - \frac{2\alpha_3 h \Delta t}{h_y^2} - \frac{2\alpha_1 \Delta t}{h} u_{i,j,k} \tag{277}$$

$$d_{i,j,k} = -1 - 4a + \frac{\alpha_1 \Delta t}{h} u_{i,j-1,k} \tag{278}$$

$$e_{i,j,k} = -au_{i,j+2,k} + (1+4a)u_{i,j+1,k} + (\frac{2\alpha_3 h\Delta t}{h_y^2} - 6a)u_{i,j,k} + (-1+4a)u_{i,j-1,k}$$

$$-au_{i,j-2,k} - \frac{2\alpha_3 h\Delta t}{h_y^2}(u_{i+1,j,k+1} + u_{i+1,j,k-1} + u_{i,j,k+1} + u_{i,j,k-1})$$

The algorithm (274) represents l sets of linear algebraic equations. As long as each one of these sets are self-consistent, they must be solved iteratively, until the answer converges to the solution $(U^{i+1})^f$ after f iterations. To stop the iterations, we can use the criterion [80]:

$$\frac{\|(U^{i+1})^f - (U^{i+1})^{f-1}\|}{\|(U^{i+1})^{f-1}\|} < \varepsilon \tag{279}$$

In what concerns the stability of this numerical method, the conservation of some quantities such as:

$$P(t) = \int_{-\infty}^{\infty}\int_{-\infty}^{\infty} u dxdy \tag{280}$$

and

$$E(t) = \frac{1}{2}\int_{-\infty}^{\infty}\int_{-\infty}^{\infty} u^2 dxdy \tag{281}$$

are frequently used as a criterion to verify its reliability. In refs. [81] and [82], this criterion is used to show analytically that these numerical methods based on the Crank-Nicolson scheme are unconditionally stable.

Acknowledgments

This work has been partially funded by FAPESP, CNPq, CAPES and USP (NAP program). The authors are grateful to J. Noronha, Y. Hama, F. Grassi, S. B. Duarte, T. Kodama and M. Munhoz for stimulating discussions.

References

[1] *"An Introduction to Quantum Field Theory"*, M. Peskin and D. Schroeder, Addison-Wesley Publishing Company, (1995); *"Quarks and Leptons: an Introductory Course in Modern Particle Physics"*, F. Halzen and A.D. Martin, John Wiley and Sons, (1984); *"Introduction to Elementary Particles"*, D. Griffiths, John Wiley and Sons, (1987).

[2] *"Foundations of Quantum Chromodynamics"*, T. Muta, World Scientific, (1987).

[3] B. Betz, J. Noronha, G. Torrieri, M. Gyulassy and D. H. Rischke, *Phys. Rev. Lett.* 105, 222301 (2010) and references therein.

[4] P. Staig and E. Shuryak, *Phys. Rev. C* 84, 044912 (2011); J. Casalderrey-Solana, E. V. Shuryak and D. Teaney, arXiv:0602183 [hep-ph].

[5] H. Washimi and T. Taniuti, *Phys. Rev. Lett.* 17, 996 (1966).

[6] R.C. Davidson, *"Methods in Nonlinear Plasma Theory"*, Academic Press, New York an London, (1972).

[7] For a recent review and a historical account see: H. Leblond, *J. Phys. B: At. Mol. Opt. Phys.* 41, 043001 (2008).

[8] Lokenath Debnath, *"Nonlinear Partial Differential Equations for Scientists and Engineers"*, third edition, Birkhäuser, USA, (2011).

[9] G. N. Fowler, S. Raha, N. Stelte and R.M. Weiner, *Phys. Lett. B* 115, 286 (1982); S. Raha, K. Wehrberger and R.M. Weiner, *Nucl. Phys. A* 433, 427 (1984); E.F. Hefter, S. Raha and R.M. Weiner, *Phys. Rev. C* 32, 2201 (1985).

[10] B. B. Kadomtsev and V. I. Petviashivili, *Sov. Phys. Dokl.* 15, 539 (1970).

[11] G. C. Das and K. M. Sen, *Chaos, Solitons & Fractals*, 3, 551 (1993).

[12] Wen-shan Duan, *Chaos, Solitons & Fractals*, 14, 503 (2002).

[13] Ju-Kui Xue, *Phys. Plasmas* 10, 3430 (2003).

[14] S. K. El-Labany, Waleed M. Moslem, W. F. El-Taibany and M. Mahmoud, *Physica Scripta* 70, 317 (2004).

[15] Mai-mai Lin and Wen-shan Duan, *Chaos, Solitons & Fractals*, 23, 929 (2005).

[16] Yue-yue Wang and Jie-fang Zhang, *Phys. Lett. A* 352, 155 (2006).

[17] Yunliang et al. , *Phys. Lett. A* 355, 386 (2006).

[18] Yunliang Wang, Zhongxiang Zhou et al., *Phys. Plasmas* 13, 052307 (2006).

[19] A. Mushtaq, *Phys. Plasmas* 14, 113701 (2007).

[20] Yue-yue Wang and Jie-fang Zhang, *Phys. Lett. A* 372, 3707 (2008).

[21] Guang-jun He, Wen-shan Duan and Duo-xiang Tian, *Phys. Plasmas* 15, 043702 (2008).

[22] W.M. Moslem, U.M. Abdelsalam, R. Sabry, E.F.El-Shamy and S.K. El-Labany, *J. Plasma Phys.* 76, 453 (2010).

[23] W. M. Moslem, R. Sabry and P. K. Shukla, *Physics of Plasmas* 17, 032305 (2010).

[24] D.A. Fogaça and F.S. Navarra, *Phys. Lett. B* 639, 629 (2006).

[25] D.A. Fogaça and F.S. Navarra, *Phys. Lett. B* 645, 408 (2007).

[26] D.A. Fogaça and F.S. Navarra, *Nucl. Phys. A* 790, 619c (2007); *Int. J. Mod. Phys. E* 16, 3019 (2007).

[27] D.A. Fogaça, L. G. Ferreira Filho and F.S. Navarra, *Nucl. Phys. A* 819, 150 (2009).

[28] D. A. Fogaça, L. G. Ferreira Filho and F. S. Navarra, *Phys. Rev. C* 81, 055211 (2010).

[29] A.Y. Abul-Magd, I. El-Taher and F.M. Khaliel, *Phys. Rev. C* 45, 448 (1992).

[30] D. A. Fogaça and F. S. Navarra, *Phys. Lett. B* 700, 236 (2011).

[31] D.A. Fogaça, F.S. Navarra and L.G. Ferreira Filho, arXiv:1202.213 [math-ph]; *Comm. Nonlin. Sci. Num. Sim.* 18, 221 (2013), to be published.

[32] S. Weinberg, *"Gravitation and Cosmology"*, New York: Wiley, (1972).

[33] L. Landau and E. Lifchitz, *"Fluid Mechanics"*, Pergamon Press, Oxford, (1987).

[34] U. Heinz, *J. Phys. G* 31, S717 (2005).

[35] Y. Hama, T. Kodama and O. Socolowski Jr., *Braz. J. Phys.* 35, 24 (2005).

[36] H. Stoecker, J.F. Hofmann, W. Scheid and W. Greiner, Fizika 9,671 (1977); T. Renk and J. Ruppert, arXiv:0509036 [hep-ph]; I. M. Dremin, arXiv:0507167 [hep-ph]; L. M. Satarov, H. Stoecker and I. N. Mishustin, *Phys. Lett. B* 627, 64 (2005).

[37] H. Stoecker, *Nucl. Phys. A* 750, 121 (2005); T. Renk and J. Ruppert, *Phys. Rev. C* 73, 011901 (2006) and references therein.

[38] J. Casalderrey-Solana, E. Shuryak, E.V. and D. Teaney, *J. Phys. Conf. Ser.* 27, 22 (2005); *Nucl. Phys. A* 774, 577 (2006).

[39] J. Ruppert and B. Muller, *Phys. Lett. B* 618, 123 (2005).

[40] A. Chaudhuri and U. Heinz, *Phys. Rev. Lett.* 97, 062301 (2006).

[41] T. Renk and J. Ruppert, *Phys. Rev. C* 73, 011901(R) (2006); *Phys. Rev. C* 76,014908 (2007); *Phys. Lett. B* 646,19 (2007).

[42] L. M. Satarov, H. Stoecker and I. N. Mishustin, Phys. Lett. B 627, 64 (2005); B. Betz, P. Rau and H. Stoecker, *Int. J. Mod. Phys. E* 16, 3082 (2008).

[43] H. Stoecker, B. Betz and P. Rau, *PoS C* POD2006, 029 (2006).

[44] R. B. Neufeld, B. Muller and J. Ruppert, *Phys. Rev. C* 78, 041901 (2008).

[45] P.M. Chesler, L.G. Yaffe, Phys. Rev. Lett. 99, 152001 (2007); *Phys. Rev. D* 78, 045013 (2008).

[46] B. Betz, J. Noronha, G. Torrieri, M. Gyulassy and D. H. Rischke, *Nucl. Phys. A* 830, 777C (2009).

[47] B. Betz, J. Noronha, G. Torrieri, M. Gyulassy, I. Mishustin and D. H. Rischke, *Phys. Rev. C* 79, 034902 (2009).

[48] J. Noronha, M. Gyulassy and G. Torrieri, *J. Phys. G* 35, 104061 (2008).

[49] Denes Molnar and Pasi Huovinen, *Nucl. Phys. A* 830, 475 (2009).

[50] Björn Schenke, Sangyong Jeon and Charles Gale, *Phys. Rev. C* 82, 014903 (2010).

[51] Chun Shen, Ulrich Heinz, Pasi Huovinen and Huichao Song, *Phys. Rev. C* 84, 044903 (2011).

[52] Björn Schenke, Sangyong Jeon and Charles Gale, *Phys. Rev. C* 85, 024901 (2012).

[53] P. Romatschke, *Int. J. Mod. Phys.* E 19, 1 (2010).

[54] D. A. Fogaça, F.S. Navarra and L. G. Ferreira Filho, *Nucl. Phys. A* 887, 22 (2012).

[55] J.Y. Ollitrault, *Eur. J. Phys.* 29, 275 (2008); arXiv:0708.2433 [nucl-th].

[56] Nalan Antar, *International Journal of Engineering Science* 40, 1179 (2002).

[57] R. Saeed and Asif Shah, *Phys. Plasmas* 17, 032308 (2010).

[58] D. A. Fogaça, L. G. Ferreira Filho and F. S. Navarra, *Phys. Rev. D* 84, 054011 (2011).

[59] A. Lacour, J.A. Oller and U.-G. Meissner, *Annals of Physics* 326, 241 (2011).

[60] B.D. Serot and J.D. Walecka, *Advances in Nuclear Physics* 16, 1 (1986).

[61] R.J. Furnstahl, *Lect. Notes Phys.* 641, 1 (2004); B.D. Serot, *Int. J. Mod. Phys.* A19S1, 107 (2004) and references therein.

[62] N. Sasaki, O. Miyamura, S. Muroya, C. Nonaka, *Europhys. Lett.* 54, 38 (2001); *Phys. Rev.* C62, 011901 (2000).

[63] N.D. Minh, arXiv:0807.4431 [hep-th] and references therein.

[64] C.G. Bollini and J.J. Giambiagi, *Rev. Bras. Fis.*, Vol. 17, no1, 14 (1987).

[65] Konstantin G. Zloshchastiev, arXiv:0006002v3 [hep-th].

[66] J. C. Collins and M. J. Perry, *Phys. Rev. Lett.* 34, 1353 (1975).

[67] R. Reif, *"Fundamentals os statistical and thermal physics"*, New York: McGraw-Hill, (1965).

[68] A. L. Fetter and J. D. Walecka, *"Quantum Theory of Many-Particle Systems"*, New York: McGraw-Hill, (1971).

[69] L. S. Celenza and C. M. Shakin, *Phys. Rev. D* 34, 1591 (1986).

[70] X. Li and C. M. Shakin, *Phys. Rev. D* 71, 074007 (2005).

[71] P. G. Drazin and R. S. Johnson, *"Solitons: An Introduction"*, Cambridge University Press, (1989).

[72] Kenneth L. Jones, *Internat. J. Math. & Math. Sci.* 24, No. 6, 379 (2000).

[73] Talaat S. El-Danaf and Mohamed A. Ramadan, *Open Appl. Math. J.* I, 1 (2007).

[74] Gino Biondini, *Phys. Rev. Lett.* 99, 064103 (2007).

[75] Biswajit Sahu, *Phys. Plasmas* 18, 062308 (2011).

[76] B. Franzon, D. A. Fogaça, F. S. Navarra and J. E. Horvath, *Phys. Rev. D* 86, 065031 (2012).

[77] S. C. Chapra and R. P. Canalle, *"Numerical methods for Engineers"*, 6th edition, McGraw-Hill, (2009).

[78] J. Crank and P. Nicolson, *Proc. Cambridge Philos. Soc.* 43, 50 (1947).

[79] K. Djidjeli, W. G. Price, E. H. Twizell and Y. Wang, *J. Comp. Appl. Math.* 58, 307 (1995).

[80] Q. Cao, K. Djidjeli, W. G. Price and E. H. Twizell, *J. Eng. Math.* 35, 323 (1999).

[81] S. A. Teukolsky, *Phys. Rev. D* 61, 087501 (2000).

[82] T. Iitaka, *Phys. Rev. E* 49, 4684 (1994).

INDEX

A

B

C

Q

R

S